Engineering Design
and Graphics
with Autodesk Inventor® 8

Engineering Design and Graphics with Autodesk Inventor® 8

James D. Bethune
Boston University

Upper Saddle River, New Jersey
Columbus, Ohio

Library of Congress Cataloging-in-Publication Data

Bethune, James D.
    Engineering design and graphics with Autodesk Inventor 8 / James D. Bethune.
        p. cm.
    ISBN 0-13-119073-3
    1. Engineering graphics. 2. Engineering models—Data processing. 3. Autodesk Inventor
(Electronic resource) I. Title.

T353.B4538 2005
620'.0042'0285536–dc22   2004044611

Executive Editor: Debbie Yarnell
Managing Editor: Judith Casillo
Production Editor: Louise N. Sette
Production Coordination: Lisa S. Garboski, bookworks
Design Coordinator: Diane Ernsberger
Cover Designer: Kristi Holmes
Production Manager: Deidra Schwartz
Marketing Manager: Jimmy Stephens

This book was set in Times and Arial by *The GTS Companies*/York, PA, Campus. It was
printed and bound by Courier Kendallville, Inc. The cover was printed by Phoenix Color
Corp.

Pearson Education Ltd.   Pearson Education Australia Pty. Limited
Pearson Education Singapore Pte. Ltd.   Pearson Education North Asia Ltd.
Pearson Education Canada, Ltd.   Pearson Educación de Mexico, S. A. de C.V.
Pearson Education—Japan   Pearson Education Malaysia Pte. Ltd.

10 9 8 7 6 5 4 3 2 1
ISBN: 0-13-119073-3

Preface

This book introduces Autodesk Inventor® 8 and shows how to use Inventor to create and document designs. The content of the book goes beyond the material normally presented in an engineering drawing text associated with CAD software to include an introduction to designing simple mechanisms. The design portion of the book culminates with 10 design projects that serve as practical applications for all the material in the book.

All topics are presented using a step-by-step format so that the reader can work directly from the text to the screen. The book contains many sample problems that demonstrate the subject being discussed. Each chapter contains a variety of exercise problems that serve to reinforce the material just presented and allow the reader to practice the techniques described.

Chapters 1 and 2 present 2D sketching commands and then show how to use the Extrude command to create simple models of uniform thickness.

Chapter 3 demonstrates the commands needed to create 3D models, including the Shell, Rib, Split, Loft, Sweep, and Coil commands. Work points, work axes, and work planes are explained and demonstrated.

Chapter 4 shows how to create orthographic views from given 3D models. The creation of isometric views, sectional views, and auxiliary views is also covered.

Chapter 5 shows how to create assembly drawings using both the bottom-up and top-down processes. There is an extensive example of how to create an animated assembly, that is, a drawing that moves on screen. Drawing documentation, including presentation drawings, exploded assembly drawings, and parts lists, is also included.

Chapter 6 covers threads and fasteners. Drawing conventions and callouts are defined for both inch and metric threads. The chapter shows how to calculate thread lengths and how to choose the appropriate fastener from the Standard Parts library. Nuts, washers, setscrews, and rivets are also included.

Chapter 7 shows how to apply dimensions to drawings. Both the ANSI and ISO standards are demonstrated. Different styles of dimensioning, including ordinate and baseline, are covered as well as Inventor's Hole Table command.

Chapter 8 is an extensive discussion of tolerancing, including geometric tolerancing. The chapter first shows how to calculate appropriate tolerances, then shows how to use Autodesk Inventor to apply the tolerances to drawings. The chapter also includes several design problems that demonstrate how to calculate and apply positional tolerances to both fixed and floating conditions.

Chapter 9 presents bearings and shafts. The chapter shows how to calculate clearances for shafts and bearings and how to select bearings from manufacturers' catalogs and from the Web. Shear and bending diagrams are introduced (algebra only), and the results of the diagrams are used to calculate minimum shaft diameters and critical speeds. The chapter then demonstrates how to apply the results to selecting and supporting a shaft in bearings.

Chapter 10 introduces gears. Gear ratios, gear trains, the Lewis equation, and forces in gears are covered. Center distances and backlash are included as part of an explanation of how to design gear boxes. Examples are given demonstrating how to select gears from manufacturers' catalogs and from the Web. The chapter also shows how to animate Autodesk Inventor gear drawings so that the gears rotate on the screen.

Chapter 11 presents cams, springs, and keys. The chapter shows how to create a displacement diagram and then how to convert the displacement diagram information into a cam profile. The chapter shows how to design a spring, how to select an appropriate spring from the Web, and how to draw a spring. Various type of keys are presented along with their tolerance requirements. The chapter ends with a demonstration of cam animation.

Chapter 12 introduces sheet metal drawing and weldments. The chapter discusses the various sheet metal commands and shows how to create sheet metal drawings. The chapter also shows how to redesign existing parts into weldments. Only fillet welds are presented.

Chapter 13 demonstrates the design process, including how to manage the process. Concept sketches, evaluation matrices, team calendars, responsibility charts, and Gantt charts are presented using a design problem. The design problem starts with a problem statement and ends with a drawing of one possible solution.

Chapter 14 presents 10 design projects. The idea is for students to apply what they have learned in the previous chapters. The projects require students first to design a solution to the problem, then build and test a prototype, then document the solution using engineering drawings. Each of the projects is rated as to difficulty.

Thanks to the editors, Debbie Yarnell and Judy Casillo, and thanks to my family and especially to Cheryl.

James D. Bethune

Contents

Chapter 9—Bearings and Shafts

Chapter 10—Gears

Chapter 11—Cams, Springs, and Keys

Chapter 12—Sheet Metal and Weldments

Chapter 13—The Design Process

Chapter 14—Design Projects

Appendix

Index 509

1

Getting Started

1-1 INTRODUCTION

This chapter presents a step-by-step introduction to Inventor 8.0. When the program is first accessed, the New File dialog box will appear. See Figure 1-1. If the drawing screen does not look like Figure 1-1, click the New tool in the What To Do box.

There are seven options that will create drawings using four different types of files. The files are categorized using four different extensions. The extensions are defined as follows.

.ipt: part files for either 3D model drawings or sheet metal drawings. These files are for individual parts.

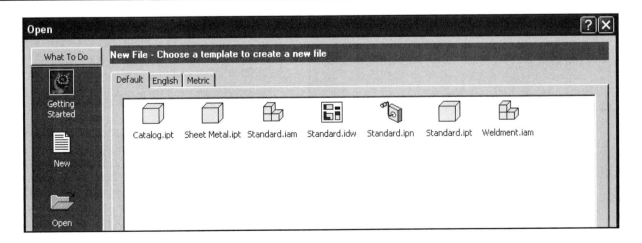

Figure 1-1

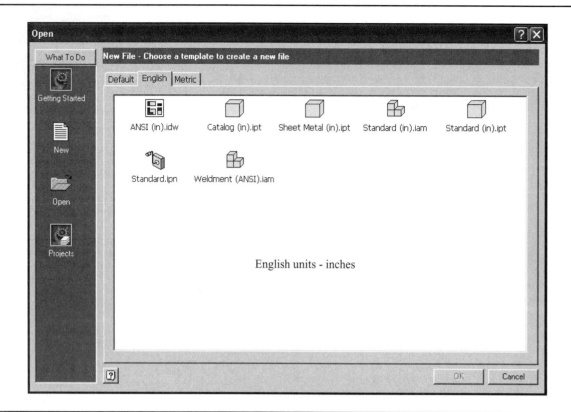

Figure 1-2

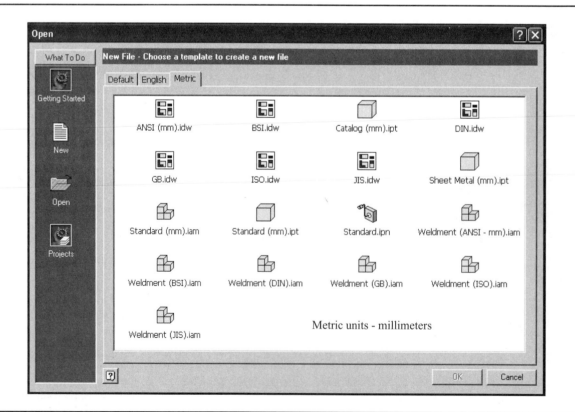

Figure 1-3

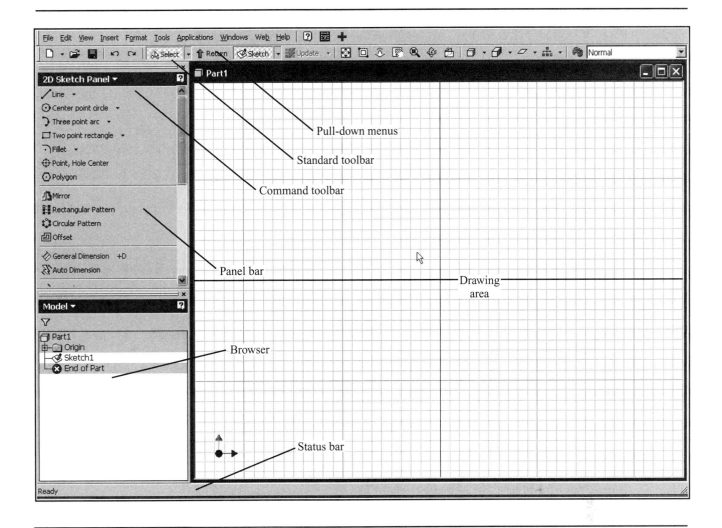

Figure 1-4

.iam: assembly drawings and weldments. Assembly drawings are formed by combining .ipt files.

.ipn: presentation files. These files are used to create exploded assembly drawings.

.idw: drawing layout files. These files are used to create orthographic views from already created assembly and presentation files.

Figure 1-2 shows the New File – English option. This option is used to create drawings in English units (inches) that conform to ANSI (American National Standards Institute) standards.

Figure 1-3 shows the New File – Metric option. This option is used to create metric drawings (millimeters).

1-2 CREATING A FIRST SKETCH

This section shows how to set up, create, and save a first drawing. The intent is to walk through a simple drawing in order to start to understand how inventor functions.

1. Select the Metric tab from the New File dialog box.
2. Select the Standard (mm) .ipt tool, then OK.

The drawing screen should change and look like the screen shown in Figure 1-4. The Inventor drawing screen includes a set of pull-down menus, the Standard toolbar, and the Command toolbar at the top of the screen. The browser area is at the lower left of the screen and contains a running list of how the drawing was created. The browser

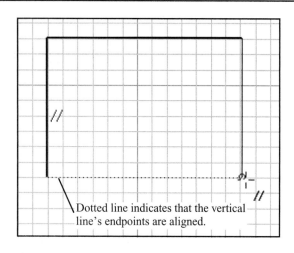

Figure 1-5

area information is used to edit models both during their creation and after they are created.

The panel bar contains command tools used to create drawings. The tool listing will change according to the operating mode selected.

To sketch a 30 × 40 rectangle

1. Select the Line tool on the 2D Sketch Panel bar.

Inventor does not use command line prompts, and there is no coordinate value input or axis reference. All work is done on the drawing screen. Each model generates its own set of reference values.

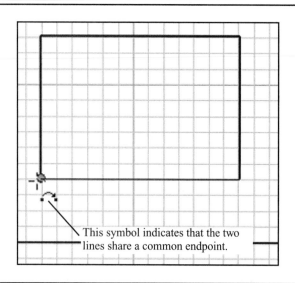

Figure 1-6

Lines are first sketched, that is, drawn without dimensions, and then modified to the required size.

2. Sketch a vertical line anywhere on the screen.

As the line is drawn, if it is vertical, a small symbol will appear next to the line indicating that the line is vertical.

3. Left-click the mouse and continue, sketching a horizontal line.

As the line is sketched a perpendicular symbol will appear if the horizontal line is perpendicular to the vertical line.

4. Left-click the mouse and continue, sketching a second vertical line.

As the second vertical line is sketched two parallel symbols will appear, one next to the line being sketched and the second next to the first vertical line, indicating that the lines are parallel.

When the endpoint of the second vertical line is aligned with the starting point of the first vertical line a broken line will appear. See Figure 1-5.

5. Sketch the second vertical line equal in length to the first vertical line.
6. Sketch a second horizontal line and locate its endpoint on the starting point of the first vertical line.

When the two points are aligned the cursor dot will change its color, and a small arclike symbol will appear. See Figure 1-6.

7. Right-click the mouse and select the Done option.

To delete lines

Lines and other objects may be deleted from a sketch.

1. Select the line to be deleted.
2. Right-click the mouse.

A dialog box will appear on the screen. See Figure 1-7.

3. Select the Delete option.

The line will disappear.

To undo a command

The Undo command will undo the last command entered.

1. Click on the Undo tool located at the top of the screen on the Standard toolbar.

The line will reappear.

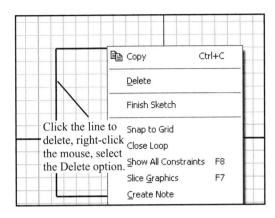

Figure 1-7

To size the rectangle

1. Select the General Dimension tool from the 2D Sketch Panel bar.
2. Select the left vertical line, then move the created dimension to the left of the object and click the left mouse button.

A small dialog box will appear containing the distance on the sketched line. See Figure 1-8.

3. Press the Delete key to remove the value, and type in 30, the required length.
4. Click on the check mark on the dialog box.

The line will change length.

5. Repeat the procedure for one of the horizontal lines, changing the sketched value to 40.
6. Right-click the mouse button and select the Done option.

Figure 1-9 shows the resulting 30 × 40 rectangle.

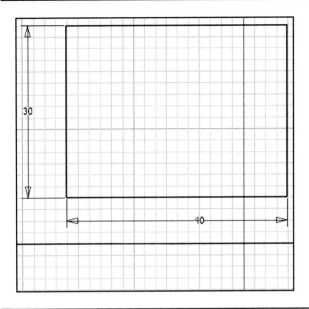

Figure 1-9

1-3 CREATING A SOLID MODEL

The 30 × 40 rectangle sketched in Section 1-2 will now be used to create a 3D solid model.

To change to an isometric view

1. Right-click the mouse and select the Isometric View option.

See Figure 1-10.

The screen will rotate into an isometric view orientation. Use the center mouse button to zoom the sketch to an acceptable size on the screen. See Figure 1-11.

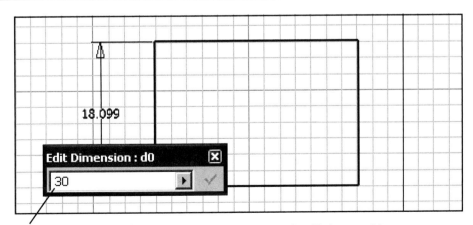

Enter the desired 30 value. The vertical sides of the rectangle will change to 30 mm.

Figure 1-8

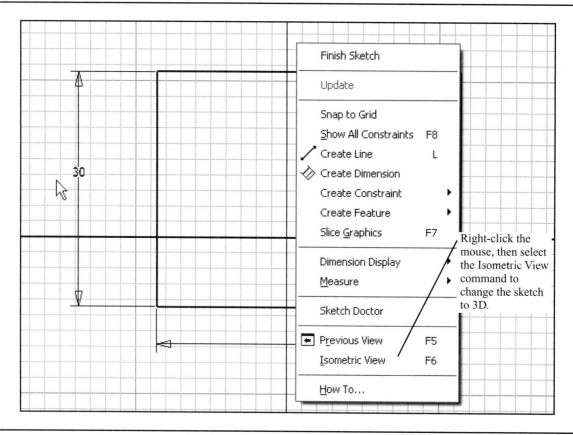

Figure 1-10

To create a solid model

1. Click the Return tool located on the Command toolbar or right-click the mouse and select the Finish Sketch option.

   The panel bar will change to a listing of Part Features tools. See Figure 1-12.

2. Select the Extrude tool from the Panel bar.

   The Extrude dialog box will appear. See Figure 1-13.

3. Change the Extents value to 15, then select OK.

   Figure 1-14 shows the results.

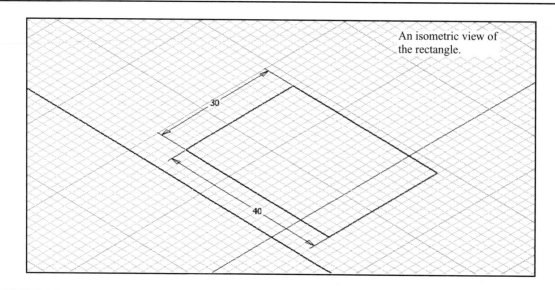

Figure 1-11

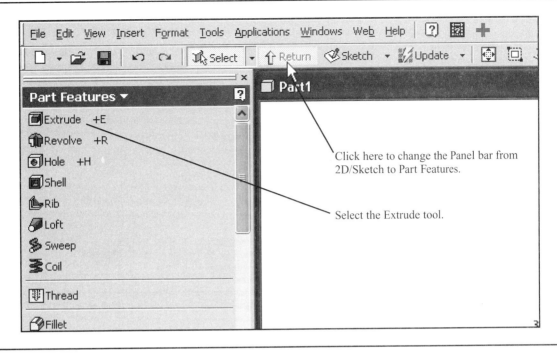

Figure 1-12

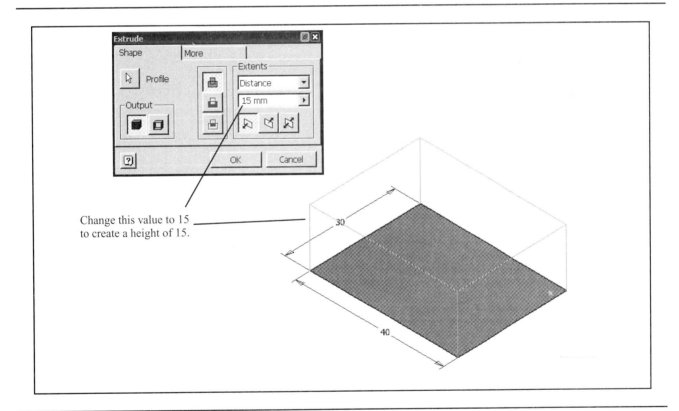

Figure 1-13

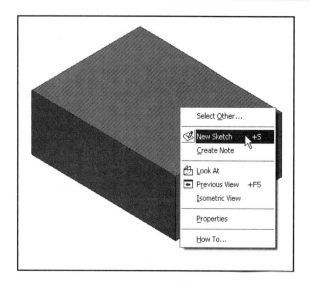

Figure 1-14

To locate the center of a hole

This section explains how to locate a Ø15 hole in the center of the top surface of the model.

1. Click on the top surface of the solid model.

The surface will change color, confirming that it has been selected. The top surface is a new sketch plane.

2. Right-click the mouse and select the New Sketch command.

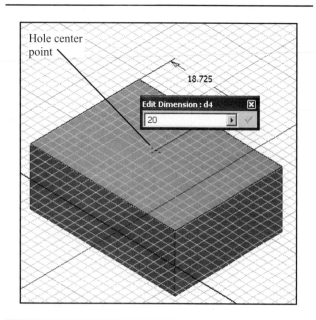

Figure 1-15

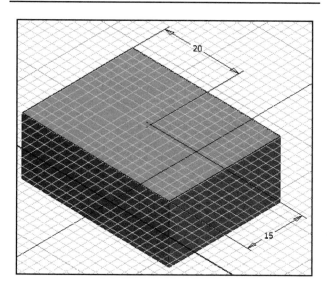

Figure 1-16

The panel bar will change to 2D Sketch tools, and the screen grid will move to the top surface of the model.

3. Select the Point, Hole Center tool from the Sketch Panel bar.
4. Locate the point near the center of the top surface, then right-click the mouse and select the Done option.
5. Select the General Dimension tool from the panel bar.
6. Select the left edge line, then the hole's center point, and change the dimension value to 20.

See Figure 1-15.

7. Repeat the procedure for the required vertical distance of 15.
8. Right-click the mouse and select the Done option.

See Figure 1-16.

To draw a Ø15 hole

1. Click the Return tool on the Command toolbar or right-click the mouse and select the Finish Sketch option.

The Panel bar will change to Part Features tools.

2. Click the Hole tool.

The Holes dialog box will appear. See Figure 1-17.

3. Locate the cursor between the hole's diameter value and the mm symbol shown in the preview box on the Holes dialog box, backspace out the value, and type in 15.

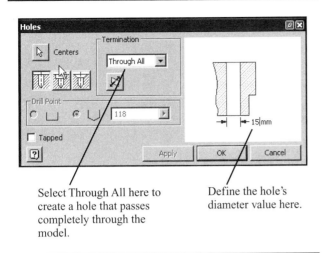

Select Through All here to create a hole that passes completely through the model.

Define the hole's diameter value here.

Figure 1-17

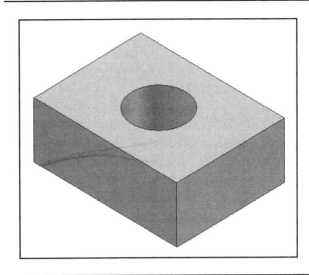

Figure 1-18

4. Set the Termination for Through All, then click the OK box.

Figure 1-18 shows the resulting model.

To save the model

1. Click on the File pull-down menu, then select the Save Copy As... option.

The Save Copy As dialog box will appear. See Figure 1-19.

2. Select a directory and file name and save the model.

In this example the model was saved in a directory called Inventor using the file name Model-1.

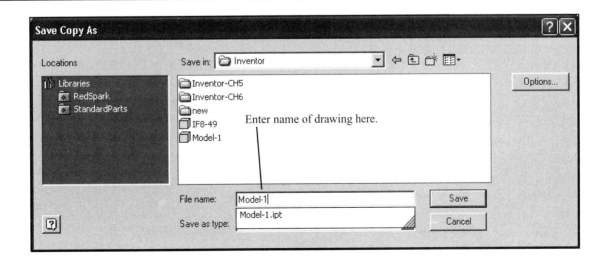

Enter name of drawing here.

Figure 1-19

1-4 EXERCISE PROBLEMS

Sketch the shapes shown in Figures EX1-1 through EX1-8, then create solid models using the specified thickness values.

EX1-1 INCHES

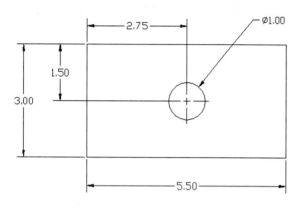

THICKNESS = 1.00

EX1-3 INCHES

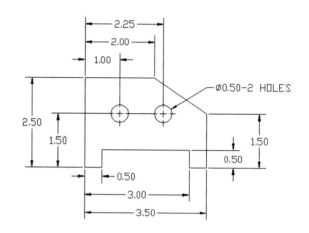

THICKNESS = 0.750

EX1-2 INCHES

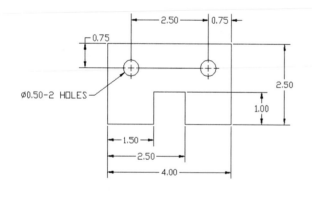

THICKNESS = 0.500

EX1-4 INCHES

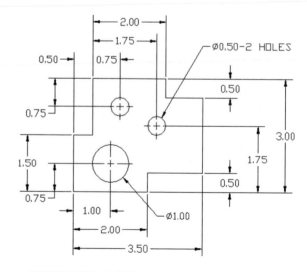

THICKNESS = 1.125

EX1-5 MILLIMETERS

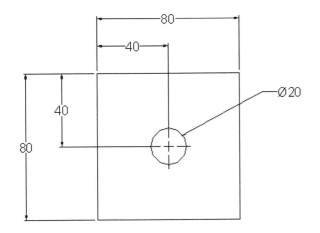

THICKNESS = 10

EX1-7 MILLIMETERS

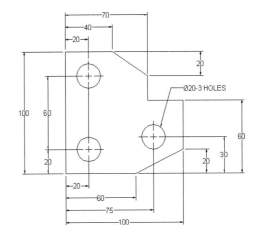

THICKNESS = 5

EX1-6 MILLIMETERS

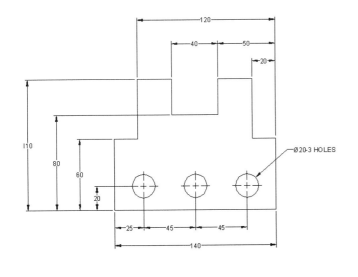

THICKNESS = 15

EX1-8 MILLIMETERS

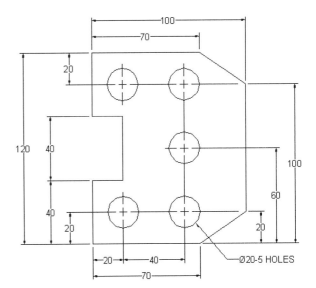

THICKNESS = 12

EX1-9 INCHES

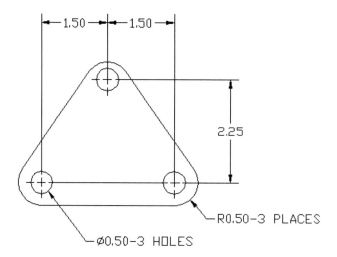

THICKNESS = .250

EX1-10 MILLIMETERS

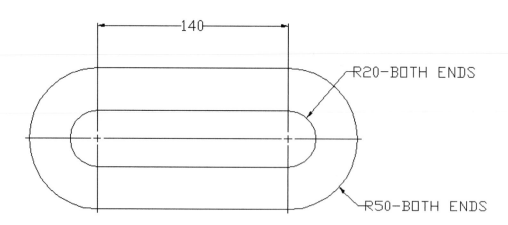

THICKNESS = 6.5

EX1-11 MILLIMETERS

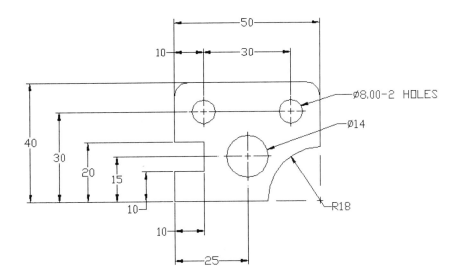

THICKNESS = 16

EX1-12 INCHES

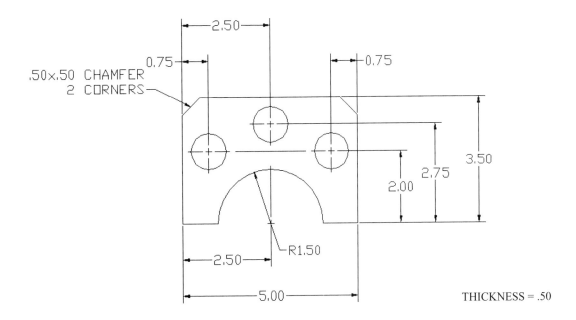

THICKNESS = .50

EX1-13 MILLIMETERS

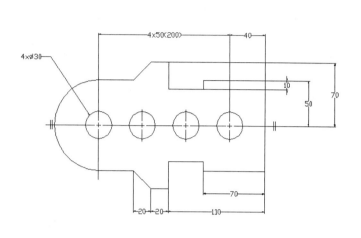

THICKNESS = 12

EX1-15 MILLIMETERS

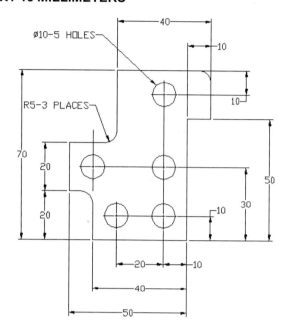

THICKNESS = 8.25

EX1-14 MILLIMETERS

EX1-16 INCHES

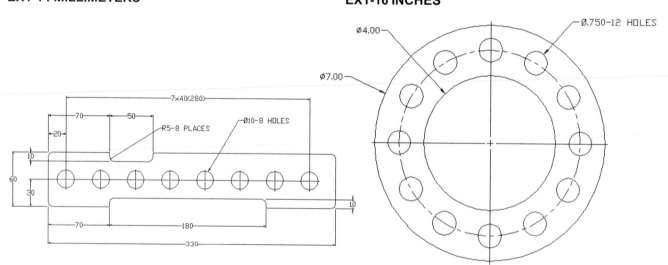

THICKNESS = 2

THICKNESS = .1875

Two-Dimensional Sketching

2-1 INTRODUCTION

This chapter introduces most of the commands found in the options of the 2D Sketch Panel bar. These commands are used to create two-dimensional (2D) sketches. Inventor models are usually based on an initial 2D sketch that is first extruded then manipulated using additional planes to develop the final model shape.

Figure 2-1 shows the sketch options available on the Sketch Panel bar.

2-2 LINE

1. Click the Line tool on the Sketch Panel bar.
2. Select any point on the screen.
3. Move the cursor around the screen.

As the cursor is moved a bar symbol will appear when the line is either horizontal or vertical.

4. Click the line when an appropriate endpoint has been located.
5. Right-click the mouse and select the Done option.

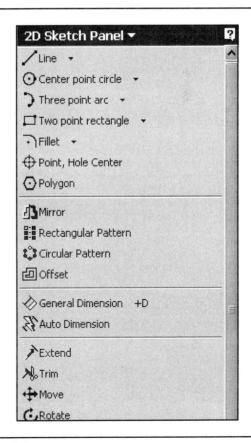

Figure 2-1

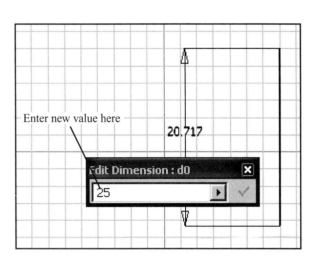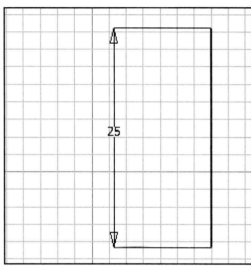

Figure 2-2

To size lines

Figure 2-2 shows a vertical line that was sketched using the Line command.

1. Click the General Dimension tool on the 2D Sketch Panel bar.
2. Click the vertical line.

A dimension indicating the length of the line will be created.

3. Change the value to 25 and click the check mark on the dialog box.

The line's length will change to the indicated distance.

To draw lines at an angle

1. Sketch a second line with its starting endpoint on the top endpoint of the vertical line so that it forms an angle to the right of the vertical line.

   See Figure 2-3.

2. Click the General Dimension tool.
3. Click the vertical line, then the angled line.

An angular dimension will appear between the lines. Move the cursor around to verify that other angular values are available.

4. Locate the angular dimension and press the left mouse button twice.

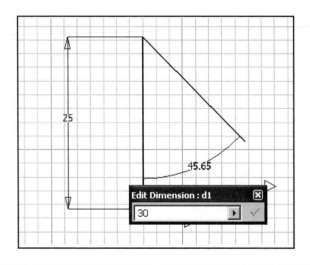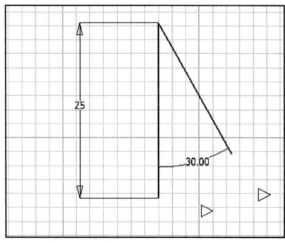

Figure 2-3

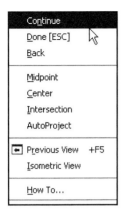 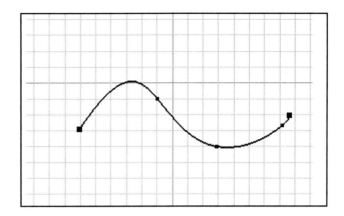

Figure 2-4

A dialog box will appear defining the existing angle between the lines.

5. Change the value to 30 and click the check mark on the dialog box.

The angle between the lines will change to the entered value. The length of the angle line can also be changed using the General Dimension tool.

2-3 SPLINE

The Spline tool command is a flyout from the Line command. An *open spline* is a curved line. A *closed spline* is an enclosed curved line on which the start and end points are the same point.

1. Click on the Spline tool on the Sketch Panel bar.
2. Select four random points, then press the right mouse button.

   A dialog box will appear.

3. Select the Continue option.

   See Figure 2-4.

To edit a spline

A spline may be edited with the General Dimension tool.

1. Click the General Dimension tool and select the first two defining points of the spline.

   See Figure 2-5.

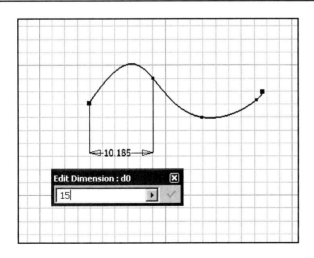 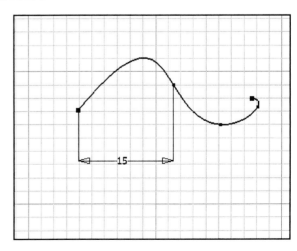

Figure 2-5

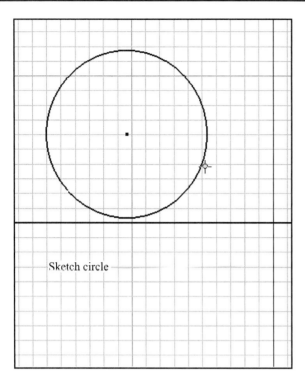

Sketch circle

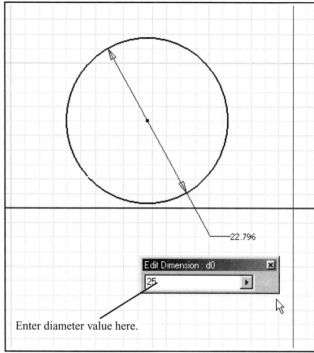

22.796

Edit Dimension : d0

25

Enter diameter value here.

Figure 2-6

2. Change the dimension value to 15 and click the check mark.

The distance will change.

2-4 CIRCLE

There are two ways to sketch a circle using Inventor: select a center point, then define a diameter; or define three tangent points.

To use the center point option

1. Select the Center point circle tool from the Sketch Panel bar.
2. Select a point on the screen.
3. Move the cursor away from the center point and left-click when an approximate diameter is created.
4. Right-click the mouse and select the Done option.

See Figure 2-6.

5. Select the General Dimension tool and use it to enter the desired diameter.

The circle will change to the defined diameter value.

To use the tangent circle option

The Tangent circle option requires that some entities already exist on the screen. This option is a flyout from the Center point circle tool. In this example a triangle was drawn using the Line tool. See Figure 2-7.

1. Select the Tangent circle option from the Sketch Panel bar.
2. Select each of the three lines of the triangle.

A circle will appear tangent to the three lines.

2-5 ELLIPSE

The Ellipse tool is a flyout from the Center point circle tool.

1. Click the Ellipse tool on the Sketch Panel bar.
2. Select a point on the screen.

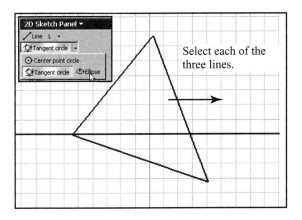
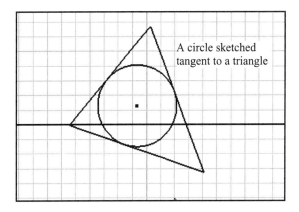

Figure 2-7

This point will become the center point of the ellipse.

3. Move the cursor away from the point and select a point.

A *centerline* (a line with a pattern of long and short dashes) will extend from the selected point through the first point and equidistant to the other side of the point. See Figure 2-8.

4. Move the cursor above the line to define the elliptical shape.
5. Select a point, then right-click the mouse and select the Done option.

To size the ellipse

1. Click the General Dimension tool.
2. Select the left edge of the ellipse, then move the cursor to a location below the existing ellipse.
3. Locate the dimension and then click the dimension.
4. Enter the desired distance value.

See Figure 2-9.

5. Click the check mark on the dialog box.

The ellipse will change shape.

6. Click the General Dimension tool again and define the vertical elliptical value.
7. Right-click the mouse and select the Done option.

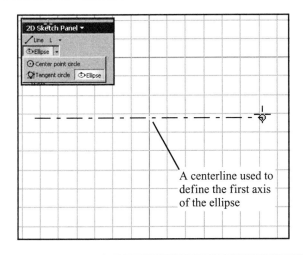
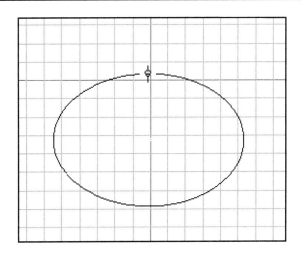

Figure 2-8

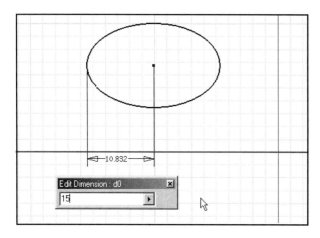
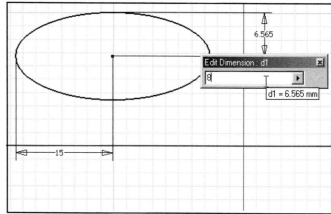
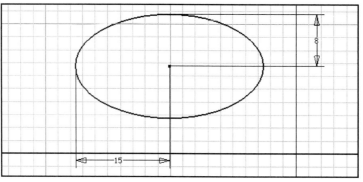

Figure 2-9

2-6 ARC

There are three ways to draw arcs using Inventor: select three points, define a tangent, or select a center point.

To sketch a three-point arc

1. Click the Three point arc tool.
2. Select a point on the screen, then move the cursor and select a second point.

    See Figure 2-10.

3. Select a third point.
4. Click the right mouse button and select the Done option.

To edit an arc

1. Click the General Dimension tool.

    Select one of the endpoints of the arc, then the center point.

2. Enter the appropriate dimensional value.

To sketch a tangent arc

The Tangent arc command requires existing entities. The Tangent arc tool is a flyout from the Center point circle tool. Figure 2-11 shows two parallel lines.

1. Click the Tangent arc tool.
2. Select the endpoint of one of the lines, then select the endpoint of the other line.
3. Right-click the mouse and select the Done option.

The arc may be edited using the General Dimension tool.

To sketch a center point arc

The Center point arc tool is a flyout from the Center point circle tool.

1. Click the Center point arc tool.
2. Select a point on the screen, then select a second point.

    See Figure 2-12. The distance between the two points will define the radius of the arc.

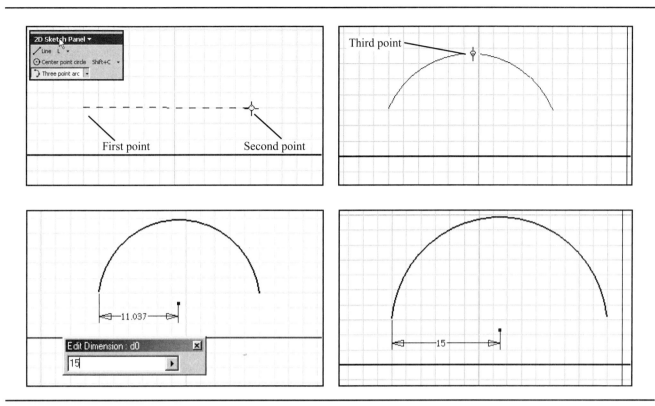

Figure 2-10

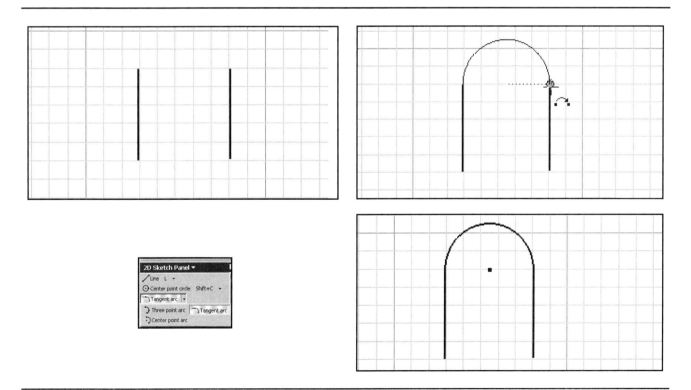

Figure 2-11

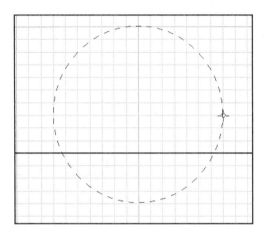
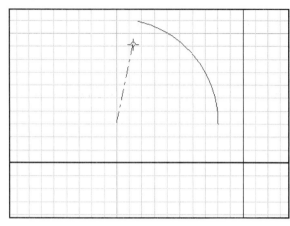

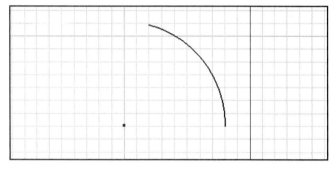

Figure 2-12

3. Select a third point.
4. Right-click the mouse and select the Done option.

The arc may be edited using the General Dimension tool.

2-7 RECTANGLE

There are two ways to sketch a rectangle using Inventor: by selecting two points and by selecting three points.

To sketch a two-point rectangle

1. Click the Two point rectangle tool.
2. Select a point on the screen.
3. Select a second point on the screen.
4. Right-click the mouse and select the Done option.

The rectangle may be edited using the General Dimension tool. See Figure 2-13.

To sketch a three-point rectangle

The Three point tool is a flyout from the Two point rectangle tool.

1. Click the Three point rectangle tool.
2. Select a point on the screen.
3. Select a second point on the screen.

The distance between the two points will define one side of the rectangle.

4. Select a third point.
5. Right-click the mouse and select the Done option.

The rectangle may be edited using the General Dimension tool.

2-8 FILLET

A fillet is a rounded edge or corner that is added to an existing entity.

1. Click the Fillet tool.

The 2D Fillet dialog box will appear. See Figure 2-14.

2. Enter the radius value for the fillet.

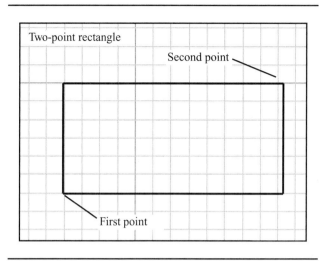

Figure 2-13

3. Select two lines on the rectangle.

    The fillet will be added to the sketch.

4. Create as many fillets as needed, then right-click the mouse and select the Done option.

2-9 CHAMFER

A *chamfer* is an angled edge or corner that is added to an existing entity. The Chamfer tool is a flyout from the Fillet tool.

Chamfers may be defined in one of three ways: with two equal distances, with two distances not equal, or with a distance and an angle.

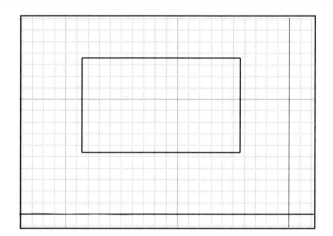

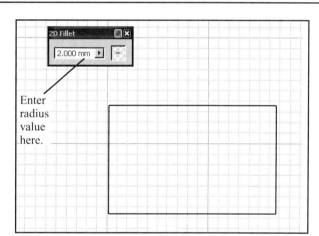

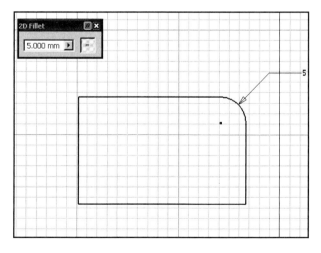

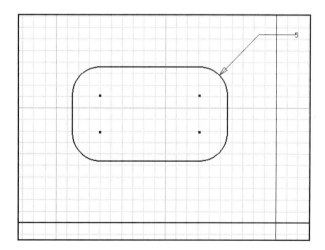

Figure 2-14

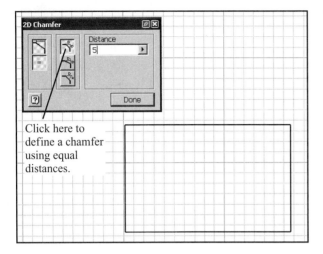
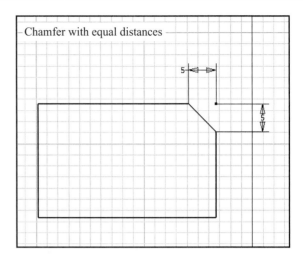

Figure 2-15

To define a chamfer with two equal distances

1. Click the Chamfer tool.

    The 2D Chamfer dialog box will appear. See Figure 2-15.

2. Select the Equal distances box.
3. Enter the chamfer distance.
4. Select two lines on the rectangle, then click the Done box.

To define a chamfer with unequal distances

1. Click the Chamfer tool.
2. Click the Unequal distances box, then click the Done box.

    The 2D Chamfer dialog box will change to allow the entry of two distances. See Figure 2-16.

3. Select two lines on the rectangle, then click the Done box.

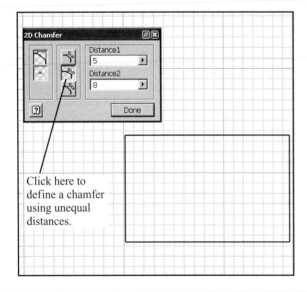
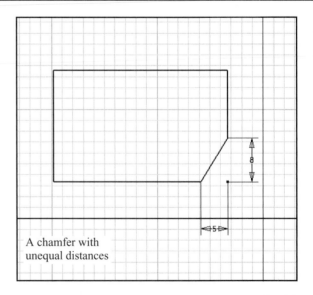

Figure 2-16

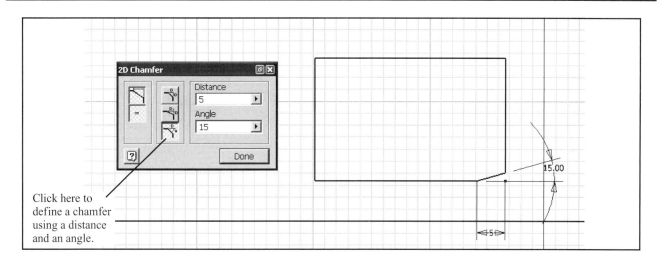

Click here to define a chamfer using a distance and an angle.

Figure 2-17

To define a chamfer using a distance and an angle

1. Click the Chamfer tool.
2. Click the Distance and angle button.

The dialog box will change to allow entry of a distance and an angle value. See Figure 2-17.

3. Select two lines on the rectangle, then click the Done box.

2-10 POLYGON

The Polygon command can be used to sketch either an inscribed or a circumscribed polygon.

To create an inscribed polygon

1. Select the Polygon tool.

The Polygon dialog box will appear. See Figure 2-18.

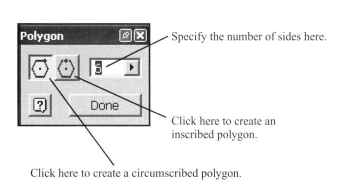

Specify the number of sides here.

Click here to create an inscribed polygon.

Click here to create a circumscribed polygon.

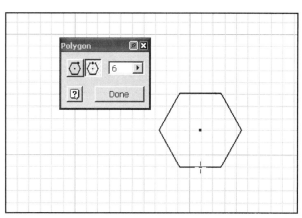

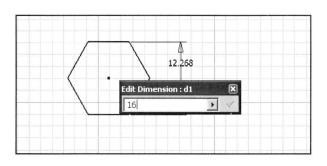

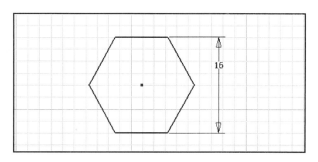

Figure 2-18

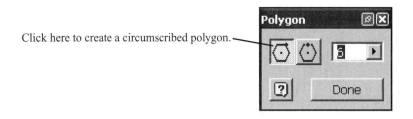

Click here to create a circumscribed polygon.

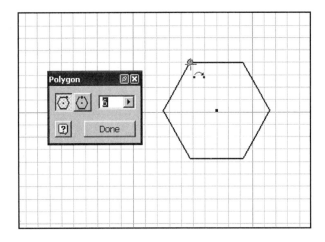

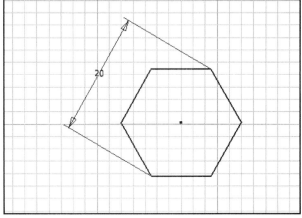

Figure 2-19

2. Select the Inscribe box and specify the number of sides for the polygon.
3. Select a point on the screen, then move the cursor away from the point, sketching the polygon.
4. Right-click the mouse and select the Done option.
5. Click the General Dimension tool and enter the appropriate dimension for the polygon.

To create a circumscribed polygon

1. Select the Polygon tool.

   The Polygon dialog box will appear. See Figure 2-19.

2. Select the Circumscribe box and specify the number of sides for the polygon.
3. Select a point on the screen, then move the cursor away from the point, sketching the polygon.
4. Right-click the mouse and select the Done option.
5. Click the General Dimension tool and enter the appropriate dimension for the polygon.

2-11 MIRROR

The Mirror command creates a reverse copy (mirror image) of an existing sketch. Mirror is very helpful for drawing symmetrical images.

Figure 2-20 shows a hexagon and a vertical line.

1. Click the Mirror tool.

The Mirror dialog box will appear. The Mirror command will automatically be in the Select mode.

2. Select the hexagon either by selecting the six individual lines or by windowing the entire object.

The hexagon will change color, indicating that it has been selected. The Mirror command will automatically switch to Mirror line.

3. Select the vertical line as the mirror line.

The line will change color, indicating that it has been selected.

4. Click Apply in the Mirror dialog box.

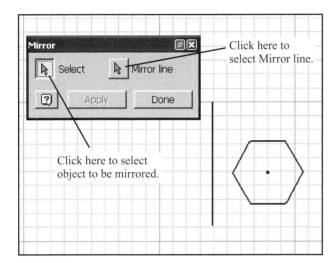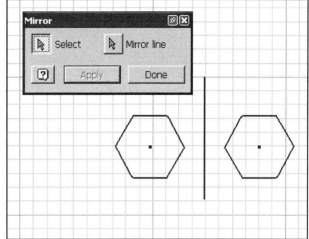

Figure 2-20

2-12 RECTANGULAR PATTERN

The Rectangular Pattern command is used to create rectangular arrays of rows and columns. Figure 2-21 shows a single rectangle. The rectangular pattern will be used to create a 3 × 4 pattern.

1. Click the Rectangular Pattern tool.

The Rectangular Pattern dialog box will appear. The Rectangular Pattern command will automatically be in the Select mode.

2. Select the rectangle, then click the right mouse button and select the Continue option.
3. Select the left vertical line of the rectangle.

This selection defines Direction 1.

4. Enter a count of 4 and a spacing of 10.
5. Click the arrow under the Direction 2 heading.
6. Select the lower horizontal line of the rectangle.
7. Change the count to 3 and the spacing to 14.
8. Click the OK button.

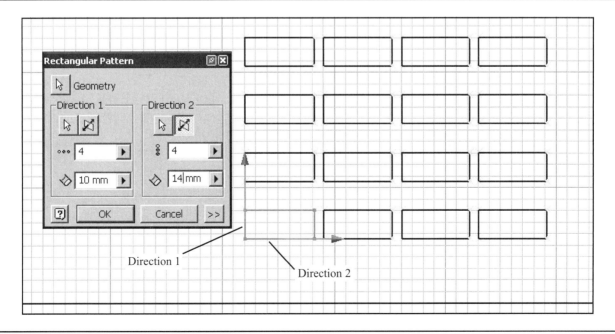

Figure 2-21

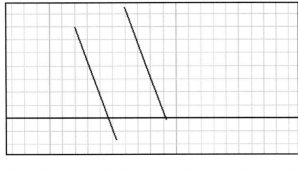

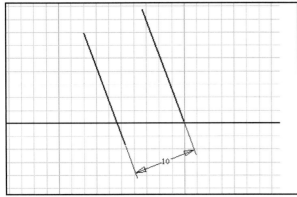

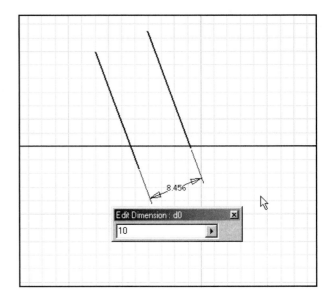

Figure 2-22

2-13 OFFSET

The Offset command is used to create lines parallel to existing lines at a specified distance. Circle and curves may also be offset. Figure 2-22 shows a line. Create a second line parallel to the existing line 10 mm away.

1. Click the Offset tool.
2. Select the line.
3. Move the cursor away from the line and select a new location for the second line, then right-click the mouse and select the Done option.
4. Click the General Dimension tool, then select the two lines.
5. Enter a dimensional value of 10 and click the check mark on the dialog box.

2-14 EXTEND

The Extend command is used to lengthen an existing line. Figure 2-23 shows two lines. The horizontal line is to be extended to the vertical line. The vertical line will serve as a boundary to the extension. The vertical line will be erased after the extension is complete.

1. Click the Extend tool.
2. Select the horizontal line.
3. Right-click the mouse and select the Done option.
4. Click the vertical line, then right-click the mouse and select the Delete option.

2-15 TRIM

The Trim command is used to shorten an existing line. Figure 2-24 shows two lines. The horizontal line will be shortened using the vertical line as a cutting line.

1. Click the Trim tool.
2. Move the cursor onto the horizontal line to the right of the vertical line.

The right portion of the horizontal line will change its color and line pattern.

3. Click the line.
4. Delete the vertical line.

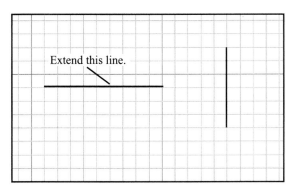

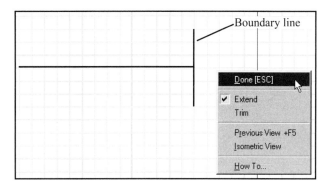

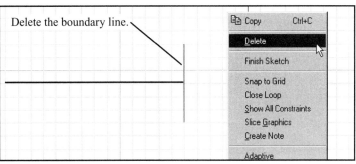

Figure 2-23

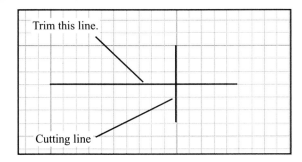

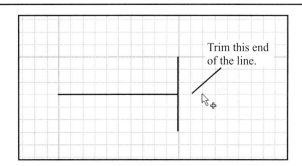

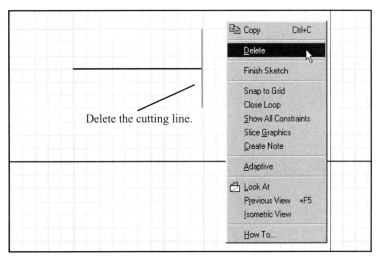

Figure 2-24

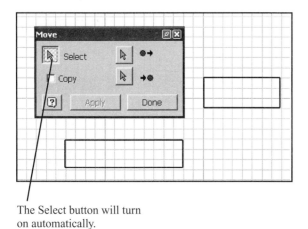

The Select button will turn
on automatically.

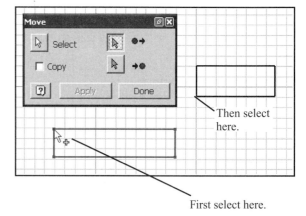

Then select
here.

First select here.

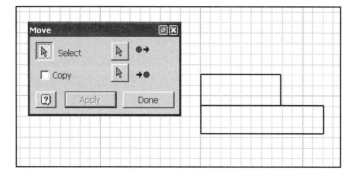

Figure 2-25

2-16 MOVE

The Move command is used to reposition an existing sketch or to copy an existing sketch. Figure 2-25 shows two rectangles. The lower rectangle is to be moved so that it is aligned with the upper rectangle.

To move a sketch

1. Click the Move tool.

The Move dialog box will appear. The Move command is automatically in the Select mode.

2. Select the lower rectangle, press the right mouse button, and select the Continue option.
3. Select the top arrow, then select the upper left corner of the lower rectangle.

The Move command will automatically shift to the second arrow box on the Move dialog box.

4. Select the lower left corner of the upper rectangle.
5. Click the Apply box, then click the Done box.

To copy a sketch

Figure 2-26 shows a rectangle and a line. The line was added to help define the location of the copy. The line will be deleted after the move.

1. Click the Move tool.
2. Click the Copy box.

A check mark will appear.

3. Select the rectangle, then right-click the mouse and select the Continue option.
4. Select the top arrow box, then select the lower left corner of the rectangle.

The Move command will automatically shift to the lower arrow.

5. Select the top endpoint of the line.
6. Click the Apply box, then the Done box.
7. Delete the line.

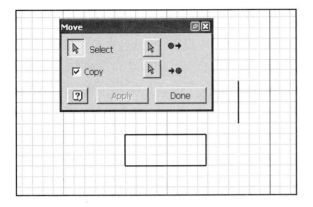
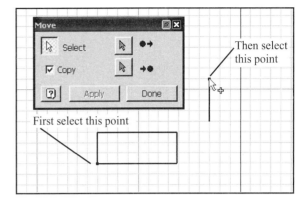
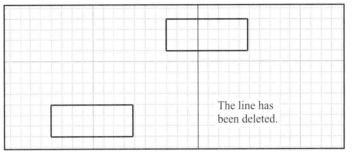

Figure 2-26

2-17 ROTATE

The Rotate command is used to rotate a sketch about a point. Figure 2-27 shows a rectangle. It is to be rotated 35° about its lower left corner.

1. Click the Rotate tool.

The Rotate dialog box will appear.

2. Select the rectangle.
3. Change the angle value to 35.
4. Click the Center Point box.
5. Click the lower left corner of the rectangle.
6. Click the Apply box, then the Done box.

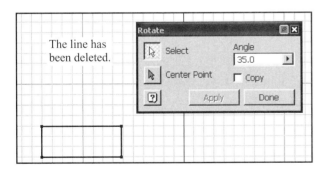
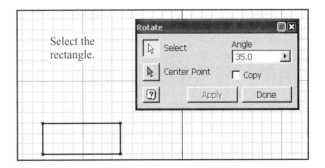
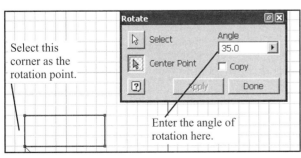
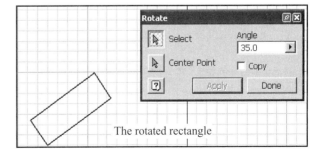

Figure 2-27

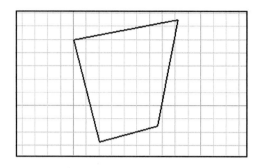
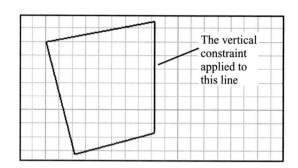
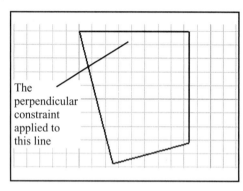
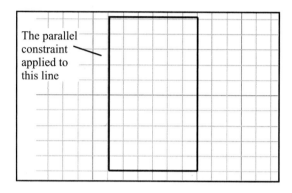

Figure 2-28

2-18 CONSTRAINTS

The Contraints tool contains 11 tools that can be used to change and define the shape of a sketch. The constraint tools are accessed by using the flyouts located above the Show Constraints tool. The 11 constraints are Perpendicular, Parallel, Tangent, Coincident, Concentric, Colinear, Horizontal, Vertical, Equal, Fix, and Symmetric.

To use horizontal, vertical, perpendicular, and parallel constraints

Figure 2-28 shows a sketch that is to be changed into a rectangle.

1. Click the Vertical constraint, then click the right line of the sketch.

    The line will become vertical.

2. Click the Perpendicular contraint, then click the top line of the sketch, then the vertical line.

    The top line will be made perpendicular to the vertical line.

3. Click the Parallel constraint, then click the top horizontal line, then the lower line.

    The lower line will become horizontal or parallel to the top line. The Horizontal contraint could also have been used.

To make two circles concentric

Figure 2-29 shows two circles. They are not concentric.

1. Click the Concentric constraint tool, then each of the circles.

    The circles will become concentric.

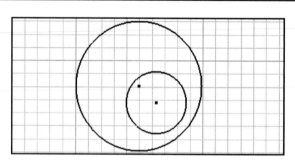
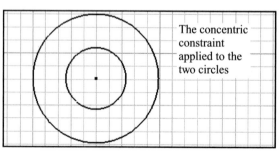

Figure 2-29

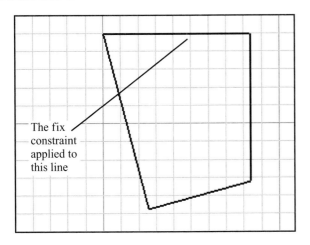 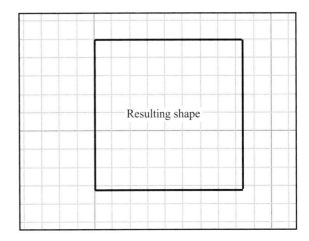

Figure 2-30

To fix a point

Figure 2-30 shows the same sketch presented in Figure 2-28. When the parallel constraint was applied to the lower line in Figure 2-28 the top horizontal line became shortened. This horizontal line can be fixed so that it will remain the same length.

1. Click the Fix constraint.
2. Click the endpoints of the top horizontal line, then right-click the mouse and select the Done option.

   The points will be fixed in their current location.

3. Click the Perpendicular constraint, then the left slanted line, then the fixed horizontal line.

   Note the difference between the sketches in Figures 2-28 and 2-30.

To make a circle tangent to a line

Figure 2-31 shows a line and a circle. The circle will be made tangent to the line.

1. Click the Tangent constraint.
2. Click the line first, then click the circle.

   The circle will move to a point tangent to the line.

2-19 SHOW CONSTRAINTS

The Show Constraints command will show the constraints applied to a sketch. This command is helpful when unknown constraints interfere with the sketching process.

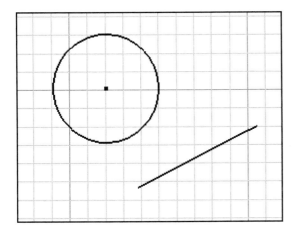 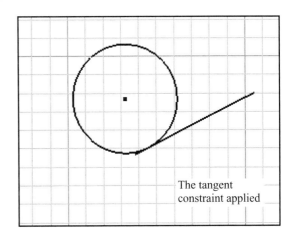

Figure 2-31

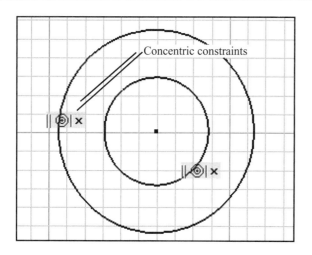

Figure 2-32

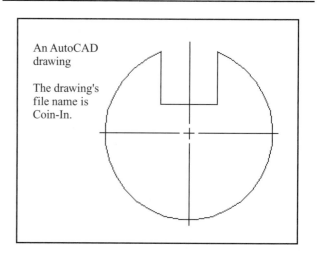

Figure 2-33

Figure 2-32 shows two circles that were constrained to be concentric.

1.   Click the Show Constraints tool.

   The concentric constraint symbols will appear on the screen.

2-20 INSERT AUTOCAD FILE

Sketches created in AutoCAD may be transferred to Inventor. Figure 2-33 shows a drawing created using AutoCAD 2002. The drawing was given the file name Coin-In.

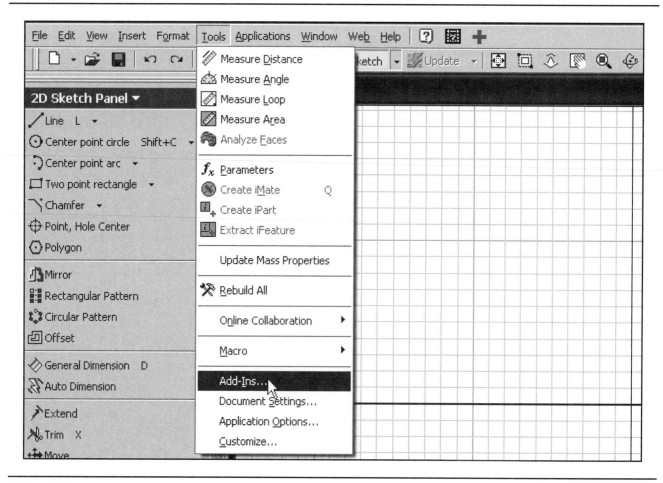

Figure 2-34

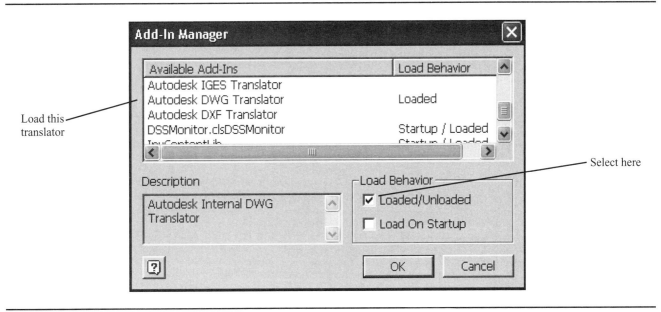

Figure 2-35

The drawing was drawn using inch dimensions.

1. Click the Tools pull-down menu and select the Add-Ins... option.

   See Figure 2-34.
   The Add-In Manager dialog box will appear. See Figure 2-35.

2. Select the Autodesk DWG Translator option and click the Loaded/Unloaded box, then click OK.

3. Click the Open file tool, and access the AutoCAD .dwg files.

   The Open dialog box will appear. See Figure 2-36.

4. Select the appropriate AutoCAD file, then click the Options box.

   The DWG File Import Options dialog box will appear. See Figure 2-37.

5. Set the dialog box as shown in Figure 2-37, then click the Next box.

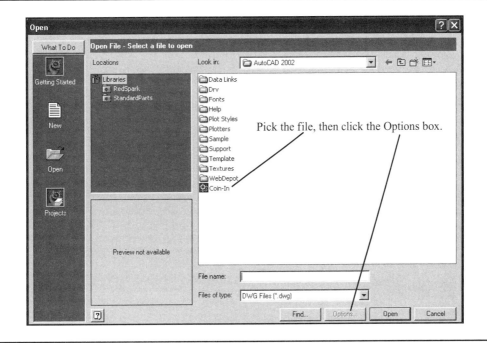

Figure 2-36

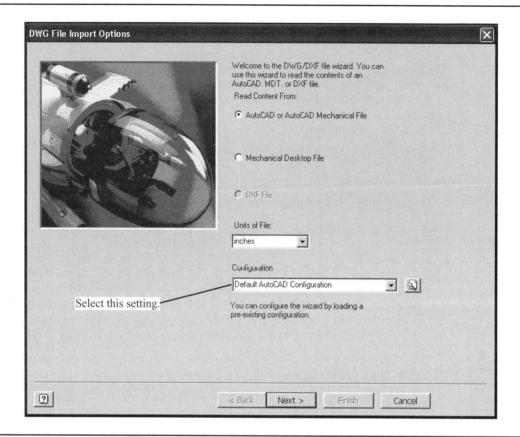

Figure 2-37

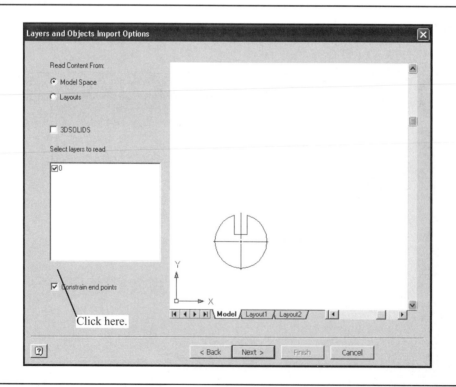

Figure 2-38

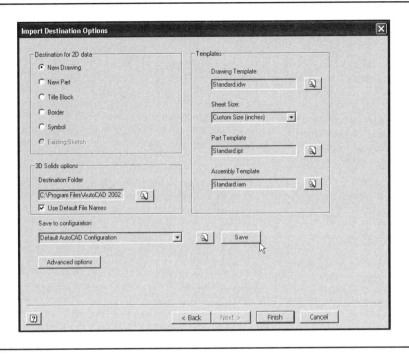

Figure 2-39

The Layers and Objects Import Options dialog box will appear. See Figure 2-38.

6. Click the Constrain end points box, then click the Next box.

The Import Destination Options dialog box will appear. See Figure 2-39.

7. Click the Save box, then click the Finish box.

The Inventor Open dialog box will reappear. See Figure 2-40.

8. Click the Open box.

The Coin-In drawing will appear on the screen. See Figure 2-41.

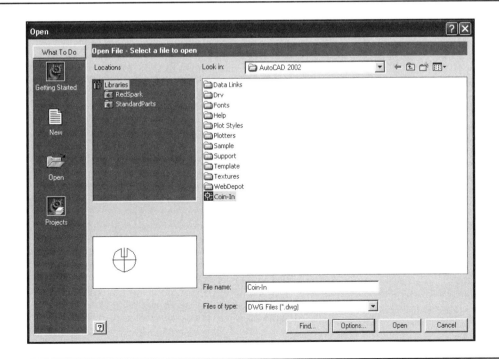

Figure 2-40

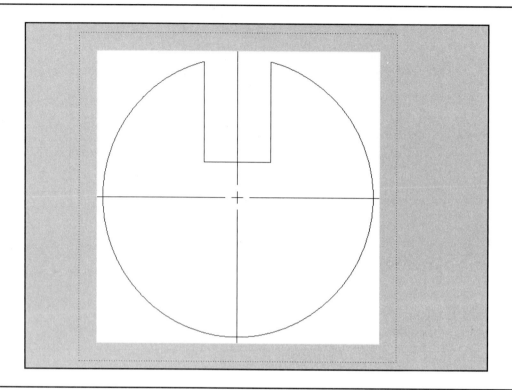

Figure 2-41

2-21 EXERCISE PROBLEMS

Redraw the following objects using the given dimensions. Create solid models of the objects using the specified thicknesses.

EX2-1 INCHES

GUIDE PLATE

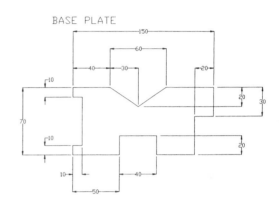

THICKNESS = 1.00

EX2-3 MILLIMETERS

BASE PLATE

THICKNESS = 12

EX2-2 INCHES

TOP GASKET

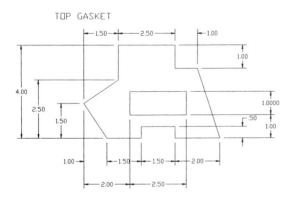

THICKNESS = .625

EX2-4 MILLIMETERS

GASKET

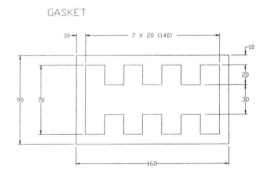

THICKNESS = 2

EX2-5 INCHES

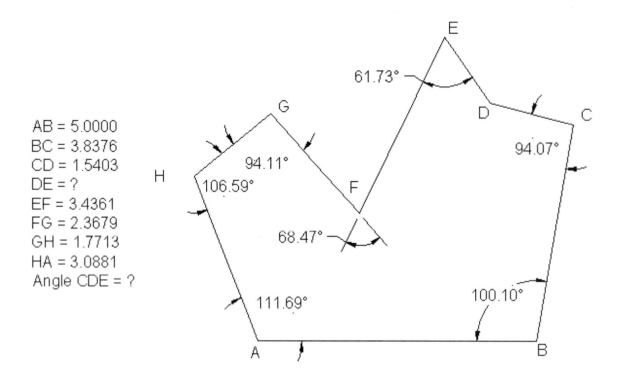

AB = 5.0000
BC = 3.8376
CD = 1.5403
DE = ?
EF = 3.4361
FG = 2.3679
GH = 1.7713
HA = 3.0881
Angle CDE = ?

EX2-6 MILLIMETERS

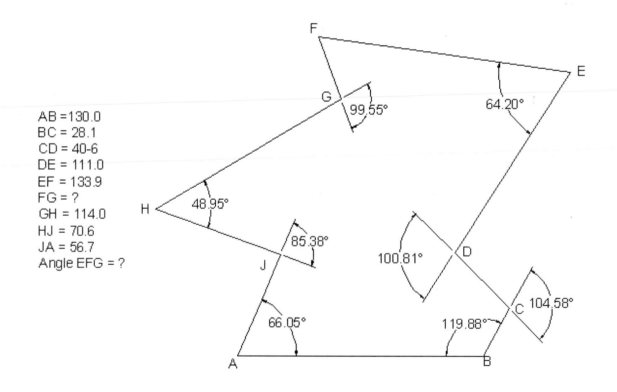

AB = 130.0
BC = 28.1
CD = 40-6
DE = 111.0
EF = 133.9
FG = ?
GH = 114.0
HJ = 70.6
JA = 56.7
Angle EFG = ?

EX2-7 INCHES

SIDE BRACKET

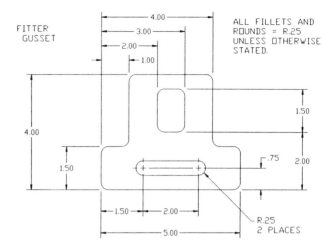

THICKNESS = 1.25

EX2-9 INCHES

THICKNESS = .375

EX2-8 MILLIMETERS

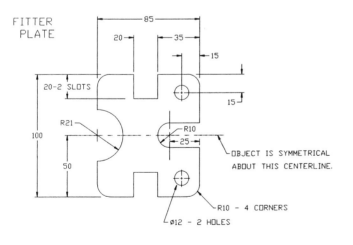

THICKNESS = 8

EX2-10 MILLIMETERS

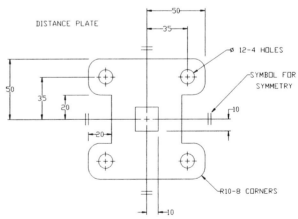

THICKNESS = 16

EX2-11 MILLIMETERS

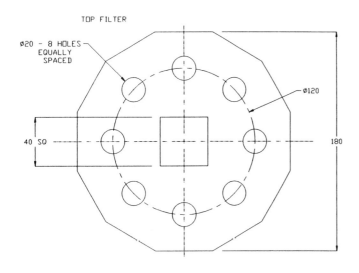

THICKNESS = 6

EX2-13 INCHES

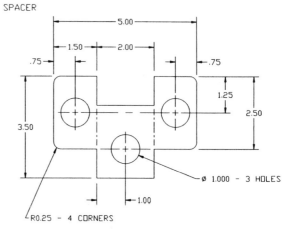

THICKNESS = .75

EX2-12 MILLIMETERS

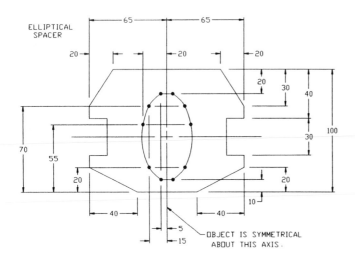

THICKNESS = 10

EX2-14 MILLIMETERS

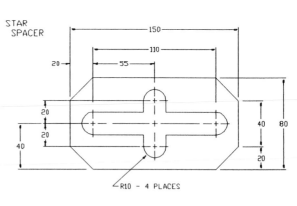

THICKNESS = 7.5

EX2-15 MILLIMETERS

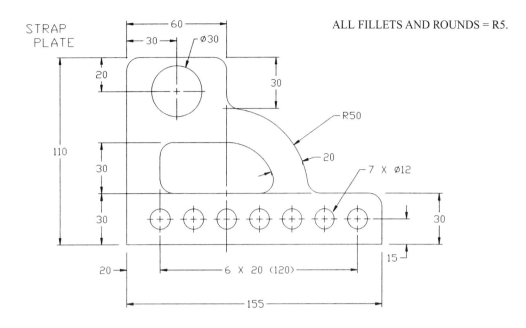

ALL FILLETS AND ROUNDS = R5.

STRAP PLATE

THICKNESS = 5

EX2-16 MILLIMETERS

LACE GASKET

THICKNESS = 6; central area is 4 thick

EX2-17 INCHES

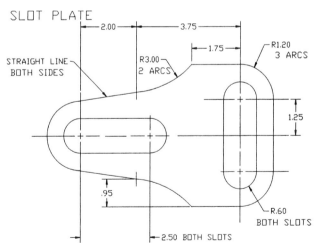

SLOT PLATE

NOTE: Object is symmetrical about its horizontal centerline.

THICKNESS = .875

EX2-18 MILLIMETERS

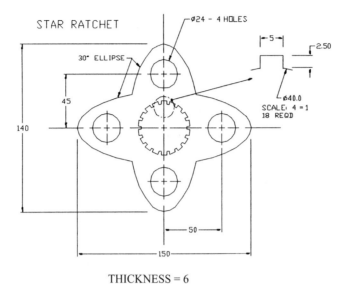

THICKNESS = 6

EX2-19 MILLIMETERS

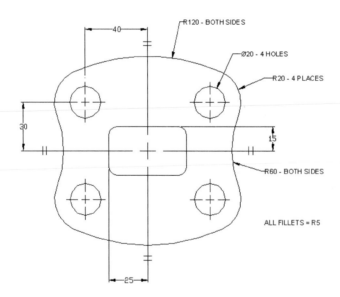

THICKNESS = 7.25

EX2-20 INCHES

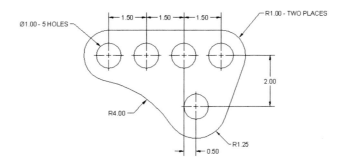

THICKNESS = .25

EX2-21 MILLIMETERS

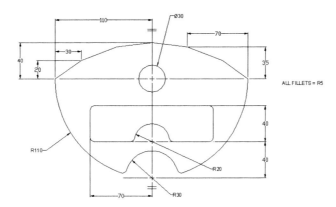

THICKNESS = 5

EX2-22 MILLIMETERS

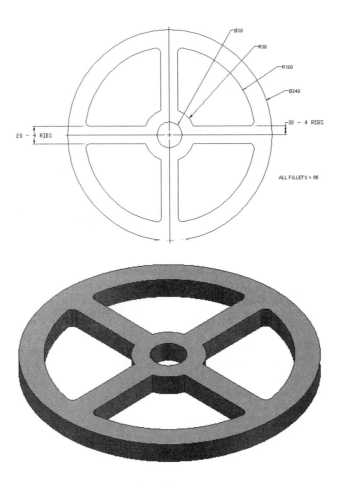

THICKNESS = 10

C H A P T E R **3**

Three-Dimensional Models

3-1 INTRODUCTION

This chapter introduces and demonstrates how to create 3D models using the commands in the Part Features panel bar. These commands are used to convert 2D sketches into 3D solid models and to modify existing models.

The first part of the chapter demonstrates some feature-modifying commands. The second part of the chapter introduces sketch and work planes and shows how they are used to alter and refine 3D models.

To access the Part Features panel bar

1. Click on the Return heading on the Command toolbar.

The Part Features panel bar will appear. See Figure 3-1.

3-2 EXTRUDE

The Extrude command is used to convert 2D sketches into solid models.

Figure 3-2 shows a 12 mm × 30 mm rectangle created using the Standard.ipt format and the 2D Sketch Panel Two point rectangle command.

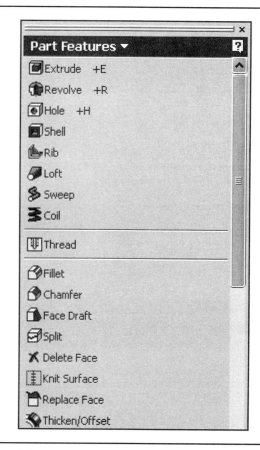

Figure 3-1

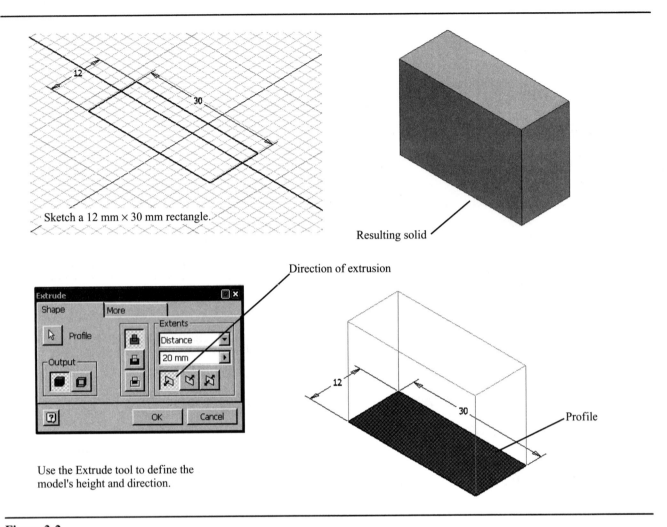

Sketch a 12 mm × 30 mm rectangle.

Resulting solid

Direction of extrusion

Use the Extrude tool to define the
model's height and direction.

Profile

Figure 3-2

1. Click the Return heading, then the Extrude tool on the Part Features panel bar.

   The Extrude dialog box will appear. See Figure 3-2.

2. Change the height distance to 20, then click OK.

   Figure 3-2 shows the extruded rectangle.

To create a taper

1. Click the Extrude tool.
2. Click the More tab on the Extrude dialog box, then set the Taper angle for 15.

3. Click OK.

   Figure 3-3 shows the resulting tapered shape.

To control the direction of the taper

1. Click the Extrude tool.
2. Set the Taper for 15 and click the middle button in the Extents area as shown.
3. Click OK.

   Figure 3-4 shows the resulting tapered shape.

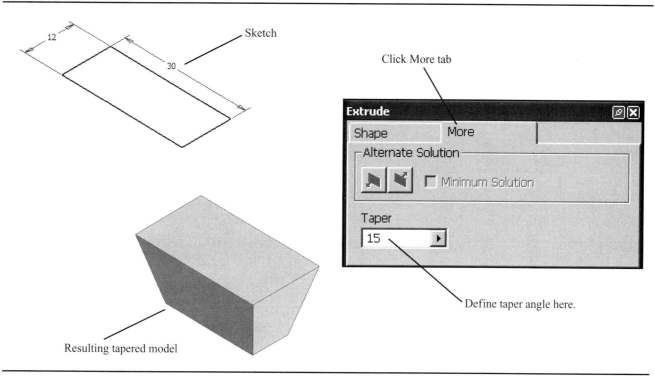

Figure 3-3

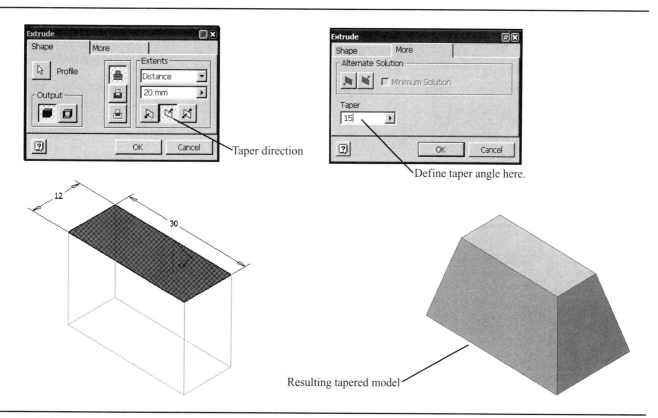

Figure 3-4

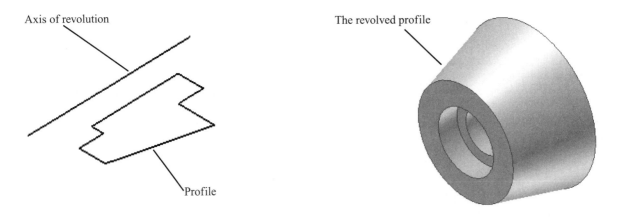

Axis of revolution

Profile

The revolved profile

Figure 3-5

3-3 REVOLVE

Figure 3-5 shows a 2D sketch and a straight line. The sketch will be revolved about the straight line to create a model.

1. Access the Part Features panel bar, then click the Revolve tool.

The Revolve dialog box will appear. The Revolve command will automatically select the 2D shape as the profile to be extruded. If it does not, click the Profile box and window the 2D profile.

2. Select the straight line as the axis.
3. Click OK.

Figure 3-5 shows the resulting revolved model.

The 2D sketch may be revolved through any angle. Figure 3-6 shows the same 2D sketch revolved through 180°.

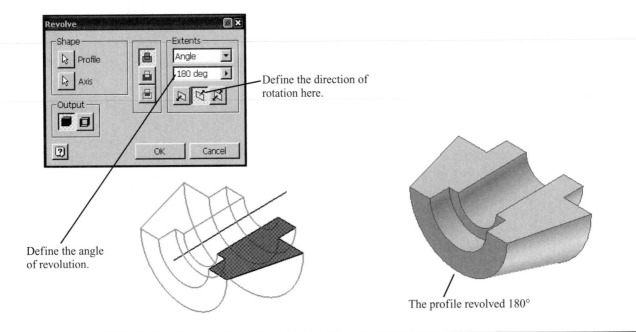

Define the direction of rotation here.

Define the angle of revolution.

The profile revolved 180°

Figure 3-6

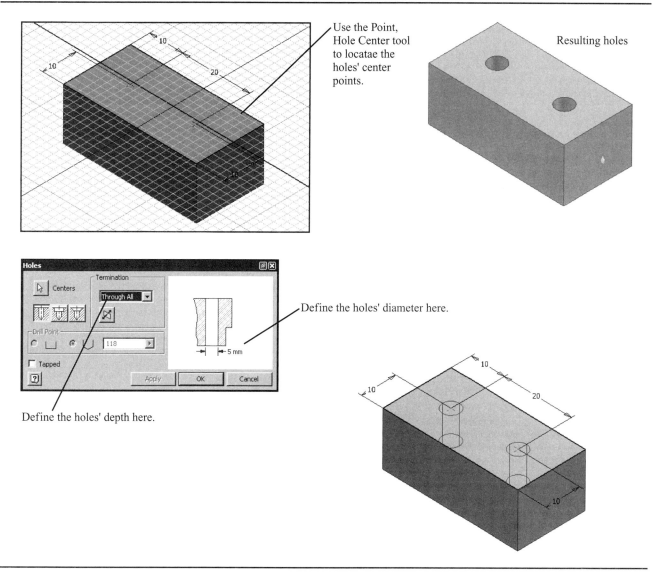

Use the Point, Hole Center tool to locatae the holes' center points.

Resulting holes

Define the holes' diameter here.

Define the holes' depth here.

Figure 3-7

3-4 HOLES

Holes may be added to a 3D solid model by first defining a sketch plane for the hole and then locating the hole's center point. Figure 3-7 shows a model. Two small holes will be located in the top surface, and a large hole will be located in the front surface.

1. Click on the top surface of the model, right-click the mouse, and select the New Sketch option.

   A grid will appear aligned with the top surface. The top surface is now the current sketch plane.

2. Click the Point, Hole Center tool.

3. Locate two hole center points on the top surface, then right-click the mouse and select the Done option.

4. Click the General Dimension tool and accurately locate the two hole center points. See Figure 3-7.

5. Click the Return heading on the Command toolbar to access the Part Features panel bar, then click the Hole tool.

   The Holes dialog box will appear. See Figure 3-7.

6. Set the Termination for Through All and set the holes' diameter for 5 mm, then click OK.

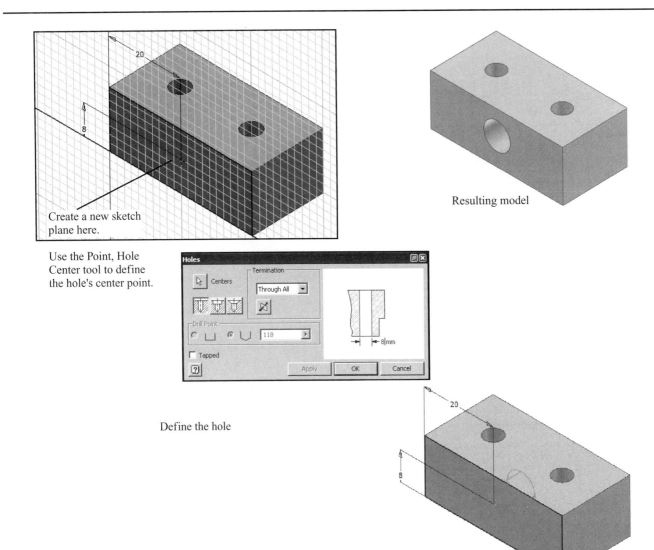

Create a new sketch plane here.

Use the Point, Hole Center tool to define the hole's center point.

Define the hole

Resulting model

Figure 3-8

To locate a hole in the front plane

1. Click the front surface, right-click the mouse, and select the New Sketch option.
2. Click the Sketch heading on the Command tool-bar.

   The 2D Sketch Panel bar will appear. See Figure 3-8.

3. Use the Point, Hole Center and General Dimension tools and locate a center point in the center of the surface.
4. Click the Sketch heading on the Command tool-bar and return the Part Features tools.
5. Click the Hole tool and add a Ø8 hole through the model.
6. Click OK.

3-5 SHELL

The Shell command is used to create thin-walled objects from existing models. Figure 3-9 shows a 12 × 30 × 20 model.

1. Click the Shell tool on the Part Features panel bar.

   The Shell dialog box will appear. There are three different ways to define a shell, which are accessed by the three boxes on the right side of the Shell dialog box. The options are as follows:

   Inside: The external wall of the existing model will become the external wall of the shell.

   Outside: The external wall of the existing model will become the internal wall of the shell.

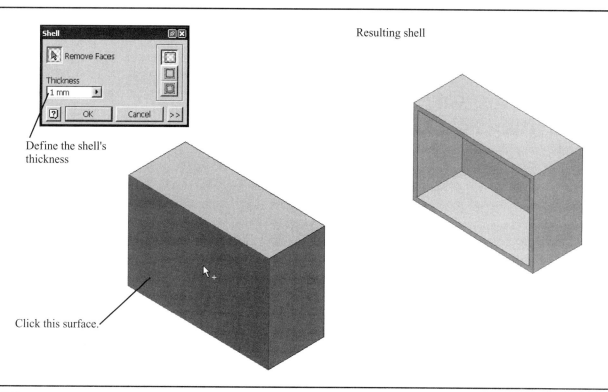

Define the shell's thickness

Resulting shell

Click this surface.

Figure 3-9

Both sides: The existing outside wall will become the center of the shell; half the thickness will be added to the outside and half to the inside.

2. Click on the front surface of the model, then click OK.

Shells may be created from any shaped model. Figure 3-10 shows a cone that has been used to create a hollow thin-walled cone.

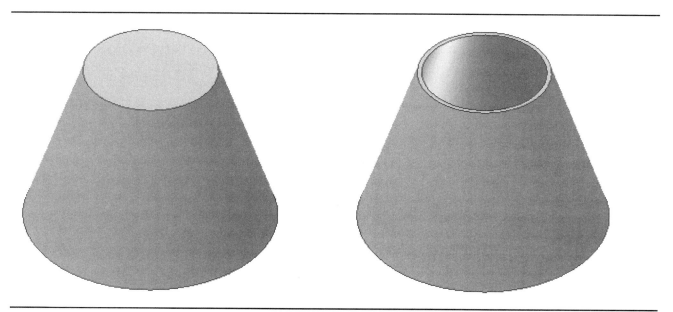

Figure 3-10

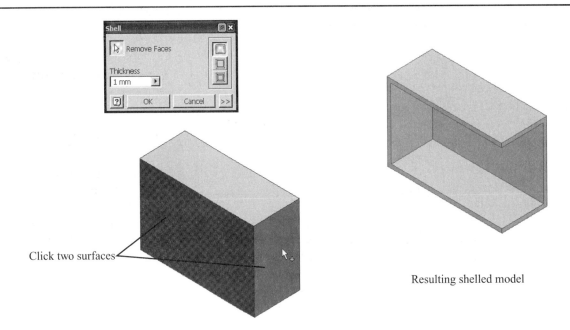

Click two surfaces

Resulting shelled model

Figure 3-11

To remove more than one surface

1. Click the Shell tool.
2. Click the surfaces to be removed.
3. Select the Inside option.
4. Click OK.

   See Figure 3-11.

3-6 FILLET

The Fillet command is used to create rounded edges. See Figure 3-12.

1. Click the Fillet tool on the Part Features panel bar.

   The Fillet dialog box will appear.

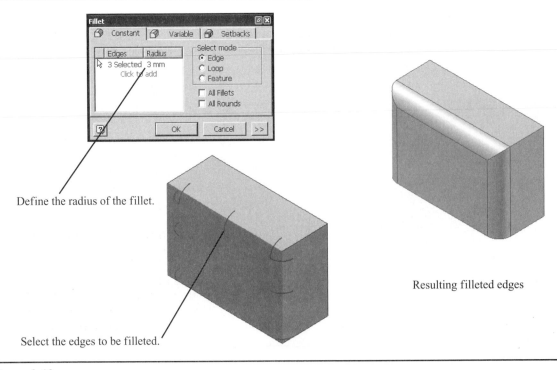

Define the radius of the fillet.

Resulting filleted edges

Select the edges to be filleted.

Figure 3-12

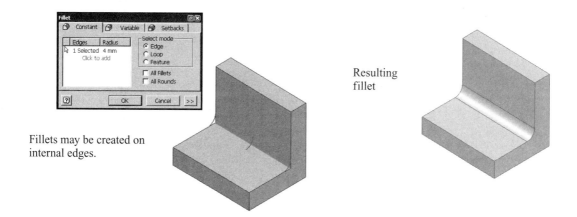

Fillets may be created on internal edges.

Resulting fillet

Figure 3-13

2. Change the Radius value to 3.
3. Click the Edges option and select the edges to be filleted.
4. Click OK.

A fillet may be added to an internal edge such as shown in Figure 3-13. Internal fillets are called *rounds*.

3-7 CHAMFER

The Chamfer command is used to create beveled edges. See Figure 3-14. Chamfers are defined by specifying linear setback distances or by specifying a setback distance and an angle. Most chamfers have equal setback distances or an angle of 45°.

1. Click the Chamfer tool.

The Chamfer dialog box will appear. The first option box on the left side of the Chamfer dialog box is used to create chamfers with equal distances.

2. Set the distance for 1.
3. Select the edges to be chamfered.
4. Click OK.

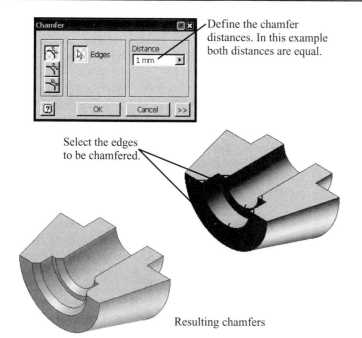

Define the chamfer distances. In this example both distances are equal.

Select the edges to be chamfered.

Resulting chamfers

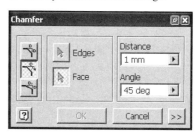

A chamfer defined by a distance and an angle

A chamfer defined by two distances.

Figure 3-14

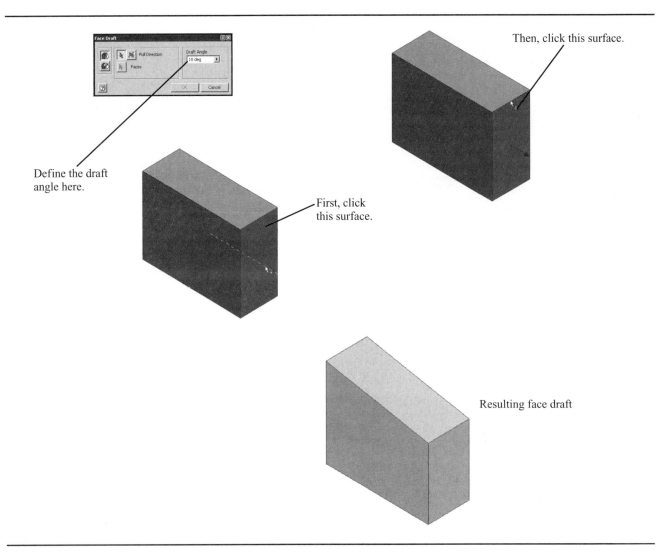

Define the draft angle here.

First, click this surface.

Then, click this surface.

Resulting face draft

Figure 3-15

3-8 FACE DRAFT

The Face Draft tool is used to create angled surfaces. See Figure 3-15.

1. Click the Face Draft tool.

   The Face Draft dialog box will appear.

2. Click the right front surface of the model.
3. Click the top surface of the model.
4. Change the Draft Angle value to 10.
5. Click OK.

Several surfaces can be drafted at the same time. See Figure 3-16.

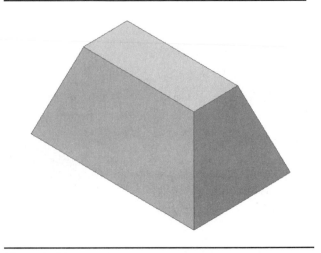

Figure 3-16

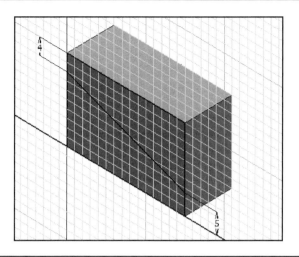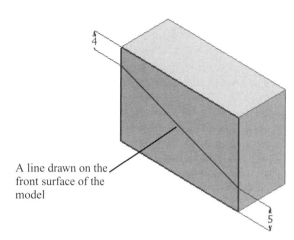

A line drawn on the front surface of the model

Figure 3-17

3-9 SPLIT

The Split command is used to trim away a portion of a model. See Figure 3-17. A sketch line is used to define the location and angle of the split.

To define the split line

1. Click the left front surface of the model.

   The surface will change color.

2. Right-click the mouse and select the New Sketch option.

   A grid will appear on the screen oriented to the selected face.

3. Click the Line tool and sketch a line across the front left surface.
4. Right-click the mouse and select the Done option.

5. Use the General Dimension tool to locate the line as needed.

To split the model

1. Click the Return box on the Command toolbar to access the Part Features panel bar, or right-click the mouse and select the finish Sketch option.

   The Split dialog box will appear. See Figure 3-18.

2. Click the Split Tool box.
3. Click the sketch line.
4. Select the Split part box under Method in the Split dialog box.
5. Use the Remove option to define which side of the model is to be removed.
6. Click OK.

   Figure 3-19 shows a split that was created using a sketched circle.

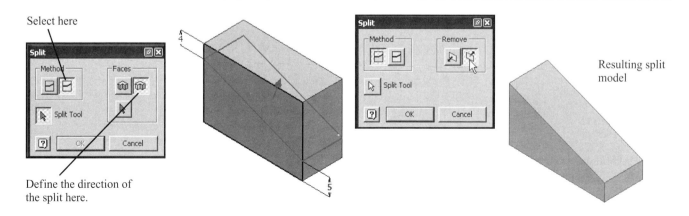

Select here

Define the direction of the split here.

Resulting split model

Figure 3-18

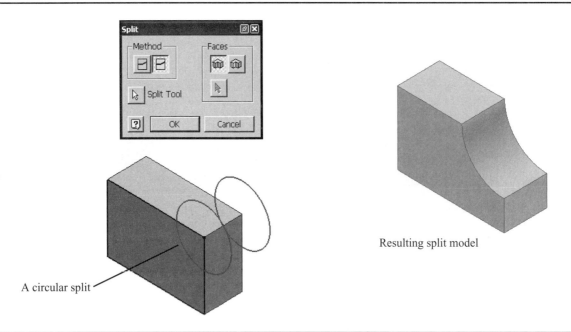

Resulting split model

A circular split

Figure 3-19

3-10 MIRROR

The Mirror command is used to create mirror images of an existing model. See Figure 3-20.

1. Click the Mirror tool.

    The Mirror Pattern dialog box will appear.

2. Click the model.
3. Click the Mirror Plane box.
4. Select one of the model's surfaces as a mirror plane.
5. Click OK.

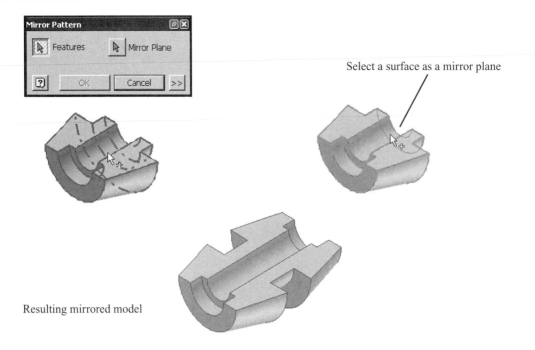

Select a surface as a mirror plane

Resulting mirrored model

Figure 3-20

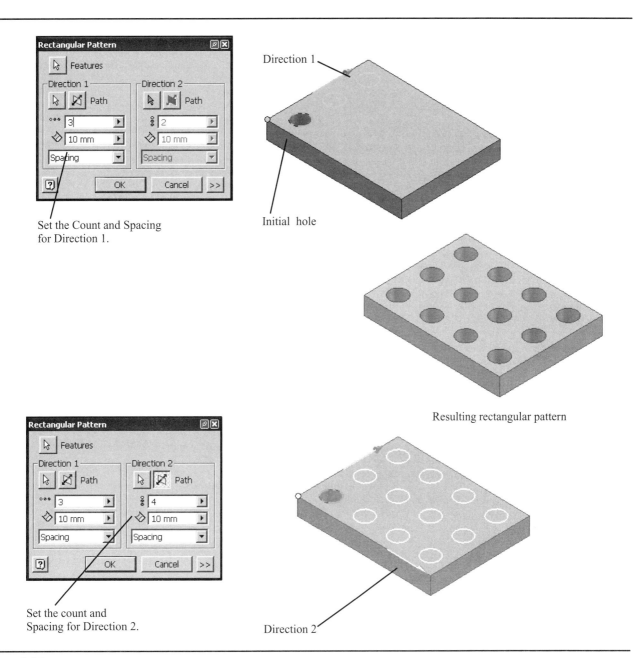

Set the Count and Spacing for Direction 1.

Direction 1

Initial hole

Resulting rectangular pattern

Set the count and Spacing for Direction 2.

Direction 2

Figure 3-21

3-11 RECTANGULAR PATTERN

The Rectangular Pattern command is used to create a rectangular array of an existing model feature. Figure 3-21 shows a 30 40 × 5 plate with a Ø5 hole located 5 mm from each edge.

1. Click the Rectangular Pattern tool located on the Part Features panel bar.

The Rectangular Pattern dialog box will appear. The Features box will automatically be active.

2. Click the hole.
3. Click the Direction 1 box, then click the back left edge of the model to define direction 1.

Use the Direction 1 box to reverse the direction if necessary.

4. Set the count value for 3 and the Spacing value for 10.
5. Click the Direction 2 box, then click the front left edge of the model to define direction 2.
6. Set the count for 4 and the Spacing for 10.
7. Click OK.

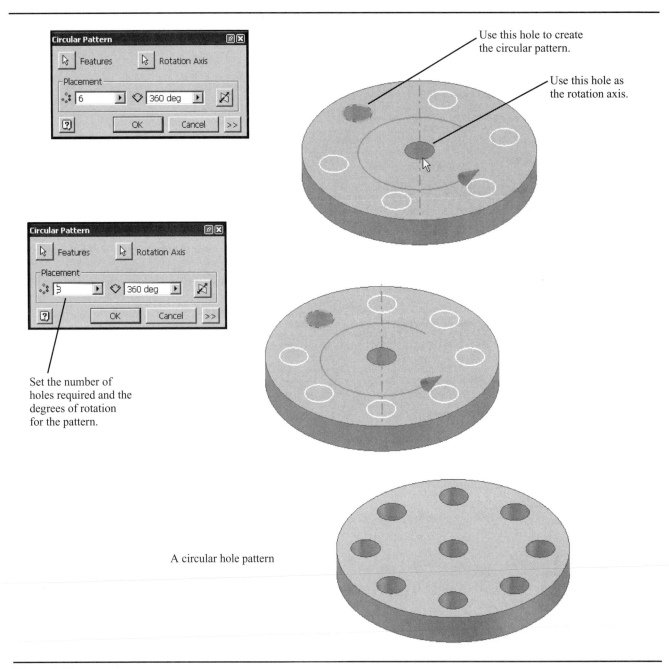

Figure 3-22

3-12 CIRCULAR PATTERN

The Circular Pattern command is used to create a polar array of an existing model feature. Figure 3-22 shows a Ø40 cylinder 5 mm high with two Ø5 holes. One hole is located in the center of the model; the second is located 15 mm from the center.

1. Click the Circular Pattern tool located on the Part Features panel bar.

The Circular Pattern dialog box will appear. The Features box will automatically be active.

2. Click the hole to be used to create the circular pattern.
3. Click the Rotation Axis button, then click the center hole.
4. Set the count value for 8 and the angle value for 360.
5. Click OK.

A model created using several
different sketch planes

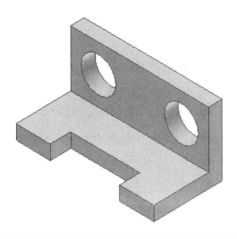

Figure 3-23

3-13 SKETCH PLANES

Sketches are created on *sketch planes*. Any surface on a model may become a sketch plane. As models become more complex they require the use of additional sketch planes.

Figure 3-23 shows a model that was created using several different sketch planes. The model is a composite of basic geometric shapes added to one another.

To create the base

1. Start a new drawing using the metric Standard (mm).ipt settings.
2. Click the Two point rectangle tool and sketch a 10 × 20 rectangle.

   See Figure 3-24. The object will automatically be drawn on a sketch plane aligned with the program's XY plane.

3. Click the right mouse button and select the Isometric View option.

   Use the mouse wheel to zoom the rectangle as necessary.

4. Click the Sketch heading on the Command toolbar, then click the Extrude tool.

   The Extrude dialog box will appear.

5. Set the extrusion height for 2 mm and click OK.

To create the vertical portion

The rectangular vertical back portion of the model will be created by first defining a new sketch plane on the top surface of the base, then sketching and extruding a rectangle that will be joined to the existing base. See Figure 3-25.

1. Click the top surface of the base.

   The surface will change color, indicating that it has been selected.

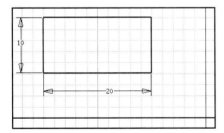

Sketch a 10 × 20 rectangle.

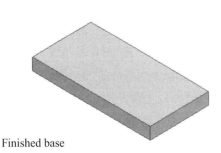

Finished base

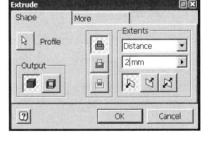

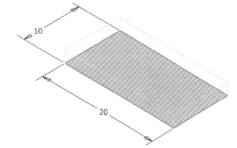

Extrude the rectangle
to a height of 2 mm.

Figure 3-24

Click the top surface.

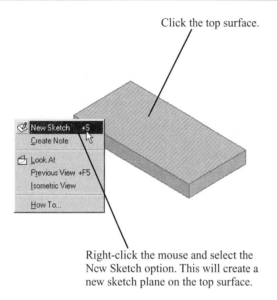

New Sketch   +S
Create Note

Look At
Previous View +F5
Isometric View

How To...

Right-click the mouse and select the
New Sketch option. This will create a
new sketch plane on the top surface.

Sketch plane on the top surface

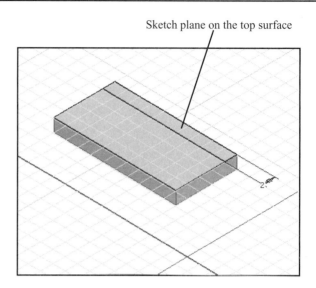

Figure 3-25

2. Right-click the mouse and select the New Sketch
   option.

   A new grid pattern will appear aligned with the top
surface of the base. This is a new sketch plane.

3. Click the Two point rectangle tool and sketch a
   2 × 20 rectangle on the top surface so that it is
   aligned with the back edge of the base.
4. Click the Sketch heading on the Command toolbar and click the Extrude tool.
5. Select the 2 × 20 rectangle and set the extrusion
   height for 8, then click OK.

   Note that the surfaces are unioned together to form
one object.

To add holes to the vertical surface

1. Click the front edge of the vertical surface.

   The surface will change color, indicating that it has
been selected.

2. Press the right mouse button and select the New
   Sketch option.

A grid will appear on the surface. This is a new sketch
plane. The holes are located 4 mm from the top edge and
from each of the side edges. See Figure Figure 3-26.

3. Use the Point, Hole Center and the General Dimension tools to locate the center points for the
   two holes.
4. Right-click the mouse and select the Done option.
5. Click the Sketch heading on the Command toolbar, then click the Hole tool.

   The Holes dialog box will appear.

6. Set the Termination for Through All and the diameter value for 5.
7. Click OK.

To create the cutout

1. Create a new sketch plane on the top surface of
   the base.

   The cutout is 3 deep with edges 5 from each end of the
model.

2. Use the Two point rectangle and General Dimension tools to define the cutout's size.

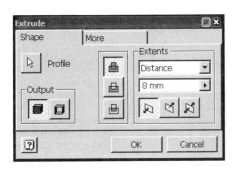

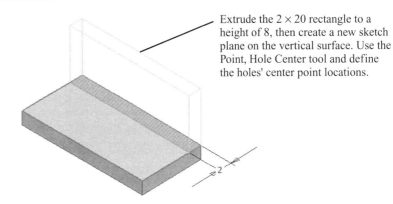

Extrude the 2×20 rectangle to a height of 8, then create a new sketch plane on the vertical surface. Use the Point, Hole Center tool and define the holes' center point locations.

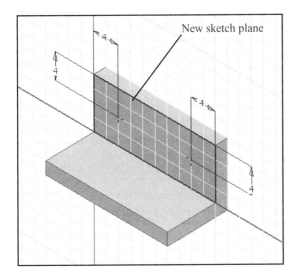

New sketch plane

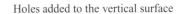

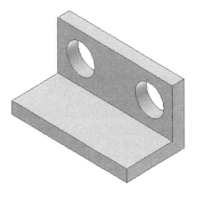

Holes added to the vertical surface

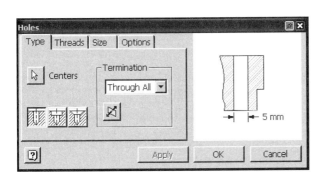

Use the Holes tool to create the holes.

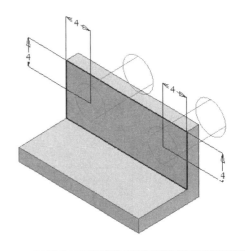

Figure 3-26

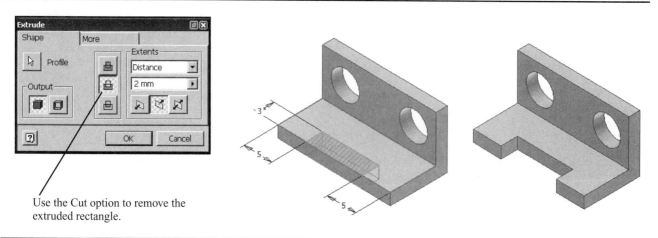

Use the Cut option to remove the extruded rectangle.

Figure 3-27

3. Click the Sketch heading on the Command toolbar, then click the Extrude tool.

   The Extrude dialog box will appear. See Figure 3-27.

4. Select the cutout rectangle and set the extrusion distance for 2, the direction arrow for a direction into the model, and select the Cut option.

3-14 DEFAULT PLANES AND AXES

Inventor includes three default planes and three default axes. The three default planes are YZ, XZ, and XY, and the three axes are X, Y, and Z. The default planes and axes tools are accessed through the browser. See Figure 3-28.

1. Click the + sign to the left of the Origin heading.

   The default planes and axes headings will cascade down.

To display the default planes and axes

Figure 3-29 shows a Ø30 × 16 cylinder that was drawn with its center point on the 0,0,0 origin. The base of the cylinder is on the XY plane. Inventor sketches are automatically created on the default XY axis.

1. Move the cursor onto the XY Plane tool.

   A plane outline will appear on the screen.

2. Click the XY Plane tool.

   The plane will be filled with color.

3. Move the cursor to the XZ Plane tool.

   The XZ plane will be outlined.

4. Move the cursor to the Z Axis tool.

   The Z axis will appear.

5. Move the cursor through all the tools and note the planes and axes that appear.

3-15 WORK PLANES

Work planes are planes used for sketching, but unlike sketch planes, work planes are not created using the surfaces of models. Work planes are created independent of the

The browser

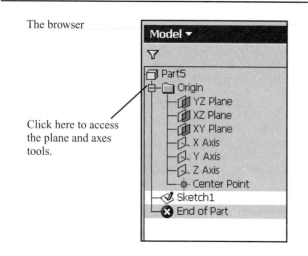

Click here to access the plane and axes tools.

Figure 3-28

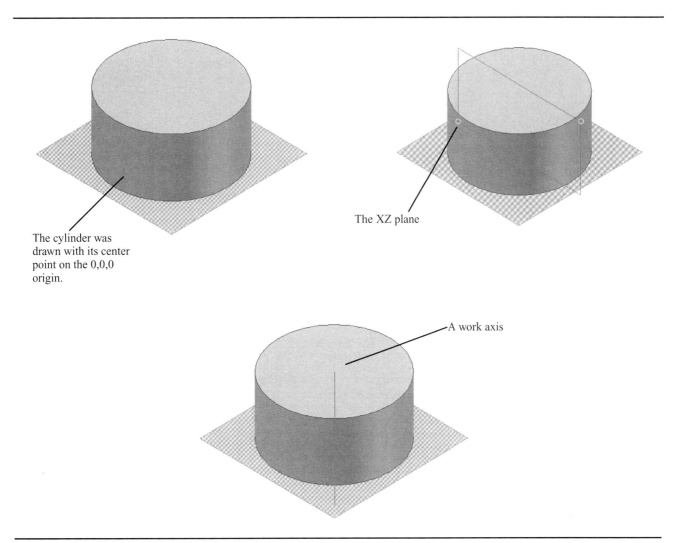

The cylinder was drawn with its center point on the 0,0,0 origin.

The XZ plane

A work axis

Figure 3-29

model. Work planes may be created outside or within the body of a model. Work planes are used when no sketch plane is available.

Work planes may be defined using the following parameters:

Angled
Edge and Face Normal
Edge and Tangent
Offset
Point and Face Normal
Point and Face Parallel
Sketch Geometry
Tangent and Face Parallel
Tangent
3-Point
2-Edge or 2-Axis
Through Line End Point Perpendicular to Line

Work plane help

If you are not sure how to create a work plane, Inventor includes an animated help feature.

1. Click the Work Plane tool, then right-click the mouse.

   A dialog box will appear. See Figure 3-30.

2. Click the How to… option.

   A Show me dialog box will appear.

3. Select the desired set of work plane parameters, and an animated video will demonstrate how to create a work plane for the selected parameters.

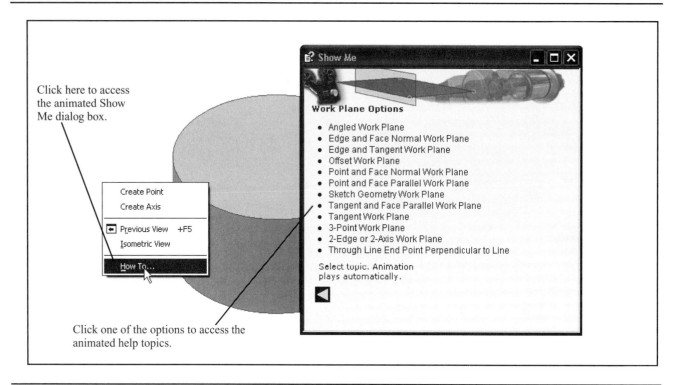

Figure 3-30

Sample problem

Figure 3-31 shows a Ø20 × 10 cylinder that was sketched but not aligned with the system's origin. The sketch was created on the default XY plane.

Create a Ø4 hole through the cylinder so that its centerline is parallel to the XY plane and 5 above the plane.

The sides of the cylinder cannot be used as a sketch plane, so a work plane is needed. Two different methods will be presented to create a work plane tangent to the edge of the cylinder.

To create a tangent work plane – Method 1

1. Click the Work Plane tool.
2. Click the YZ Plane tool in the browser area.

   A YZ plane will appear on the screen. See Figure 3-32.

3. Move the cursor and click the lower outside edge of the cylinder.

   A work plane will be created tangent to the cylinder.

4. Right-click the mouse and select the New Sketch option.
5. Move the cursor to one of the work plane's corner points.

   A small circle will appear.

6. Click the circle.

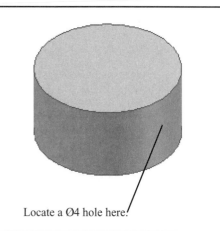

Locate a Ø4 hole here.

Figure 3-31

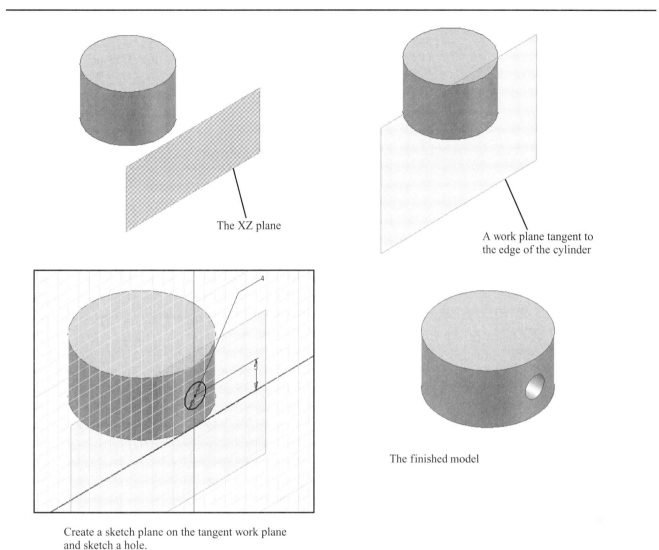

The XZ plane

A work plane tangent to the edge of the cylinder

Create a sketch plane on the tangent work plane and sketch a hole.

The finished model

Figure 3-32

A grid will appear. The grid will include intersecting horizontal and vertical lines that are darker than the other grid lines. The horizontal line is aligned with the XY plane, and the vertical line is parallel to the YZ plane tangent to the edge of the cylinder.

To create the hole through the cylinder

1. Click the Circle tool.
2. Sketch a hole with its center point located on the darker vertical line.

3. Use the General Dimension tool to create a Ø4 circle with a center point located 5 from the top surface of the cylinder.
4. Click the Sketch heading on the Command toolbar, then click the Extrude tool.

   The Extrude dialog box will appear.

5. Set the extrusion distance for 20 in a direction that passes through the cylinder, and select the Cut option.
6. Click OK.

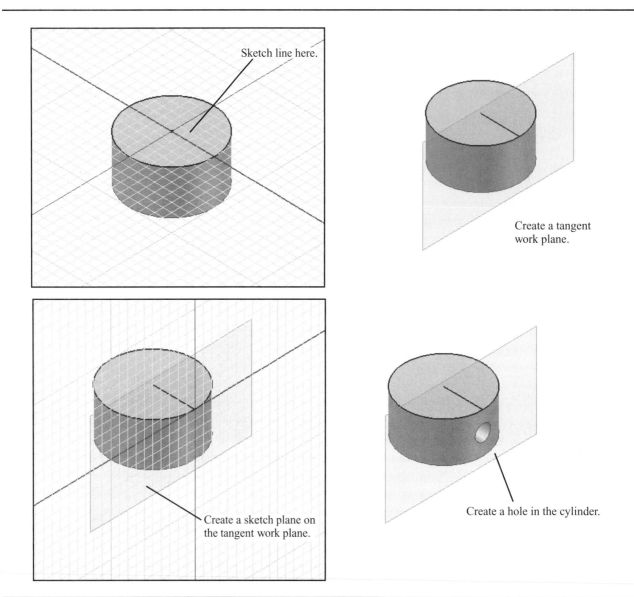

Sketch line here.

Create a tangent work plane.

Create a sketch plane on the tangent work plane.

Create a hole in the cylinder.

Figure 3-33

To create a tangent work plane – Method 2

Figure 3-33 shows a Ø20 × 10 cylinder.

1. Create a new sketch plane on the top surface of the cylinder.
2. Use the Line tool and draw a line from the cylinder's center point to its edge.
3. Click the Sketch heading on the Command toolbar to access the Work Plane tool.

4. Click the Work Plane tool, then click the endpoint of the line that is aligned with the edge of the cylinder.

A work plane will appear tangent to the outside edge of the cylinder.

5. Click one of the work plane's corner points, then right-click the mouse.
6. Select the New Sketch option.
7. Create the through hole as explained previously.

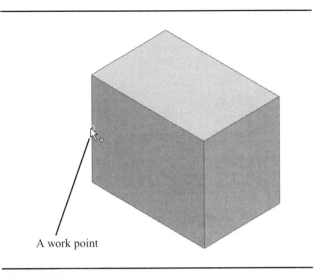

A work point

Figure 3-34

3-16 WORK POINTS

Work points are defined points on a model. They are used to help locate work planes and work axes.

To define a work point

1. Click the Work Point tool.
2. Select the location for the work point and click the mouse.

In the example shown in Figure 3-34 the midpoint of the left edge was selected. The browser box will list the work point once it is created.

To create an oblique work plane using work points

An oblique work plane may be created using work points. Figure 3-35 shows a 20 × 30 × 24 rectangular prism.

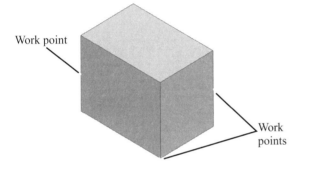

Work point

Work points

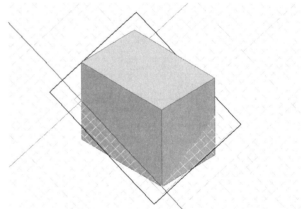

A sketch plane created on the work plane, and a rectangle sketched on the sketch plane

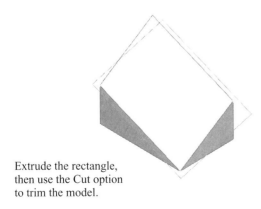

Extrude the rectangle, then use the Cut option to trim the model.

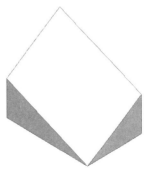

An oblique surface

Figure 3-35

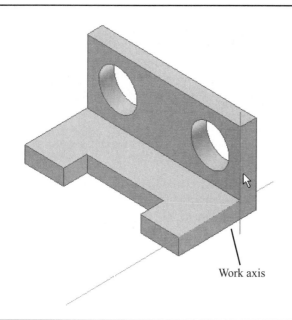

Work axis

Figure 3-36

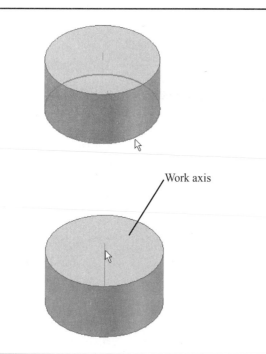

Work axis

Figure 3-37

1. Create three work points on the prism, two on the midpoints of the vertical egdes and one at the lower corner as shown in Figure 3-35.
2. Click on the Work Plane tool, then click the three work points.
3. Left-click one of the work plane's corner points, right-click the mouse, and select the New Sketch option.
4. Click the Two point rectangle tool and draw a very large rectangle on the new sketch plane.

The rectangle may be any size that exceeds the size of the prism.

5. Access the Extrude tool and cut out the top portion of the prism.

3-17 WORK AXES

A *work axis* is a defined edge. Work axes are used to help define work planes and to help define the geometric relationship between assembled models.

To create a work axis

1. Click the Work Axis tool.
2. Click the edge line that is to be defined as a work axis.

Figure 3-36 shows a model with two defined work axes.

To draw a work axis at the center of a cylinder

Figure 3-37 shows a cylinder.

1. Click the Work Axis tool.
2. Click the lower edge of the cylinder.

The words Work Axis 1 will appear in the browser.

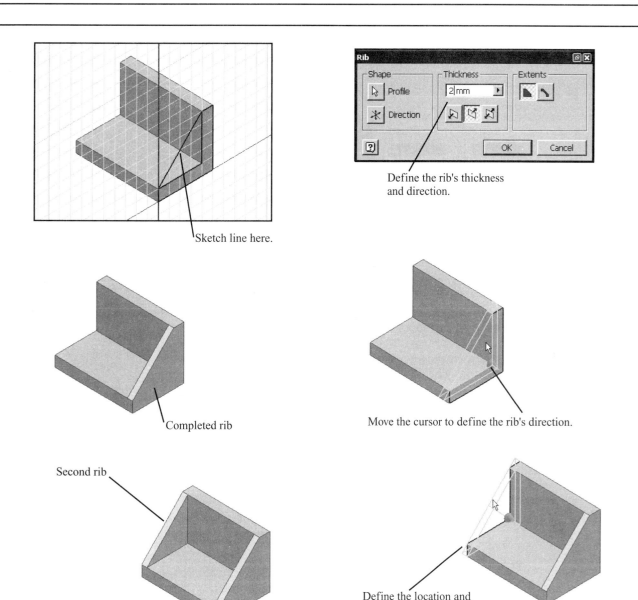

Sketch line here.

Define the rib's thickness and direction.

Completed rib

Move the cursor to define the rib's direction.

Second rib

Define the location and direction for a second rib.

Figure 3-38

3-18 RIBS (WEBS)

A *rib* is used to add strength to a model. Ribs or webs are typically used with cast or molded parts. Figure 3-38 shows an L-bracket. Ribs 1 mm thick are to be added to each end of the bracket.

1. Click the right end surface of the bracket, click the right mouse button, and select the New Sketch option.

2. Use the Line tool and add a line as shown, then access the Part Features menu.
3. Click the Rib tool.
4. Select the line as the Profile, set the thickness value for 1 mm, select the middle thickness box, click the Direction box on the Rib dialog box and define the rib's direction as shown. Click OK.

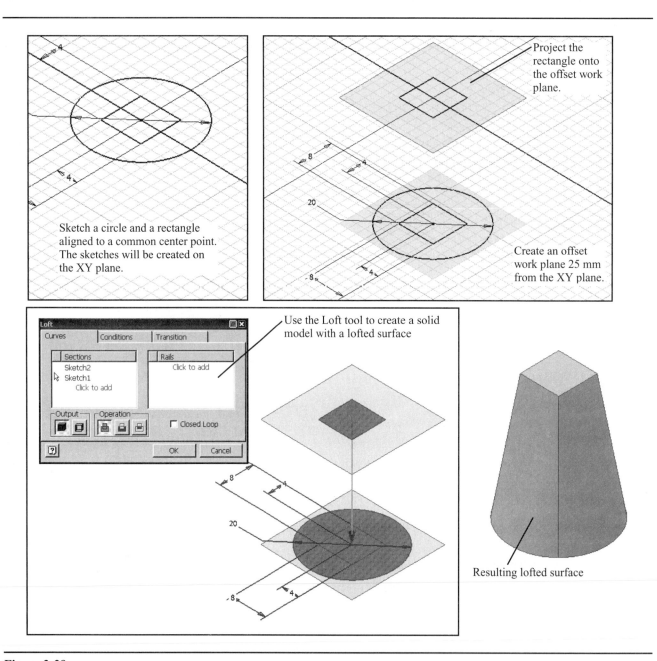

Sketch a circle and a rectangle aligned to a common center point. The sketches will be created on the XY plane.

Project the rectangle onto the offset work plane.

Create an offset work plane 25 mm from the XY plane.

Use the Loft tool to create a solid model with a lofted surface

Resulting lofted surface

Figure 3-39

3-19 LOFT

The Loft command is used to create a solid between two or more sketches. Figure 3-39 shows a loft surface created between a circle and a rectangle. Both the circle and the rectangle are first drawn on the same XY plane. This allows the General Dimension tool to be used to ensure the alignment between the two sketches. The rectangle is then projected onto another work plane. The Loft tool is then used to create a surface between the two planes.

To sketch the circle and rectangle

1. Sketch a Ø20 circle and a 6 × 8 rectangle aligned to a common center point. Use the Fix constraint if necessary.
2. Create a work plane aligned with the XY axis.
3. Create a second work plane offset 25 mm from the first XY plane.

To create an offset work plane

1. Click the Work Plane tool, then click the XY Plane tool in the browser area (see Figure 3-28).

A new plane will appear.

2. Click one of the corner points of the new plane and move the cursor upward.

An Offset dialog box will appear.

3. Set the offset distance for 25 and click the check mark on the dialog box.

Check the browser area to verify that two work planes have been created.

To project the rectangle

1. Click one of the corner points of the offset work plane, right-click the mouse and select the New Sketch option.
2. Select the Project Geometry tool on the 2D Sketch Panel.
3. Select the rectangle.

The rectangle will be projected into the offset work plane.

4. Right-click the mouse and select the Done option.

To create a loft

1. Access the Part Features panel bar and select the Loft option.

The Loft dialog box will appear.

2. Click in the Sections area of the Loft dialog box, then click the rectangle.
3. Click the Sections area of the Loft dialog box again and double-click the circle.
4. Click OK.

Hide the work planes if desired.

3-20 SWEEP

The Sweep tool is used to project a sketch along a defined path. In this example a shape is created in the XZ plane and then projected along a path drawn in the YZ plane. See Figure 3-40.

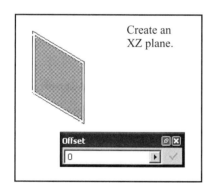

Create an XZ plane.

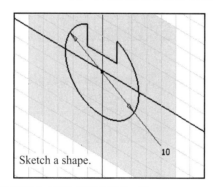

Sketch a shape.

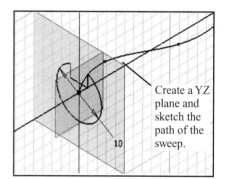

Create a YZ plane and sketch the path of the sweep.

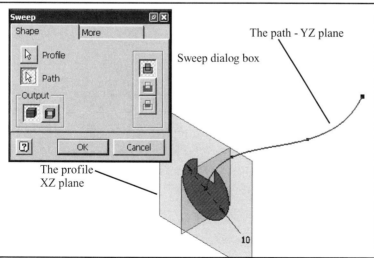

Sweep dialog box

The profile XZ plane

The path - YZ plane

Figure 3-40

To create the sketch

1. Click the Work Plane tool, then the XZ Plane tool in the browser area.

   A plane will appear.

2. Click one of the plane's corners and offset the work plane 0.000, then click the check mark on the Offset dialog box.
3. Right-click one of the plane's corner points and select the New Sketch option.
4. Sketch a Ø10 circle with a 4 × 3 keyway as shown.

To create the path

1. Click the Work Plane tool, then the YZ Plane tool in the browser.
2. Click one the plane's corner points and create a second work plane offset 0.000 mm.

3. Right-click one of the YZ work plane's corner points and select the New Sketch option.
4. Click the Spline tool and sketch a spline, click the right mouse button and select the Continue option, then right-click the mouse again and select the Done option.

   In this example a random spline was used.

To create the sweep

1. Click the Sweep tool.

   The Sweep dialog box will appear. The circular sketch will automatically be selected as the profile.

2. Click the spline to define it as the path.
3. Click OK.

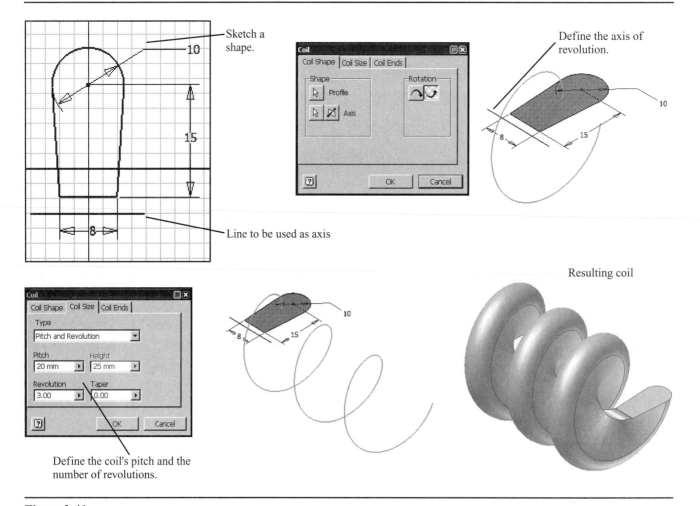

Sketch a shape.

Line to be used as axis

Define the axis of revolution.

Resulting coil

Define the coil's pitch and the number of revolutions.

Figure 3-41

3-21 COIL

A coil is similar to a sweep, but the path is a helix. A sketch is drawn, then projected along a helical-shaped path.

To create the sketch

1. Sketch the shape shown in Figure 3-41 on the XY plane.
2. Sketch a line below the shape as shown.

   This line will serve as the axis of rotation.

3. Right-click the mouse, and click the Finish Sketch option.

To create the coil

1. Click the Coil tool on the Part Features panel bar.

   The Coil dialog box will appear. The sketched profile will be selected automatically.

2. Select the sketch line as the axis.

3. Click the Coil Size tab.

   The dialog box will change.

4. Set the Type for Pitch and Revolution, the Pitch for 20, and the Revolution for 3.
5. Click OK.

   See Figure 3-41.

To create a spring

1. Sketch a Ø5 circle on the XY plane.
2. Right-click the mouse and click the Finish Sketch option.
3. Click the Coil tool on the Part Features panel bar.

   The Coil dialog box will appear. See Figure 3-42.

4. Select the X axis as the axis of revolution by clicking the X Axis tool in the browser.
5. Click the Coil Size tab and set the Type for Pitch and Revolution, the Pitch for 6, and the Revolution for 6.

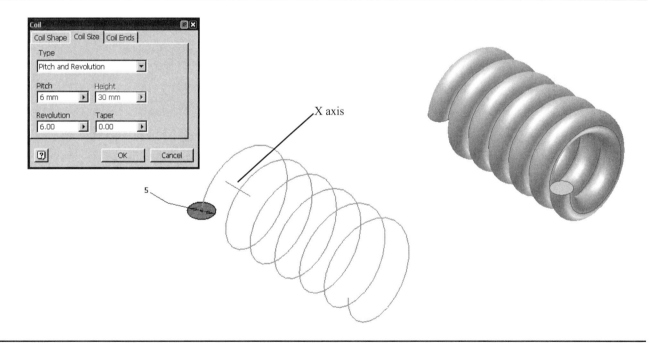

Figure 3-42

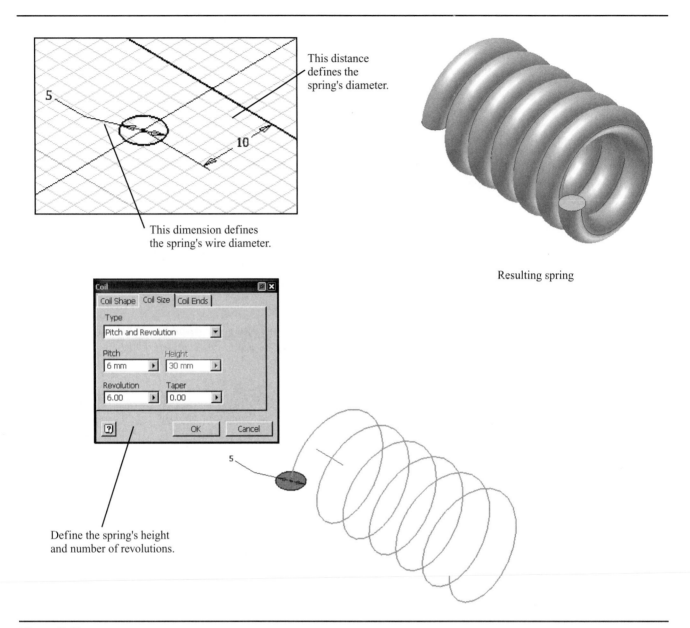

This distance defines the spring's diameter.

This dimension defines the spring's wire diameter.

Resulting spring

Define the spring's height and number of revolutions.

Figure 3-43

To specify the diameter of the spring

The diameter of the spring is controlled by the location of the center of the defining circle relative to the X axis. See Figure 3-43.

1. Sketch a Ø5 circle and sketch a line aligned with the X axis.
2. Use the General Dimension tool to locate the center point of the circle 10 mm from the line on the X axis.
3. Right-click the mouse and click the Finish Sketch option, then click the Coil tool.
4. Set the line on the X axis as the axis.

5. Click the Coil Size tab and set the Type for Revolution and Height, the Height for 30, and the Revolution for 6.
6. Click OK.

To add sections to a spring

Springs often include a straight section at their ends to facilitate assembly. These sections can be added to springs drawn using Inventor by using the Sweep command.

Figure 3-44 shows a spring that was created from a Ø5 circle located 8 mm from the X axis. A section will be added to the spring's starting point.

1. Click the Work Plane tool, then the XZ Plane tool in the browser.

   A plane will appear.

2. Click one of the corners of the plane and offset it −8.0.

   This will align the work plane with the center point of the initial Ø5 circle used to create the spring.

3. Click one of the offset work plane's corner points, click the right mouse button, and select the New Sketch option.

4. Use the Line and Tangent arc tools to create the shape as shown.

5. Click the right mouse button and select the Finish Sketch option.

6. Click the end surface of the spring, the initial circle, and create a new sketch plane.

7. Click the Sweep tool, define the end of the spring as the profile, click the right mouse button, and select the Continue option.

8. Define the line and arc as the Profile, then click OK.

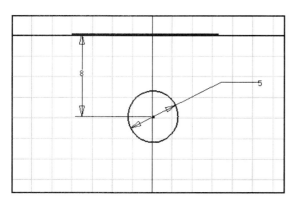

Define the spring's diameter and wire diameter.

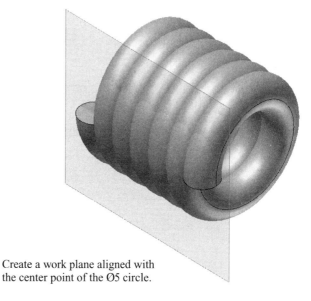

Create a work plane aligned with the center point of the Ø5 circle.

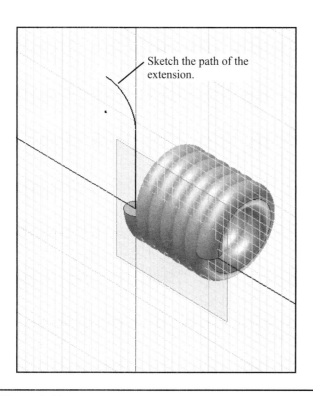

Sketch the path of the extension.

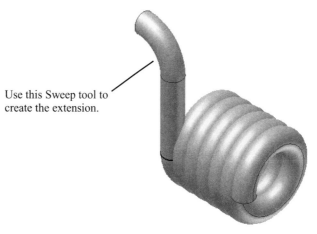

Use this Sweep tool to create the extension.

Figure 3-44

3-22 MODEL MATERIAL

A material designation may be assigned to a model. The material designation become part of the model's file and will be included on any assembly's parts list that includes the model.

To define a model's material

1. Right-click on the model's name in the Browser box and select the Properties option.

    See Figure 3-45. The Properties dialog box will appear.

2. Select the Physical tab and then the scroll arrow on the right side of the Material box.

    See Figure 3-46.

3. Select a material.

    In this example, mild steel was selected.

    Defining the material for a model will change the model's color. Different materials have different colors. Figure 3-47 shows the original model and the same model with the mild steel material designation.

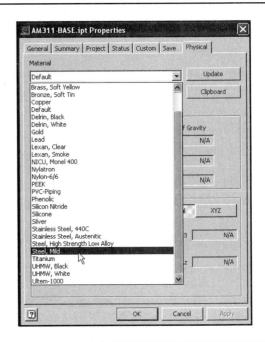

Figure 3-46

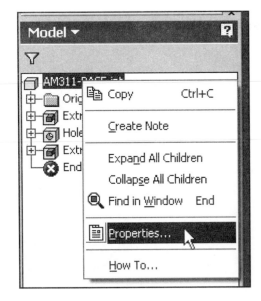

Figure 3-45

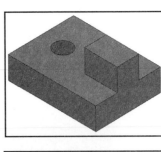

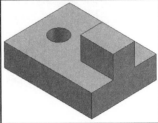

Figure 3-47

3-23 EXERCISE PROBLEMS

Redraw the following objects as solid models based on the given dimensions. Make all models from mild steel.

EX3-1 MILLIMETERS

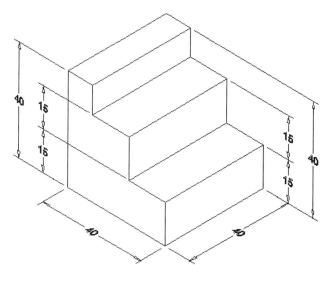

EX3-3 MILLIMETERS

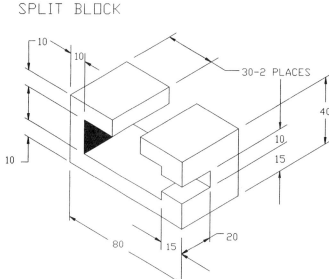

SPLIT BLOCK

EX3-2 INCHES

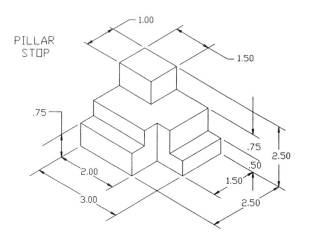

PILLAR STOP

EX3-4 MILLIMETERS

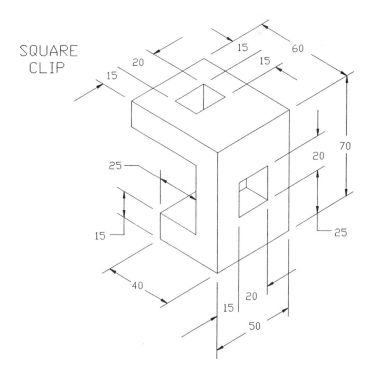

SQUARE CLIP

EX3-5 MILLIMETERS

SETTER BRACKET

50

10 - ALL AROUND

30

40 -2 PLACES

25

15

100

30 -2 PLACES

EX3-7 INCHES

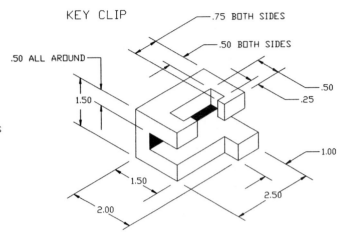

KEY CLIP

.75 BOTH SIDES

.50 BOTH SIDES

.50 ALL AROUND

1.50

.50

.25

1.50

2.00

2.50

1.00

EX3-6 MILLIMETERS

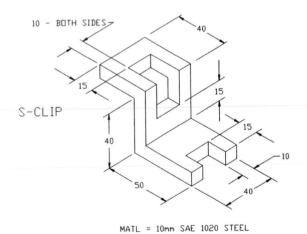

10 - BOTH SIDES

40

15

15

15

S-CLIP

40

10

50

40

MATL = 10mm SAE 1020 STEEL

EX3-8 MILLIMETERS

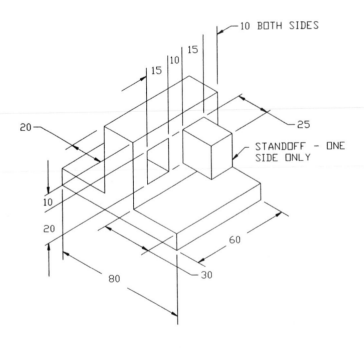

10 BOTH SIDES

15

15

10

15

20

25

STANDOFF - ONE SIDE ONLY

10

20

60

80

30

EX3-9 MILLIMETERS

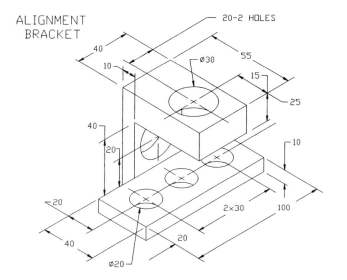

EX3-11 INCHES

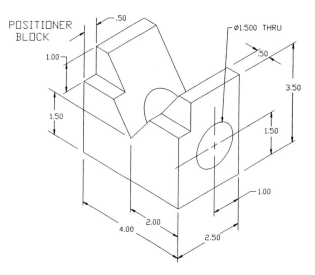

EX3-10 MILLIMETERS

EX3-12 MILLIMETERS

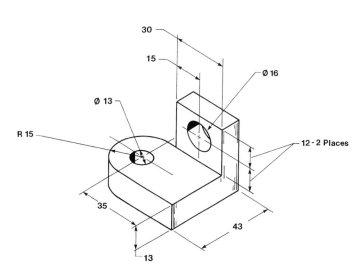

EX3-13 MILLIMETERS

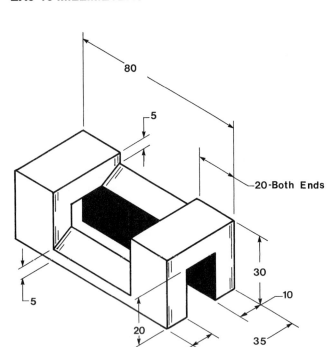

EX3-15 MILLIMETERS

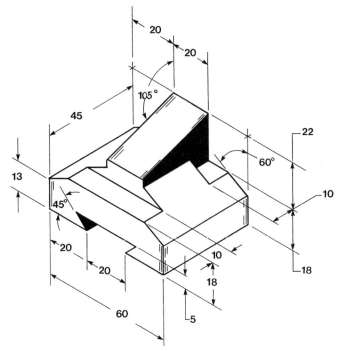

EX3-14 MILLIMETERS

EX3-16 INCHES

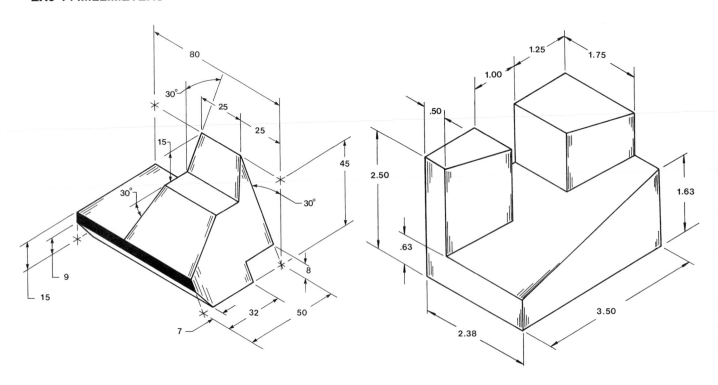

EX3-17 MILLIMETERS

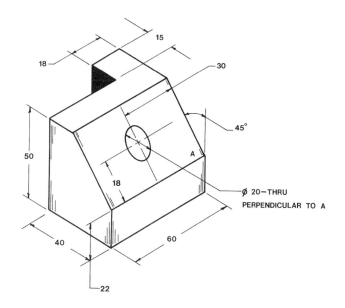

EX3-19 MILLIMETERS

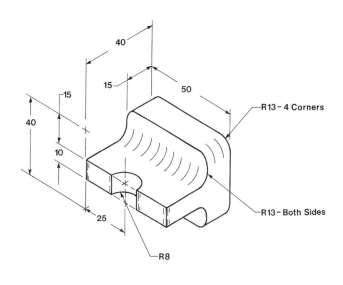

EX3-18 MILLIMETERS

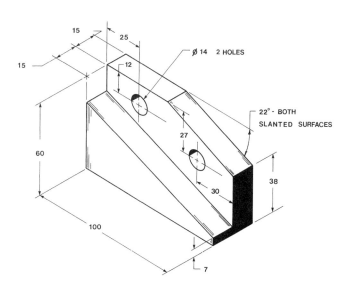

EX3-20 MILLIMETERS

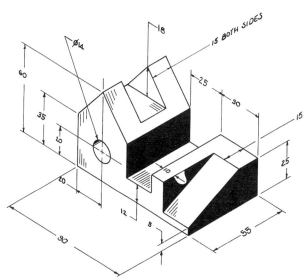

EX3-21 MILLIMETERS

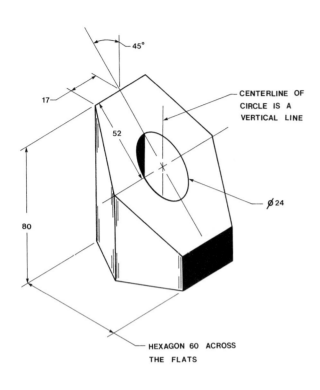

45°

17

52

80

CENTERLINE OF
CIRCLE IS A
VERTICAL LINE

⌀ 24

HEXAGON 60 ACROSS
THE FLATS

EX3-23 MILLIMETERS

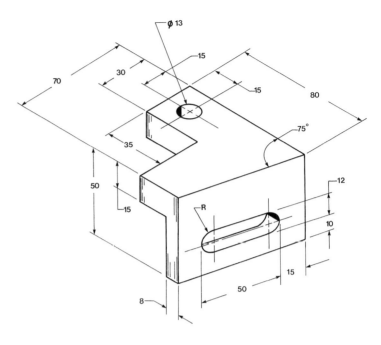

⌀ 13

70

30

15

15

80

75°

35

50

15

12

R

10

15

50

8

EX3-22 INCHES

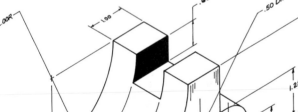
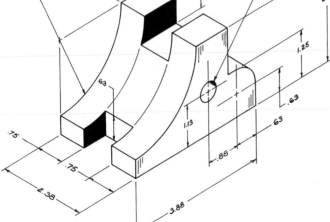

2.00R

1.90

.69

.50 DIA

2.50

1.25

.63

1.13

.63

.63

.88

.75

.75

2.38

3.88

EX3-24 MILLIMETERS

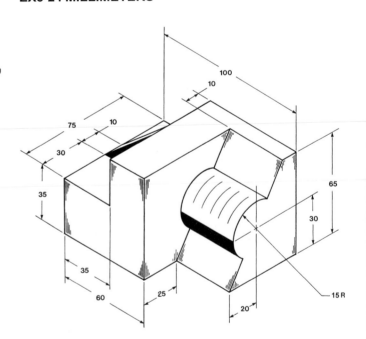

100

10

75

10

30

35

65

30

35

60

25

20

15 R

EX3-25 INCHES (SCALE: 4=1)

EX3-27 MILLIMETERS

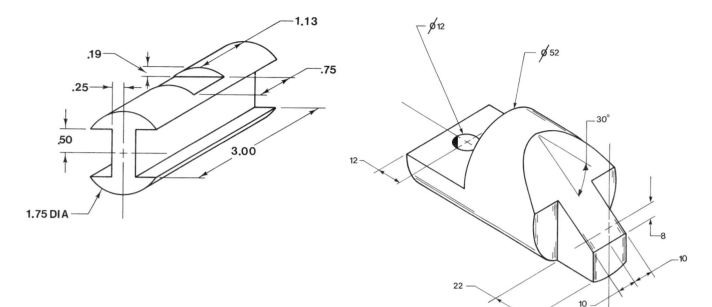

EX3-26 MILLIMETERS

EX3-28 MILLIMETERS

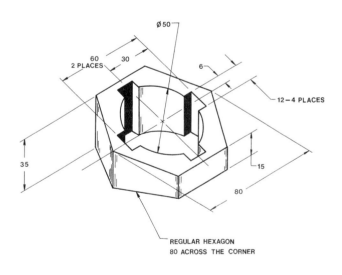

EX3-29 INCHES (SCALE: 4=1)

EX3-31 MILLIMETERS

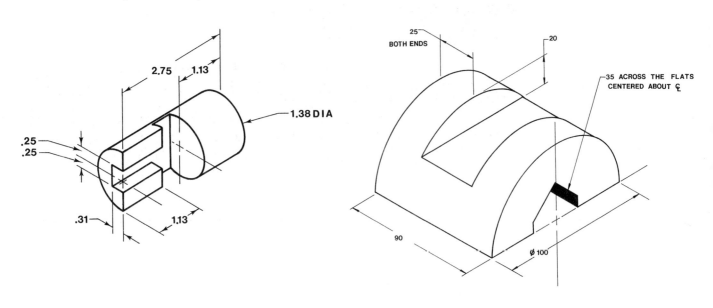

EX3-30 MILLIMETERS (SCALE: 2=1)

EX3-32 MILLIMETERS

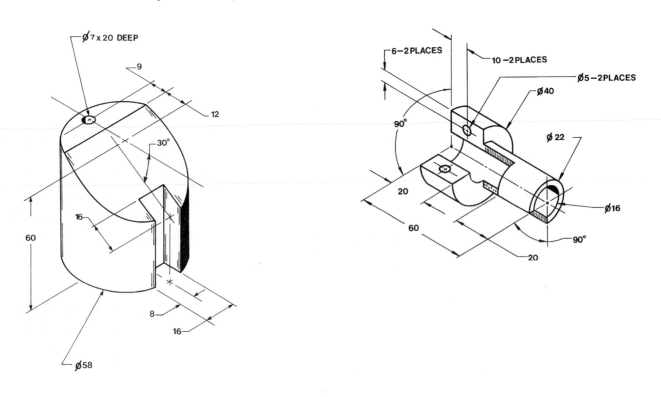

EX3-33 MILLIMETERS

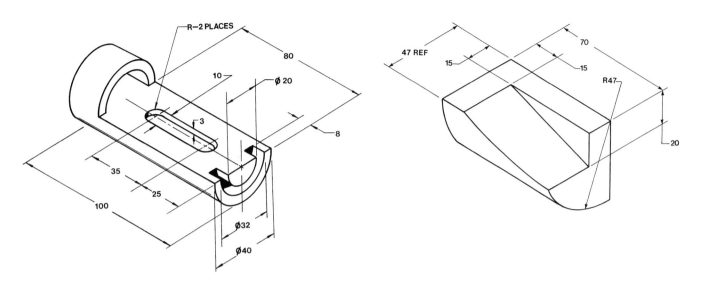

EX3-35 MILLIMETERS

EX3-34 MILLIMETERS

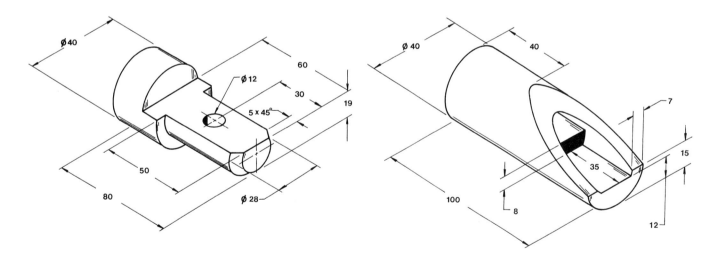

EX3-36 MILLIMETERS

EX3-37 MILLIMETERS

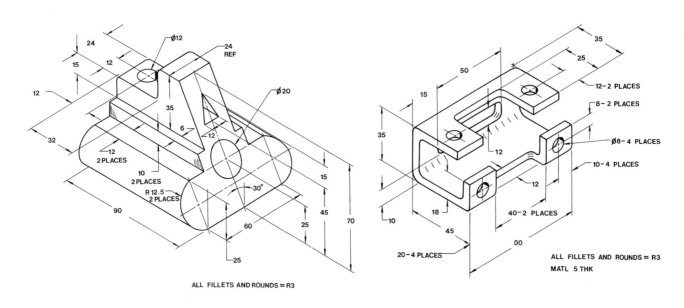

ALL FILLETS AND ROUNDS = R3

EX3-39 MILLIMETERS (CONSIDER A SHELL)

ALL FILLETS AND ROUNDS = R3
MATL 5 THK

EX3-38 MILLIMETERS

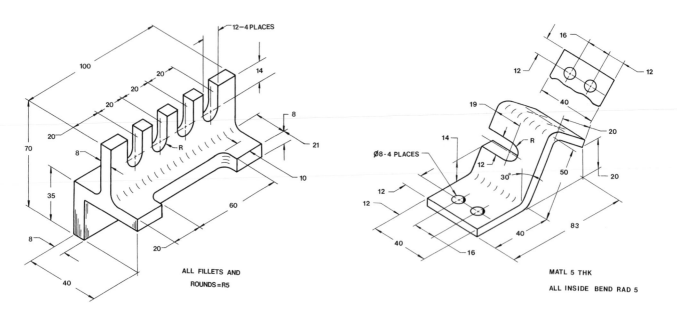

ALL FILLETS AND
ROUNDS = R5

EX3-40 MILLIMETERS

MATL 5 THK

ALL INSIDE BEND RAD 5

EX3-41

Draw a spring with a Ø5 mm wire diameter, 25 mm overall diameter, 6 revolutions, and a height of 30 mm.

EX3-42

Draw a spring with a Ø2 mm wire diameter, 10 mm overall diameter, 8 revolutions, and a height of 25 mm.

EX3-43

Draw a spring with a Ø.25 in. wire diameter, 3.00 in. overall diameter, 5 revolutions, and a height of 3.5 in.

EX3-44

Draw a spring with a Ø3 mm wire diameter, 12 mm overall diameter, 10 revolutions, and a height of 30 mm.

Add a hook-shaped extension, as dimensioned below, to both ends of the spring.

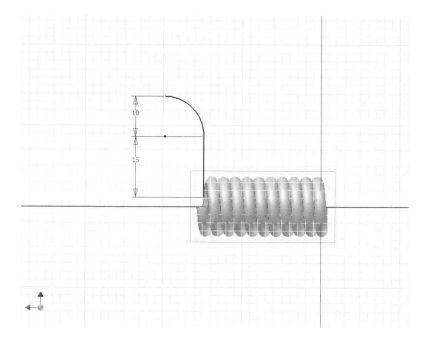

Orthographic Views

4-1 INTRODUCTION

Orthographic views may be created directly from 3D Inventor models. *Orthographic views* are two-dimensional views used to define a three-dimensional model. Unless the model is of uniform thickness, more than one orthographic view is necessary to define the model's shape. Standard practice calls for three orthographic views: a front, top, and right side view, although more or fewer views may be used as needed.

Modern machines can work directly from the information generated when a solid 3D model is created, so the need for orthographic views—blueprints—is not as critical as it once was; however, there are still many drawings in existence that are used for production and reference. The ability to create and read orthographic views remains an important engineering skill.

This chapter presents orthographic views using third-angle projection in accordance with ANSI standards. ISO first-angle projects are also presented.

4-2 ORTHOGRAPHIC VIEWS

Figure 4-1 shows an object with its front, top, and right side orthographic views projected from the object.

The views are two-dimensional, so they show no depth. Note that in the projected right plane there are three rectangles. There is no way to determine which of the three is closest and which is farthest away if only the right side view is considered. All views must be studied to analyze the shape of the object.

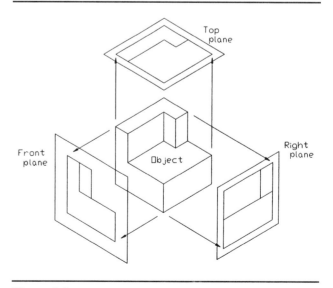

Figure 4-1

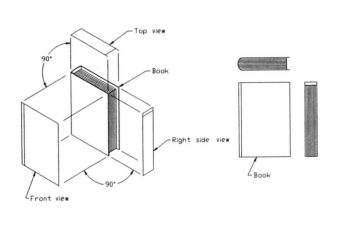

Figure 4-2

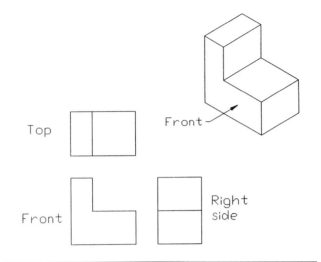

Figure 4-3

Figure 4-2 shows three orthographic views of a book. After the views are projected they are positioned as shown. The positioning of views relative to one another is critical. The views must be aligned and positioned as shown.

Normal surfaces

Normal surfaces are surfaces that are at 90° to each other. Figures 4-3, 4-4, and 4-5 show objects that include only normal surfaces and their orthographic views.

Hidden lines

Hidden lines are used to show surfaces that are not directly visible. All surfaces must be shown in all views. If an edge or surface is blocked from view by another feature, it is drawn using a hidden line. Figures 4-6 and 4-7 show objects that require hidden lines in their orthographic views.

Figure 4-8 shows an object that contains an edge line, A-B. In the top view, line A-B is partially hidden and partially visible. The hidden portion of the line is drawn using a hidden line pattern, and the visible portion of the line is drawn using a solid line.

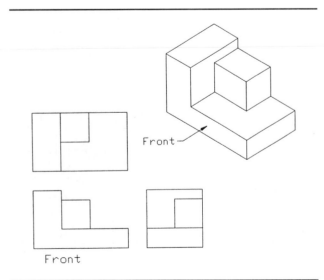

Figure 4-4

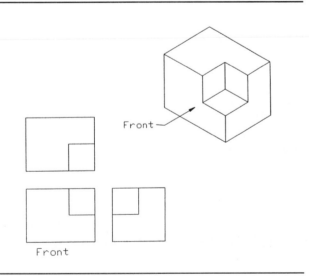

Figure 4-5

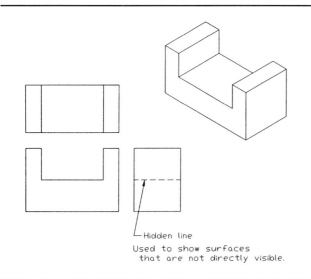

Hidden line
Used to show surfaces
that are not directly visible.

Figure 4-6

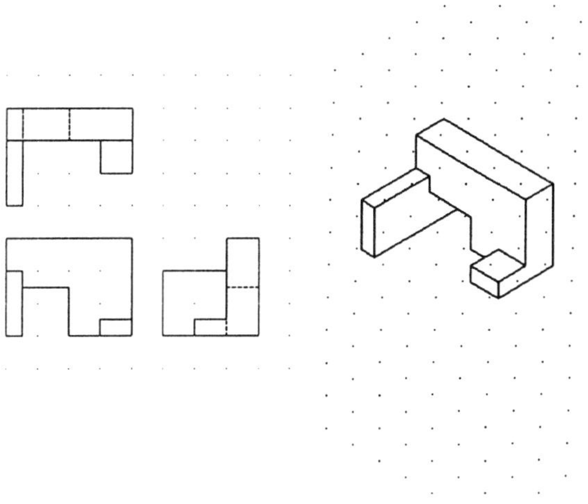

Figure 4-7

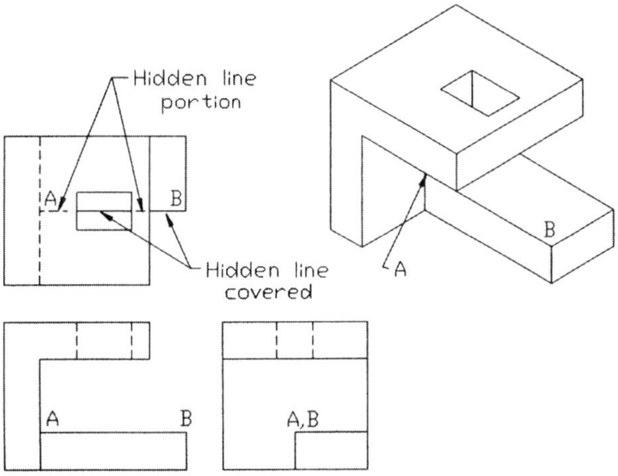

Figures 4-9 and 4-10 show objects that require hidden lines in their orthographic views.

Precedence of lines

It is not unusual for one type of line to be drawn over another type of line. Figure 4-11 shows two examples of overlap by different types of lines. Lines are shown on the views in a prescribed order of precedence. A solid line (object or continuous) takes precedence over a hidden line, and a hidden line takes precedence over a centerline.

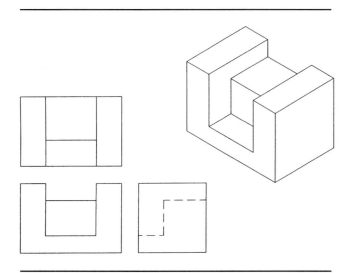

Hidden line portion

Hidden line covered

Figure 4-8

Figure 4-9

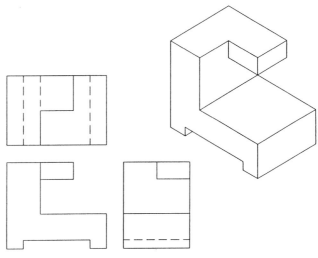

Figure 4-10

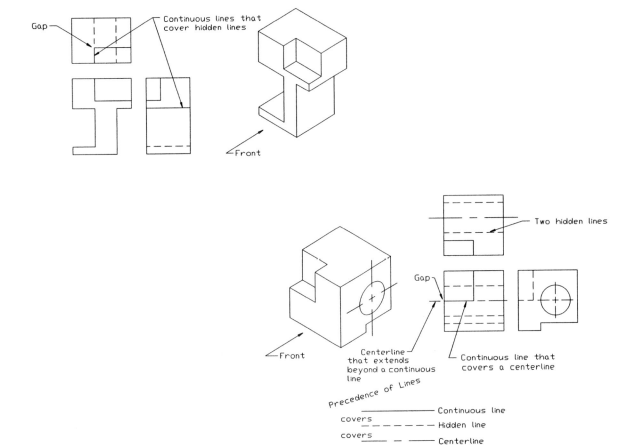

Figure 4-11

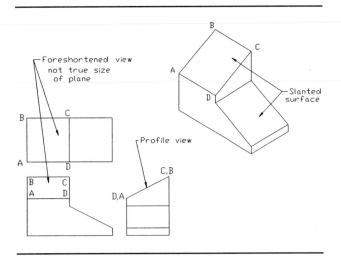

Figure 4-12

Slanted surfaces

Slanted surfaces are surfaces drawn at an angle to each other. Figure 4-12 shows an object that contains two slanted surfaces. Surface ABCD appears as a rectangle in both the top and front views. Neither rectangle represents the true shape of the surface. Each is smaller that the actual surface. Also, none of the views show enough of the object to enable the viewer to accurately define the shape of the object. The views must be used together for a correct understanding of the object's shape.

Figures 4-13 and 4-14 show objects that include slanted surfaces. Projection lines have been included to emphasize the importance of correct view location. Information is projected between the front and top views using vertical lines and between the front and side views using horizontal lines.

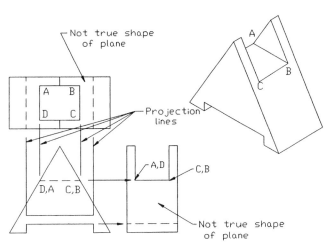

Figure 4-13

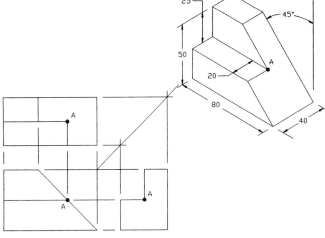

Figure 4-14

Compound lines

A *compound line* is formed when two slanted surfaces intersect. Figure 4-15 shows an object that includes a compound line.

Oblique surfaces

An *oblique surface* is a surface that is slanted in two different directions. Figures 4-16 and 4-17 show objects that include oblique surfaces.

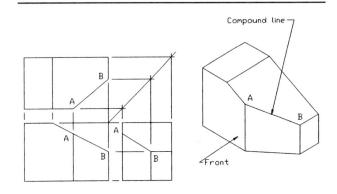

Figure 4-15

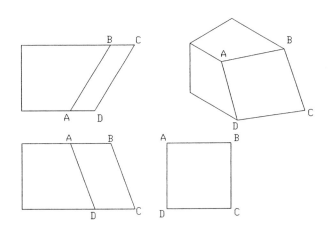

Figure 4-16

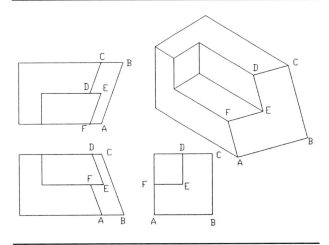

Figure 4-17

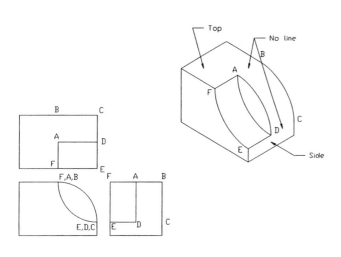

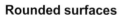

Figure 4-18

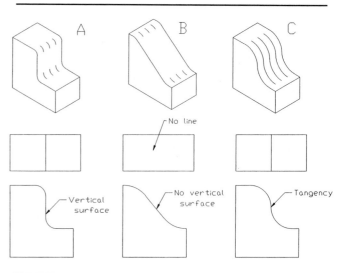

Figure 4-19

Rounded surfaces

Figure 4-18 shows an object with two rounded surfaces. Note that as with slanted surfaces, an individual view is insufficient to define the shape of a surface. More than one view is needed to accurately define the surface's shape.

Convention calls for a smooth transition between rounded and flat surfaces; that is, no lines are drawn to indicate the tangency. Inventor includes a line to indicate tangencies between surfaces in the isometric drawings created using the multiview options but does not include them in the orthographic views. Tangency lines are also not included when models are rendered.

Figure 4-19 shows the drawing conventions for including lines for rounded surfaces. If a surface includes no vertical portions or no tangency, no line is included.

Figure 4-20 shows an object that includes two tangencies. Each is represented by a line.

Figure 4-21 shows two objects with similar configurations; however, the boxlike portion of the lower object blends into the rounded portion exactly on its widest point, so no line is required.

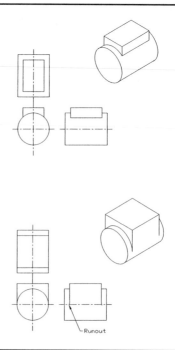

Figure 4-21

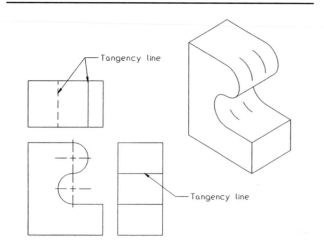

Figure 4-20

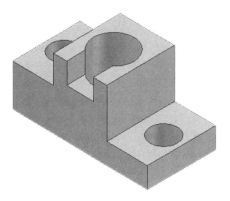

Figure 4-22

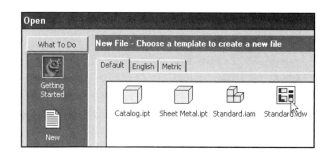

Figure 4-23

4-3 ORTHOGRAPHIC VIEWS

Inventor will create orthographic views directly from models. Figure 4-22 shows a completed three-dimensional model. It was created using an existing file, EX4-7. It will be used throughout this chapter to demonstrate orthographic presentation views.

To create an orthographic view

1. Start a new drawing and select the Standard.idw option.

   See Figure 4-23.

2. Click OK.

   The drawing management screen will appear. See Figure 4-24.

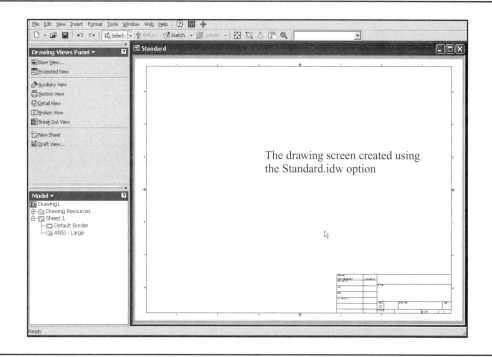

Figure 4-24

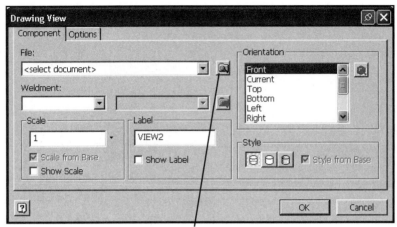

Click here to access the drawing files.

Figure 4-25

3. Click the Base View… tool in the Drawing Views Panel bar.

The Drawing View dialog box will appear. See Figure 4-25.

4. Click the Explore directories button.

The Open dialog box will appear. See Figure 4-26.

5. Select the desired model. In this example the model's file name is EX4-7.

The Create View dialog box will appear. See Figure 4-27.

6. Select the Bottom option, locate the view on the drawing screen, and click the location.

Figure 4-28 shows the resulting orthographic view.

Figure 4-26

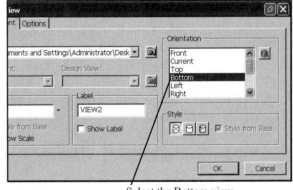

Select the Bottom view.

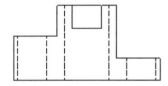

Resulting orthographic view

Figure 4-27

Figure 4-28

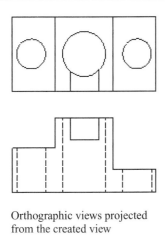

Orthographic views projected
from the created view

Figure 4-29

The screen will include a border and a title block. The lettering in the title block may appear illegible. This is normal. The text will be legible when printed. The section on title blocks will explain how to work with title blocks.

To create other orthographic views

1. Click the Projected View tool on the Drawing Management toolbox on the panel bar.
2. Click the view already on the drawing screen.
3. Move the cursor upward from the view.

   A second view will appear.

4. Select a location, click the left mouse button to place the view, then click the right mouse button and select the Create option.

Figure 4-29 shows the resulting two orthographic views. The initial view was created using the Bottom option. This is a relative term based on the way the model was drawn. The initial view can be defined as the front view, and the second view created from that front view is also, by definition, the top view.

To add centerlines

Convention calls for all holes to be defined using centerlines. The views in Figure 4-29 do not include centerlines.

1. Move the cursor into the Panel bar area and right-click the mouse.

   A small dialog box will appear.

2. Select the Drawing Annotation option.

   The Drawing Annotation Panel bar will appear. See Figure 4-30.

3. Click the Center Mark tool.

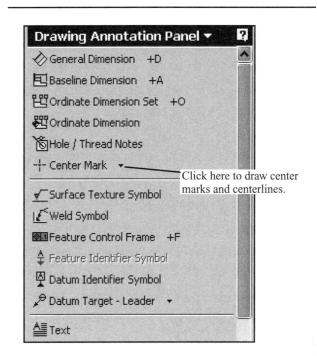

Click here to draw center marks and centerlines.

Figure 4-30

4. Move the cursor into the drawing screen and click the edges of the holes in the top view.

   See Figure 4-31.

5. Click the Centerline bisector tool.

   The Centerline bisector tool is a flyout from the Center Mark tool.

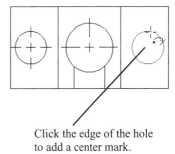

Click the edge of the hole
to add a center mark.

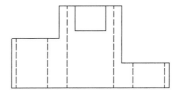

Figure 4-31

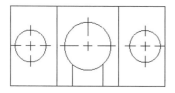

Click the two hidden lines to
add a vertical centerline.

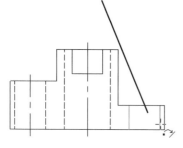

Figure 4-32

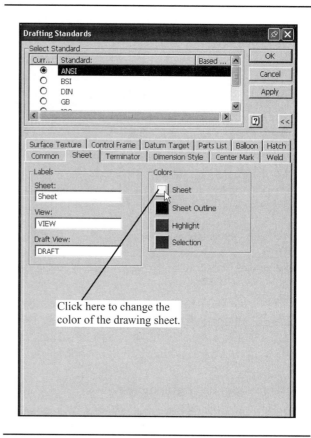

Click here to change the
color of the drawing sheet.

Figure 4-33

6. Click each side of the hole projections in the front view.

Vertical centerlines will appear. See Figure 4-32.

To change the background color of the drawing screen

1. Click the Format heading at the top of the screen.
2. Select the Standards option.

The Drafting Standards dialog box will appear. See Figure 4-33.

3. Click the Sheet tab, then the heading Sheet in the Colors box.

The Color dialog box will appear.

4. Click the desired color, then OK.

The Drafting Standards dialog box will appear.

5. Click the Apply box, then OK.

The sheet's background color will be changed.

4-4 ISOMETRIC VIEWS

An isometric view may be created from any view on the screen. The resulting orientation will vary according to the view selected. In this example the front view is selected.

1. Click the Projected View tool.
2. Click the Front view.
3. Move the cursor to the right of the front view and select a location for the isometric view.
4. Move the cursor slightly and click the right mouse button.
5. Select the Create option.

Figure 4-34 shows the resulting isometric view.

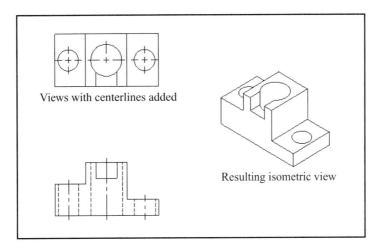

Figure 4-34

4-5 SECTION VIEWS

Some objects have internal surfaces that are not directly visible in normal orthographic views. *Section views* are used to expose these surfaces. Section views do not include hidden lines.

Any material cut when a section view is defined is hatched using section lines. There are many different styles of hatching, but the general style is evenly spaced 45° lines. This style is defined as ANSI 31 and will be applied automatically by Inventor.

Figure 4-35 shows a three-dimensional view of an object. The object is cut by a cutting plane. *Cutting planes* are used to define the location of the sectional view. Material to one side of the cutting plane is removed, exposing the sectional view.

Figure 4-36 shows the same object presented using two dimensions. The cutting plane is represented by a cutting plane line. The cutting plane line is defined as A-A, and the section view is defined as view A-A.

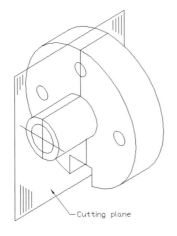
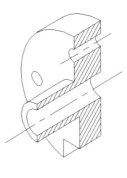

Figure 4-35

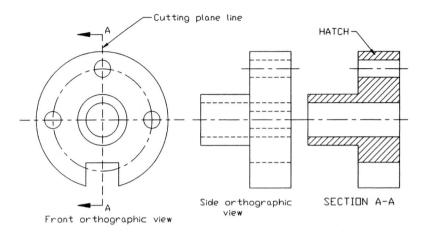

Figure 4-36

All surfaces directly visible must be shown in a section view. In Figure 4-37 the back portion of the object is not affected by the section view and is directly visible from the cutting plane. The section view must include these surfaces. Note how the rectangular section blocks out part of the large hole. No hidden lines are used to show the hidden portion of the large hole.

To draw a section view

Figure 4-38 shows the front and top views of the object defined in EX3-10. A section view will be created by first

defining the cutting plane line in the top view, then projecting the section view below the front view.

1. Click the Section View tool, then click the top view.

   The cursor will change to a +-like shape.

2. Define the cutting plane by defining two points on the top view.

   See Figure 4-38.

3. Right-click the mouse and select the Continue option.

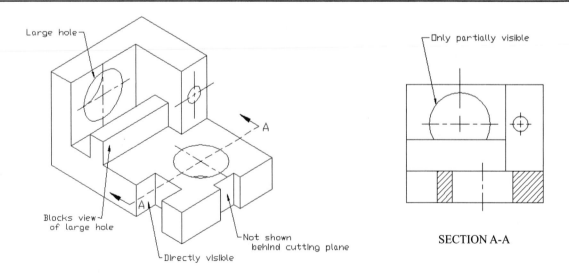

SECTION A-A

Figure 4-37

Define the cutting plane

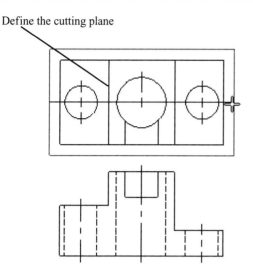

Figure 4-38

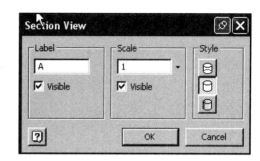

Figure 4-39

The Section View dialog box will appear. See Figure 4-39.

4. Set the Label letter for A and the Scale for 1.
5. Move the cursor so as to position the section view below the front view.
6. Click the section view location.
7. Add the appropriate centerlines using the Centerline bisector tool.

Figure 4-40 shows the resulting section view. Notice that the section view is defined as A-A, and the scale is specified. The arrows of the cutting plane line are directed away from the section view. The section view is located behind the arrows.

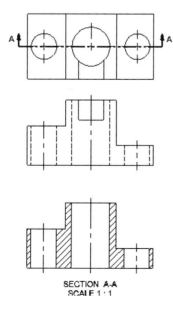

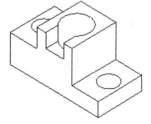

SECTION A-A
SCALE 1 : 1

Figure 4-40

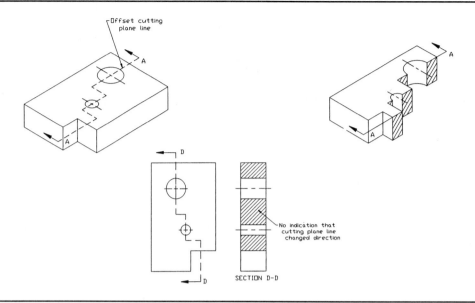

Figure 4-41

4-6 OFFSET SECTION VIEWS

Cutting plane lines need not pass directly across an object but may be offset to include several features. Figure 4-41 shows an object that has been cut using an offset cutting plane line.

To create an offset cutting plane

Figure 4-42 shows the front and top views of an object. The views were created using the Create View, Project View, and Centerline tools.

1. Click the Section View tool, then click the top view.

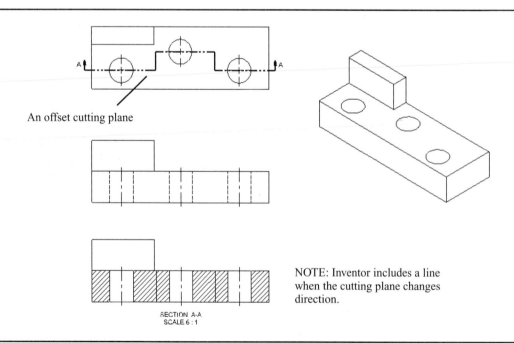

An offset cutting plane

SECTION A-A
SCALE 6 : 1

NOTE: Inventor includes a line when the cutting plane changes direction.

Figure 4-42

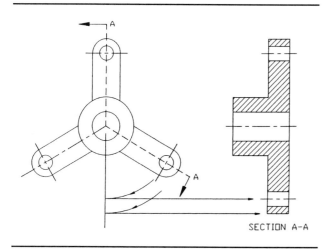

Figure 4-43

2. Draw a cutting plane across the top view through the centers of each of the three holes.
3. Locate the section view below the front view.

4-7 ALIGNED SECTION VIEWS

Figure 4-43 shows an example of an aligned section view. Aligned section views are most often used on circular objects and use an angled cutting plane line to include more features in the section view, like an offset cutting plane line.

An aligned section view is drawn as if the cutting plane line runs straight across the object. The cutting plane line is rotated into a straight position, and the section view is projected.

Figure 4-44 shows an aligned section view created using Inventor.

4-8 DETAIL VIEWS

Detail views are used to enlarge portions of an existing drawing. The enlargements are usually made of areas that could be confusing because of many crossing or hidden lines.

To create a detail view

1. Click the Detail View tool, then click the view to be enlarged.

The Detail View dialog box will appear. See Figure 4-45.

2. Set the Label letter to A and the Scale to 2, then pick a point on the view.
3. Move the cursor, creating a circle.

The circle will be used to define the area of the detail view.

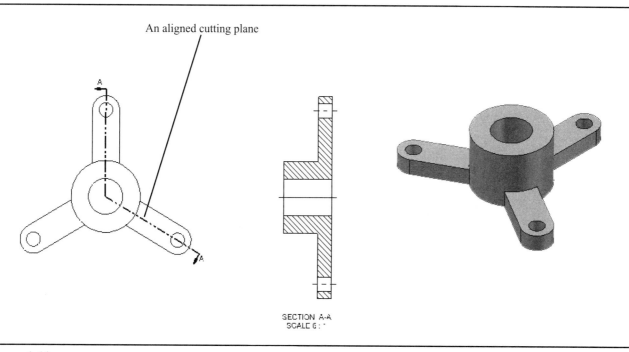

Figure 4-44

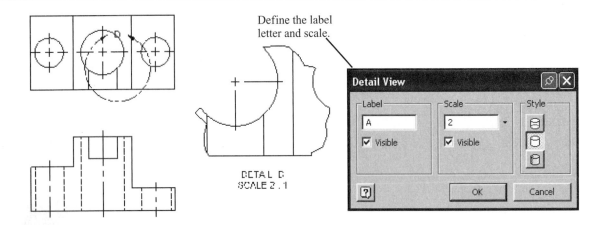

Define the label
letter and scale.

Figure 4-45

4. When the circle is of appropriate diameter click the left mouse button and move the cursor away from the view.
5. Locate the detail view and click the location.

4-9 BROKEN VIEWS

It is often convenient to break long continuous shapes so that they take up less drawing space. Figure 4-46 shows a long L-bracket that has a continuous shape; that is, its shape is constant throughout its length. Figure 4-47 shows an orthographic view of the same L-bracket.

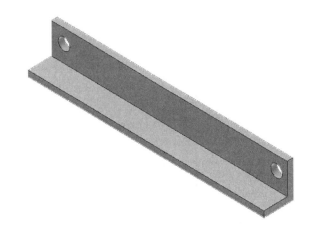

Figure 4-46

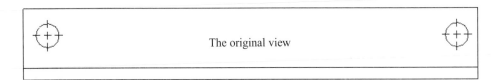

The original view

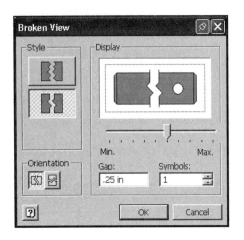

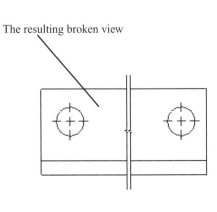

The resulting broken view

Figure 4-47

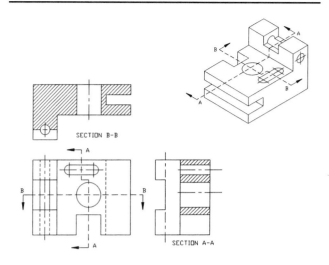

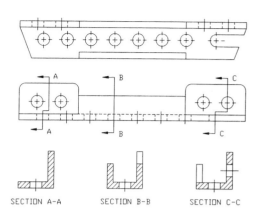

Figure 4-48

Figure 4-49

To create a broken view

1. Click the Broken View tool, then the orthographic view.

   The Broken View dialog box will appear.

2. Select the orientation of the break and the gap distance between the two portions of the L-bracket.

3. Click a point near the left end of the L-bracket, then move the cursor to the right and click a second point near the right end of the L-bracket.

   Figure 4-47 shows the resulting broken view.

Multiple section views

It is acceptable to take more than one section view of the same object in order to present a more complete picture of the object. Figures 4-48 and 4-49 show objects that use more than one section view.

4-10 AUXILIARY VIEWS

Auxiliary views are orthographic views used to present true-shaped views of slanted surfaces. Figure 4-50 shows an object with a slanted surface that includes a hole

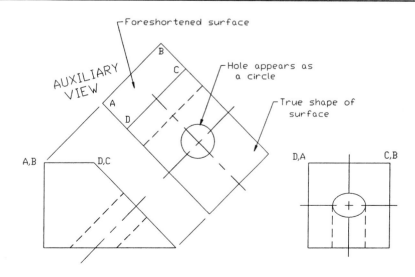

Figure 4-50

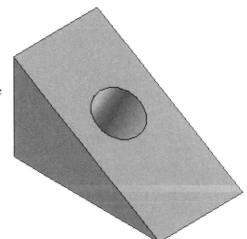

An object with a slanted surface created using Inventor

Figure 4-51

drilled perpendicular to the slanted surface. Note how the right side view shows the hole as an ellipse and that the surface A-B-C-D is foreshortened; that is, it is not shown at its true size. Surface A-B-C-D does appear at its true shape and size in the auxiliary view. The auxiliary view was projected at 90° from the slanted surface so as to generate a true-shaped view.

Figure 4-51 shows an object that includes a slanted surface and hole.

To draw an auxiliary view

1. Click the Create View and Project View tools and create front and right side views as shown in Figure 4-52.

2. Click the Auxiliary View tool, then the front view.

   The Auxiliary View dialog box will appear.

3. Enter the appropriate settings, then click the slanted edge line in the front view.

4. Move the cursor away from the front view and select a location for the auxiliary view.

5. Click the left mouse button and enter the auxiliary view.

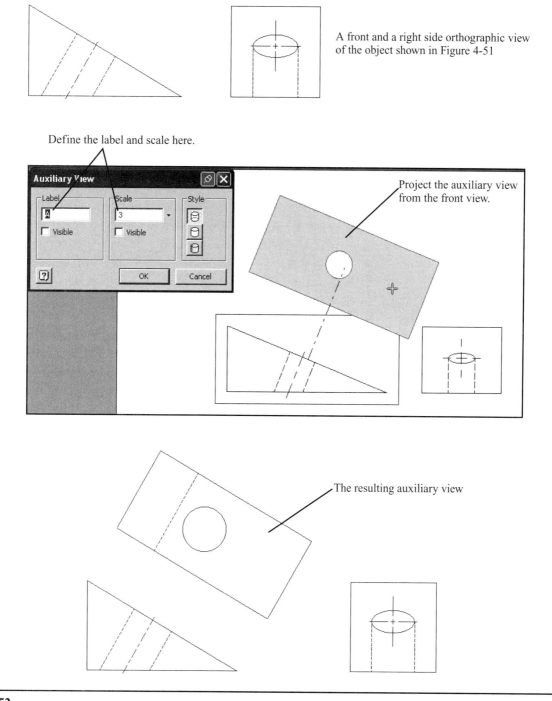

A front and a right side orthographic view of the object shown in Figure 4-51

Define the label and scale here.

Project the auxiliary view from the front view.

The resulting auxiliary view

Figure 4-52

4-11 EXERCISE PROBLEMS

Draw a front, a top, and a right side orthographic view of each of the objects in exercise problems EX4-1 through EX4-24. Make all objects from mild steel.

EX4-1 MILLIMETERS

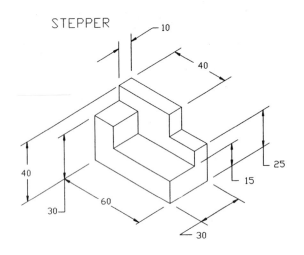

EX4-3 MILLIMETERS

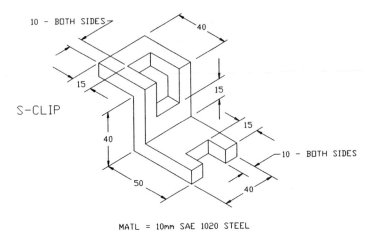

EX4-2 MILLIMETERS

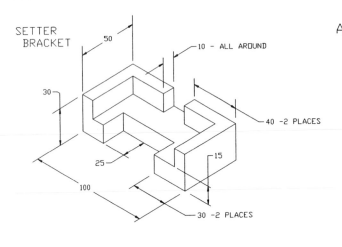

EX4-4 MILLIMETERS

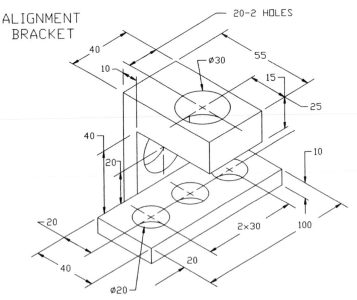

EX4-5 INCHES

KEY CLIP

.75 BOTH SIDES
.50 BOTH SIDES
.50 ALL AROUND
1.50
.50
.25
1.50
2.00
1.50
2.50
1.00

EX4-7 MILLIMETERS

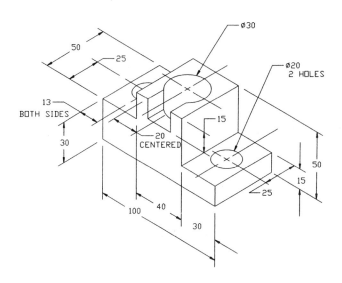

Ø30
Ø20
2 HOLES
50
25
13
BOTH SIDES
30
20
CENTERED
15
15
25
50
100
40
30

EX4-6 INCHES

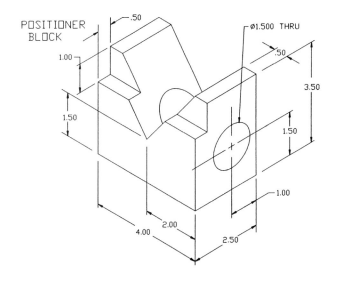

POSITIONER
BLOCK
.50
Ø1.500 THRU
.50
1.00
3.50
1.50
1.50
1.00
2.00
4.00
2.50

EX4-8 MILLIMETERS

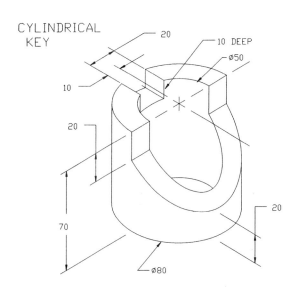

CYLINDRICAL
KEY
20
10 DEEP
Ø50
10
20
20
70
Ø80

EX4-9 MILLIMETERS

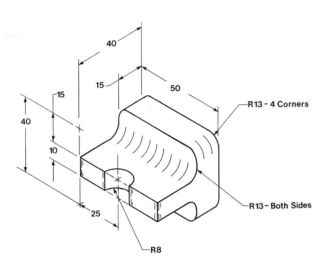

EX4-11 MILLIMETERS

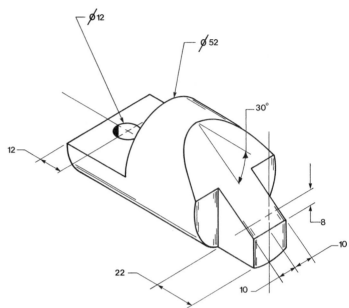

EX4-10 MILLIMETERS

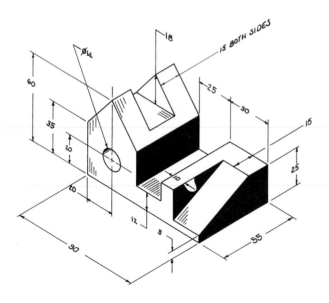

EX4-12 MILLIMETERS

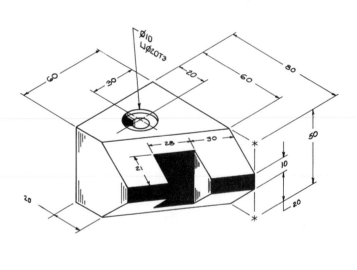

EX4-13 MILLIMETERS

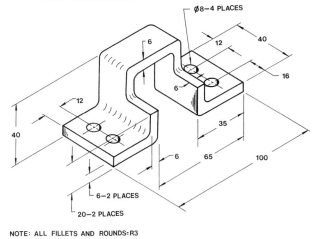

Ø8-4 PLACES

NOTE: ALL FILLETS AND ROUNDS=R3

EX4-15 MILLIMETERS

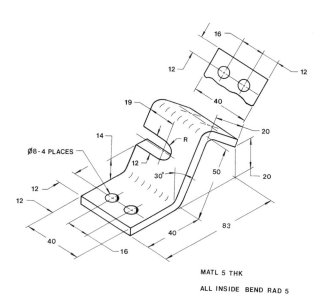

Ø8-4 PLACES

MATL 5 THK

ALL INSIDE BEND RAD 5

EX4-14 MILLIMETERS

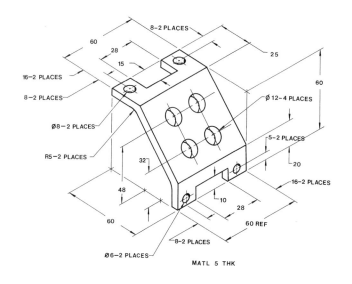

Ø12-4 PLACES
Ø8-2 PLACES
R5-2 PLACES
Ø6-2 PLACES

MATL 5 THK

EX4-16 INCHES

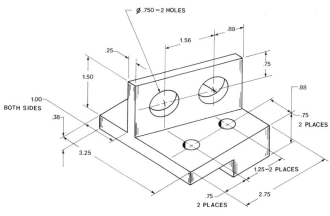

Ø.750-2 HOLES

EX4-17 MILLIMETERS

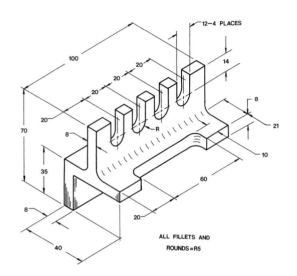

ALL FILLETS AND
ROUNDS = R5

EX4-19 MILLIMETERS

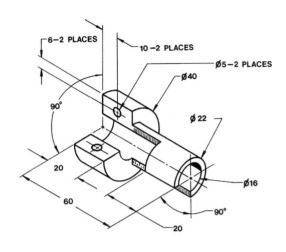

EX4-18 MILLIMETERS

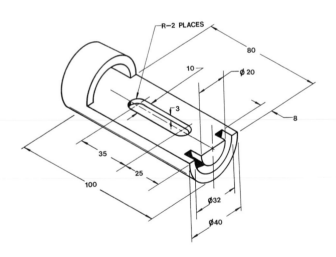

EX4-20 MILLIMETERS

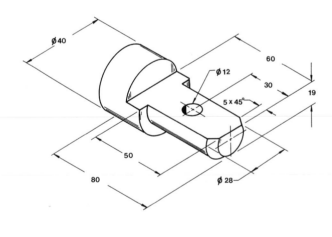

EX4-21 MILLIMETERS

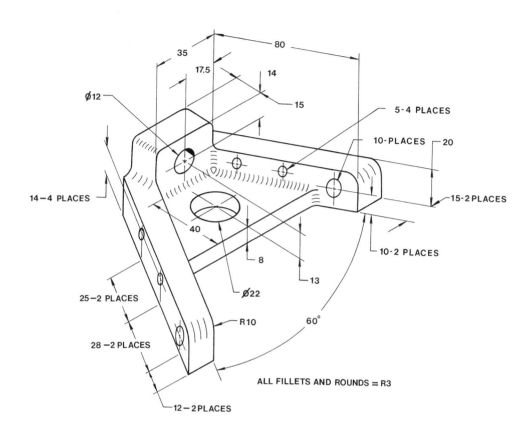

EX4-22 MILLIMETERS

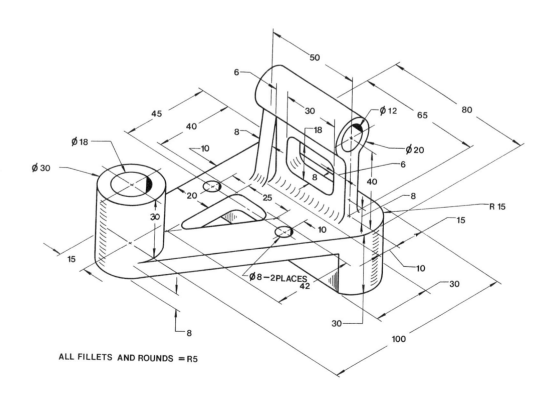

EX4-23 MILLIMETERS

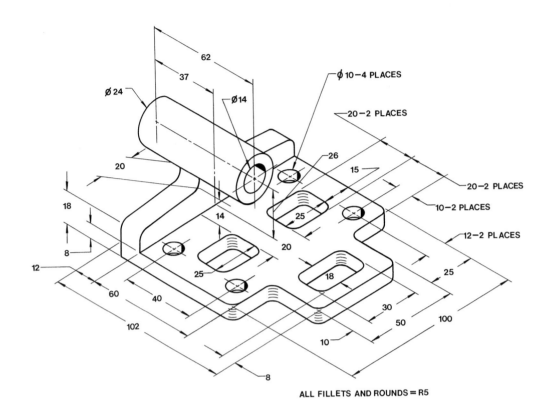

ALL FILLETS AND ROUNDS = R5

EX4-24 MILLIMETERS

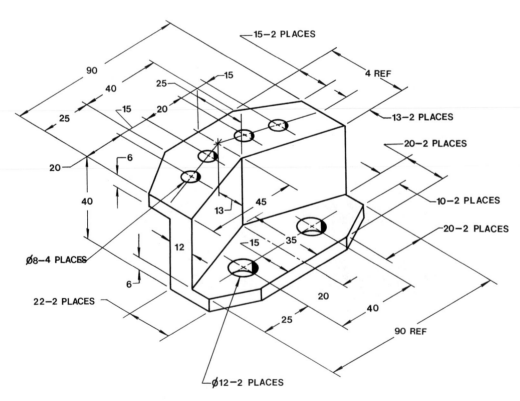

Draw at least two orthographic views and one auxiliary view of each of the objects in exercise problems EX4-25 through EX4-36.

EX4-25 MILLIMETERS

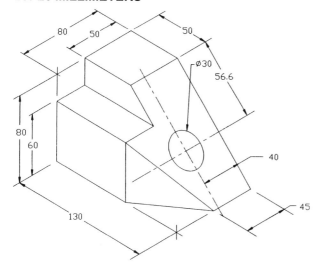

EX4-27 INCHES

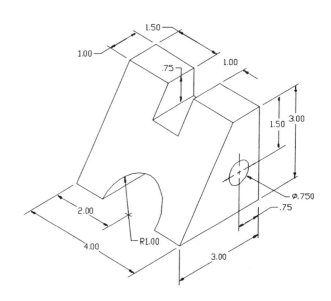

EX4-26 MILLIMETERS

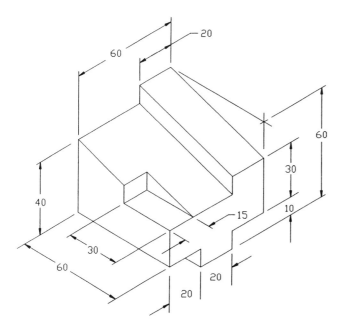

EX4-28 MILLIMETERS

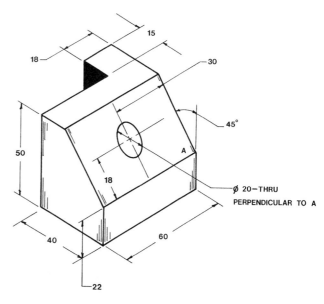

EX4-29 MILLIMETERS

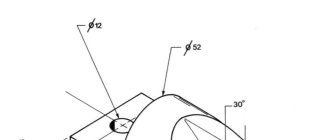

EX4-31 MILLIMETERS

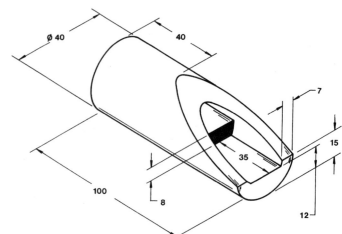

EX4-30 MILLIMETERS

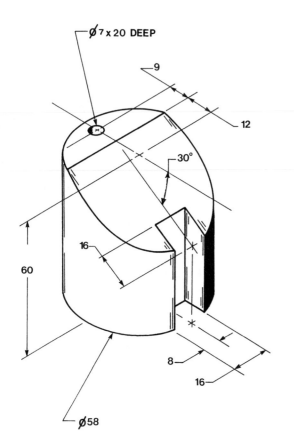

EX4-32 MILLIMETERS

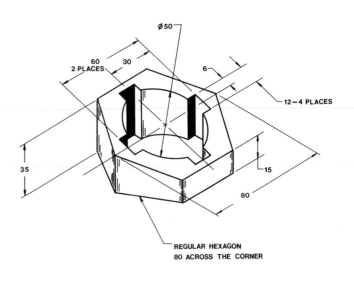

EX4-33 MILLIMETERS

EX4-35 INCHES

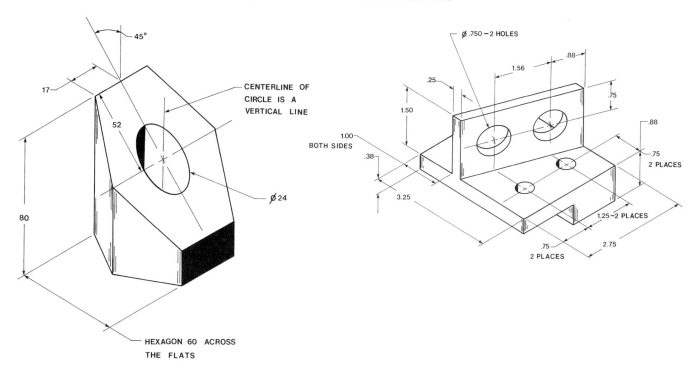

EX4-34 MILLIMETERS

EX4-36 MILLIMETERS

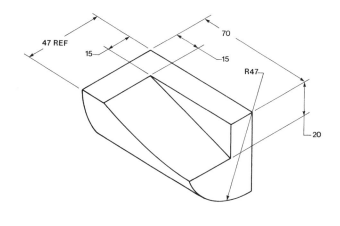

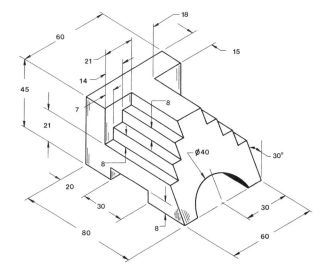

Define the true shape of the oblique surfaces in each of the objects in exercise problems EX4-37 through EX4-40.

EX4-37 INCHES

EX4-39 INCHES

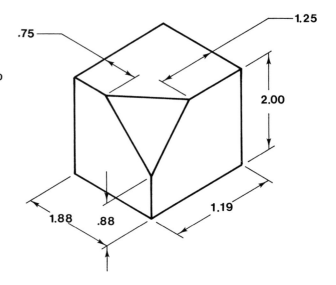

EX4-38 MILLIMETERS

EX4-40 MILLIMETERS

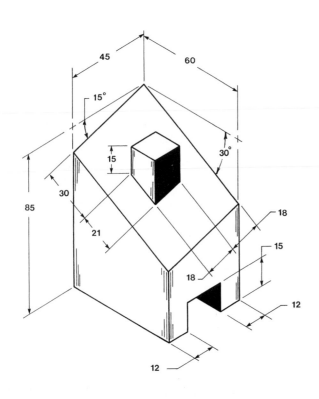

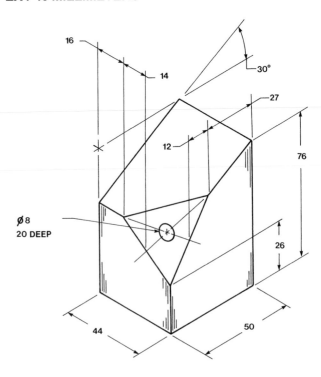

Draw each of the objects shown in exercise problems EX4-41 through EX4-44 as models, then draw a front view and an appropriate sectional view of each.

EX4-41 MILLIMETERS

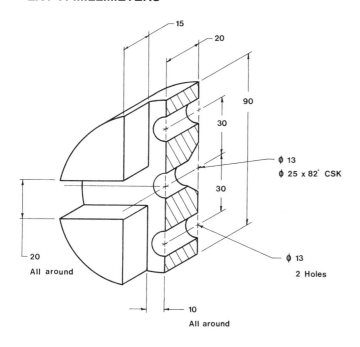

EX4-43 MILLIMETERS

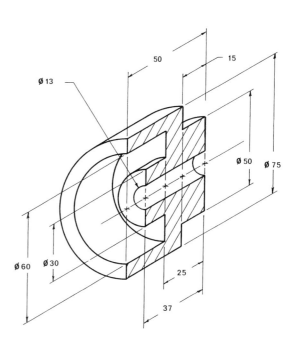

EX4-42 MILLIMETERS

EX4-44 INCHES

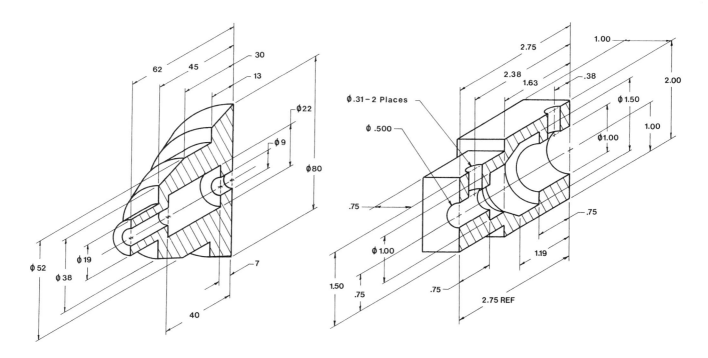

Draw at least one orthographic view and the indicated sectional view for each object in exercise problems EX4-45 through EX4-50.

EX4-45 MILLIMETERS

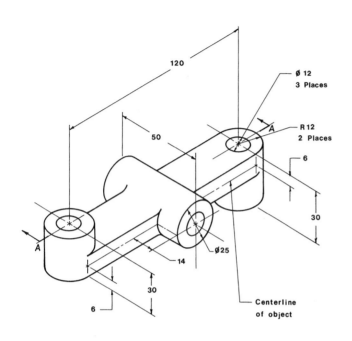

EX4-47 INCHES

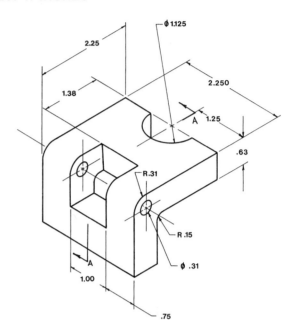

EX4-46 MILLIMETERS

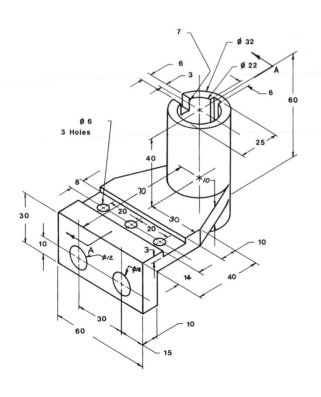

EX4-48 INCHES

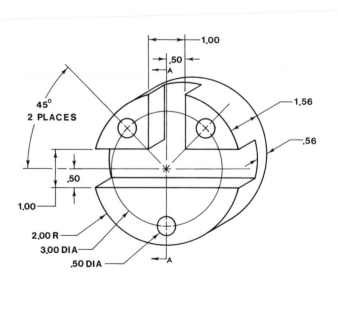

EX4-49 MILLIMETERS

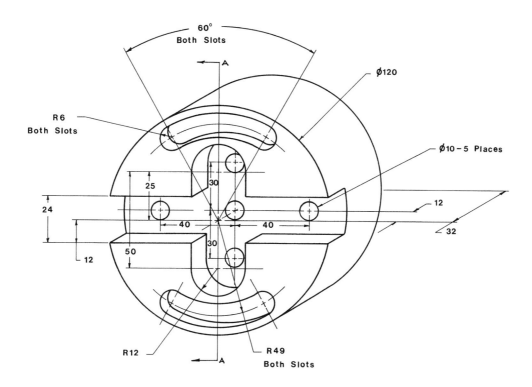

EX4-50 MILLIMETERS

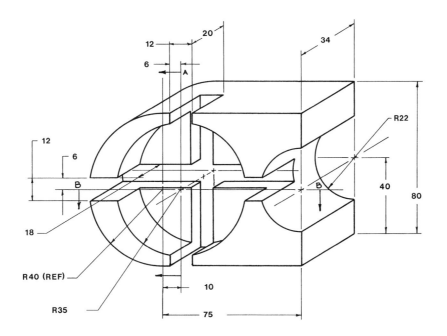

Given the orthographic views in exercise problems EX4-51 and 4-52, draw a model of each, then draw the given orthographic views and the appropriate sectional views.

EX4-51 INCHES

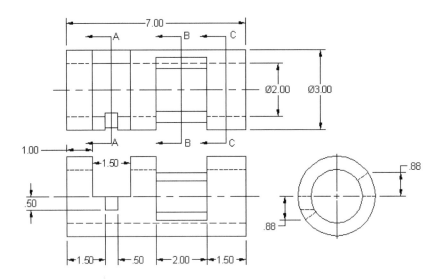

EX4-52 MILLIMETERS

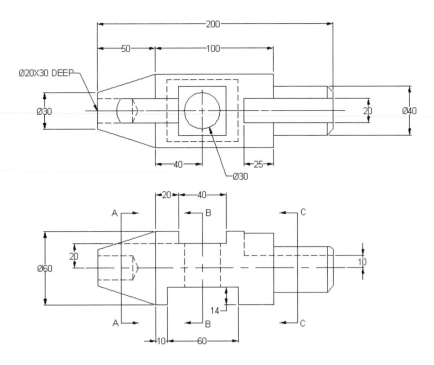

Draw a 3D model and a set of multi-views for each object shown in exercise problems EX4-53 through EX4-60.

EX4-53 INCHES

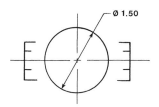

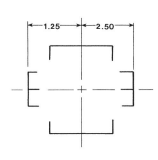

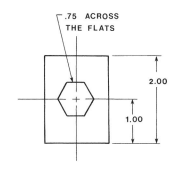

EX4-55 MILLIMETERS

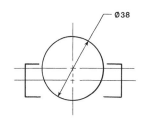

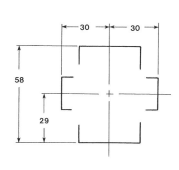

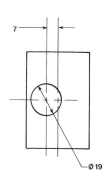

EX4-54 MILLIMETERS

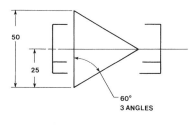

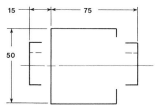

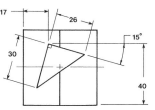

EX4-56 MILLIMETERS

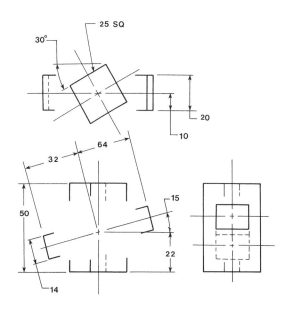

EX4-57 MILLIMETERS

Ø 50

43   86

15°

31

EQUILATERAL TRIANGLE
30 PER SIDE

62

EX4-59 MILLIMETERS

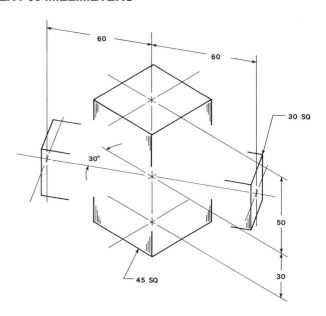

60   60

30 SQ

30°

50

45 SQ

30

EX4-58 MILLIMETERS

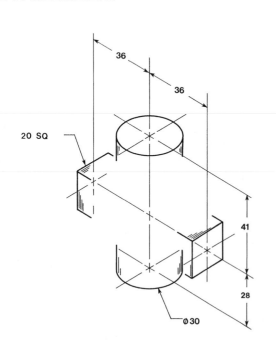

36

36

20 SQ

41

28

Ø 30

EX4-60 INCHES

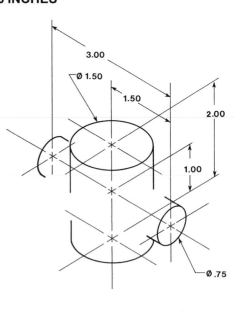

3.00

Ø 1.50

1.50

2.00

1.00

Ø .75

CHAPTER 5

Assembly Drawings

5-1 INTRODUCTION

This chapter explains how to create assembly drawings. It uses a group of relatively simple parts to demonstrate the techniques required. The idea is to learn how to create assembly drawings and then gradually apply the knowledge to more difficult assemblies. For example, the next chapter introduces threads and fasteners and includes several exercise problems that require the use of fasteners when creating assembly drawings. Assembly drawings will be included throughout the remainder of the book.

This chapter also shows how to create bills of materials, isometric assembly drawings, title blocks, and other blocks associated with assembly drawings. The chapter also shows how to animate assembly drawings.

5-2 BOTTOM-UP AND TOP-DOWN ASSEMBLIES

There are three ways to create assembly drawings: bottom up, top down, or a combination of the two. A *bottom-up* approach uses drawings that already exist. Model drawings are pulled from files and compiled to create an assembly. The *top-down* approach creates model drawings from the assembly drawing. It is also possible to pull drawings from a file and then create more drawings as needed to complete the assembly.

5-3 TO START AN ASSEMBLY DRAWING

Assembly drawings are created using the .iam format. In this example the bottom-up approach will be used. A model called SQBLOCK already exists. The SQBLOCK figure was created from a 30 mm × 30 mm × 30 mm cube with a 15 mm × 15 mm × 30 mm cutout.

To start an assembly drawing

1. Click the New tool, select the Metric tab, then Standard.iam.

   See Figure 5-1. The Assembly Panel bar will appear. See Figure 5-2.

2. Click the Place Component tool.

   The Open dialog box will appear. See Figure 5-3.

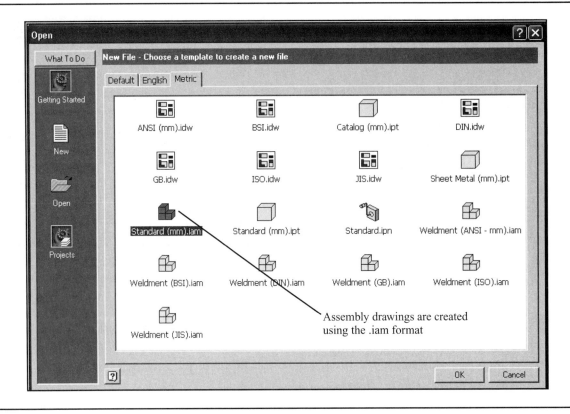

Assembly drawings are created using the .iam format

Figure 5-1

Figure 5-2

3. Click the desired file name, then OK.

   In this example the SQBLOCK file was selected. The selected model (component) will appear on the screen.

4. Zoom the component to an appropriate size, then left-click the mouse to locate the component.

   A second copy of the component will automatically appear.

5. Move the second component away from the first. Left-click the mouse to locate the second component, then right-click the mouse and select the Done option.

   See Figure 5-4.

5-4 DEGREES OF FREEDOM

Components are either free to move or they are grounded. *Grounded* components will not move when assembly tools are applied. The first component will automatically be grounded. Grounded components are identified by a push-pin icon in the browser box. See Figure 5-5.

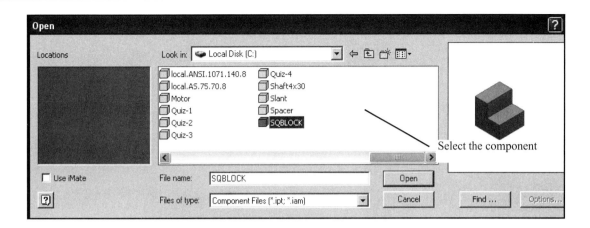

Figure 5-3

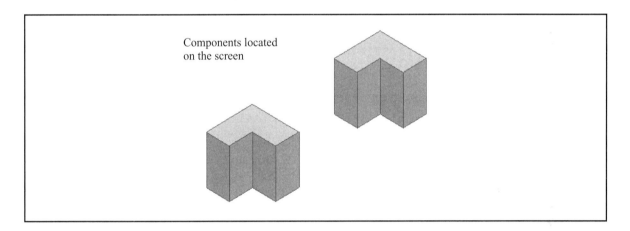

Figure 5-4

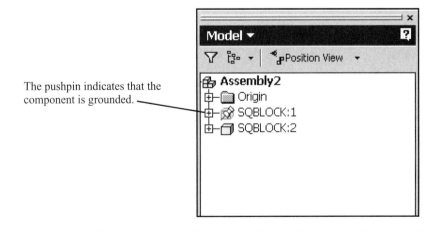

Figure 5-5

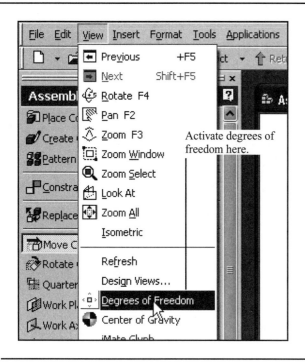

Figure 5-6

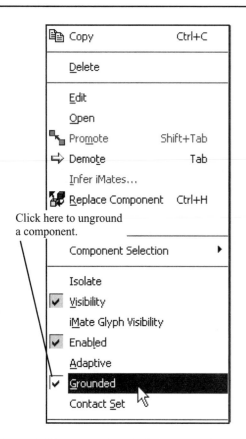

Figure 5-8

No degrees of freedom appear because the component is grounded.

Degrees of freedom available

Figure 5-7

To display the degrees of freedom

Components that are not grounded will have degrees of freedom. The available degrees of freedom for a component may be seen by using the Degrees of Freedom option.

1. Click the View heading at the top of the screen.
2. Click the Degrees of Freedom option.

See Figure 5-6. The available degrees of freedom will appear on the components. See Figure 5-7. Note that in Figure 5-7 the first component does not have any degrees of freedom; it is grounded.

To unground a component

1. Right-click on the component's heading in the browser box.

   A dialog box will appear. See Figure 5-8.

2. Click the Grounded option.

5-5 MOVE AND ROTATE

The Move and Rotate tools found on the Assembly Panel bar are used, as their names imply, to move and rotate components.

To move a component

1. Click the Move tool, then click the component to be moved.
2. Hold the left mouse button down and move the component about the screen.
3. When the desired location is reached, release the left button.
4. Right-click the mouse and select the Done option.

To rotate a component

1. Click the Rotate tool, then click the component to rotate.

A circle will appear around the component. See Figure Figure 5-9.

2. Click and hold the left mouse button outside the circle and move the cursor.

The component will rotate. It is suggested that various points outside the circle be tried to see how the component can be rotated.

3. When the desired orientation is achieved, press the right mouse button and select the Done option.

5-6 CONSTRAINT

The Constraint tool is used to locate components relative to one another. Components may be constrained using the mate, flush, angle, tangent, or insert constraint options.

To use the Mate command

1. Click the Constraint tool.

The Place Constraint dialog box will appear. See Figure 5-10. The Mate command will automatically be selected.

2. Click the front face of the left SQBLOCK as shown.
3. Click the front face of the right SQBLOCK as shown.
4. Click the Apply box on the Place Constraint dialog box or right-click the mouse and select the Apply option.

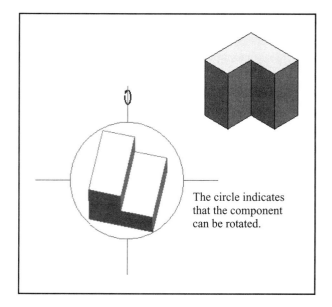

The circle indicates that the component can be rotated.

Figure 5-9

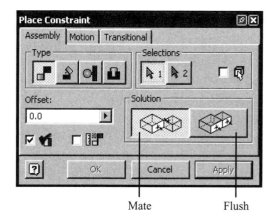

Mate   Flush

Figure 5-10

Click surfaces as indicated.

Figure 5-11

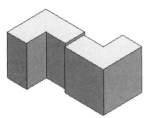

Resulting mated surfaces.

Figure 5-12

See Figure 5-11. The blocks will be joined at the selected surfaces. See Figure 5-12.

The blocks may not be perfectly aligned when assembled. This situation may be corrected using the Flush command.

To use the Flush command

1. Click on the Flush button on the Place Constraint dialog box.
2. Click the top surface of each block as shown.
3. Click the Apply button.

4. Make other surfaces flush as needed to align the two blocks.

To use the Offset option

Figure 5-13 shows two SQBLOCKs assembled using the Mate constraint with an offset value of 10 mm.

To position objects

Sometimes components are not oriented so they can be joined as desired. In these cases first rotate or move one of the components as needed, then use the Constraint com-

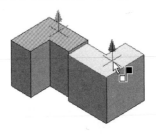

Use the Flush option to align the assemblies' surfaces.

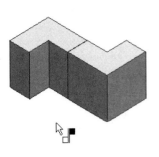

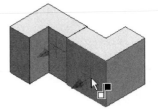

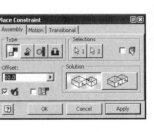

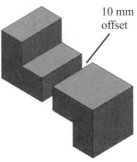

10 mm offset

Figure 5-13

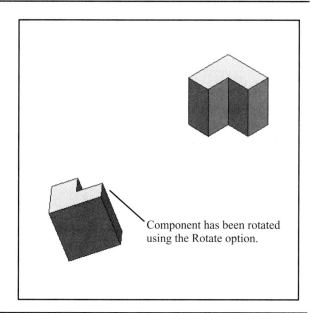

Component has been rotated using the Rotate option.

Figure 5-14

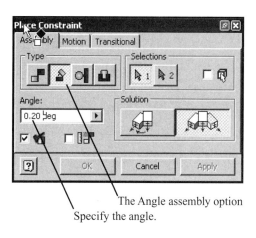

The Angle assembly option
Specify the angle.

Figure 5-16

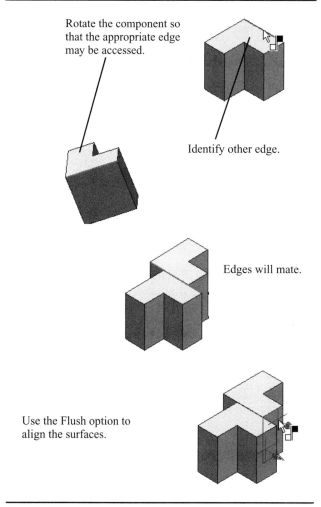

Rotate the component so that the appropriate edge may be accessed.

Identify other edge.

Edges will mate.

Use the Flush option to align the surfaces.

Figure 5-15

mands. See Figure 5-14. The left component has been rotated using the Rotate Component command.

1. Click the Place Constraint tool in the Assembly Panel bar.
2. Click the edge lines of the two components as shown.
3. Click the Apply box.

   See Figure 5-15.

4. Use the Flush command to align the components.
5. Click the Apply button.

To use the Rotate command

1. Click the Place Constraint tool.

   The Place Constraint dialog box will appear. See Figure 5-16.

2. Use the Mate constraint to align the edges of the SQBLOCKS.

   See Figure 5-17.

3. Use the Angle constraint to set an angle value of −20.00 to create an angle between the two SQBLOCKS.

   Inventor defines counterclockwise as the positive angular direction.

4. Use the Mate constraint to align the SQBLOCKS as desired.
5. Click the Apply box.

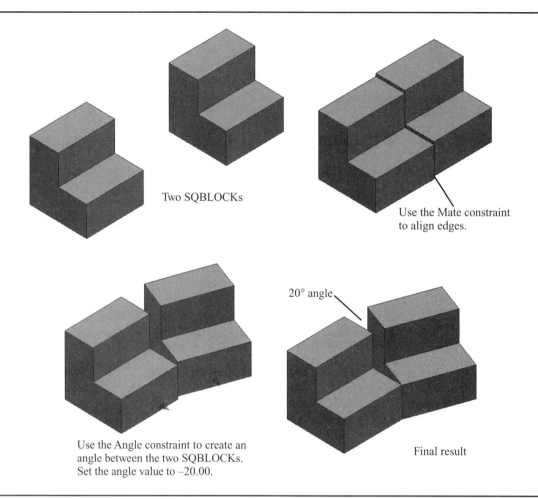

Two SQBLOCKs

Use the Mate constraint
to align edges.

Use the Angle constraint to create an
angle between the two SQBLOCKs.
Set the angle value to –20.00.

20° angle

Final result

Figure 5-17

Ø10 × 20

Ø20 × 20 with Ø10 hole through

Figure 5-18

6. Click the Mate command, then the two edge lines as shown.
7. Click the Apply box.
8. Use the Flush command to align the blocks.

To use the Tangent command

Figure 5-18 shows two cylinders. The smaller cylinder has dimensions of Ø10 × 20, and the larger cylinder has dimensions of Ø20 × 20 with a Ø10 centered longitudinal hole.

1. Click the Place Constraint tool.

The Place Constraint dialog box will appear. See Figure 5-19.

2. Click the Tangent box under the Type heading.

The Outside option will be selected automatically.

3. Select the outside edge of the large cylinder, then the outside edge of the smaller cylinder.

Figure 5-20 shows the resulting tangent constraint for the cylinders.

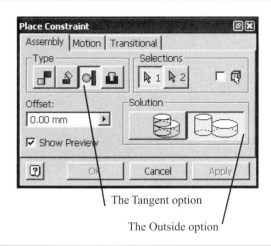

The Tangent option

The Outside option

Figure 5-19

The resulting tangent cylinders

Figure 5-20

To use the Insert command

1. Click the Place Constraint tool.

    The Place Constraint dialog box will appear. See Figure 5-21.

2. Click the Insert box under the Type heading, then click the Insert box under the Solution heading.
3. Click the top surface of each cylinder as shown.

    See Figure 5-22.

4. Click the Apply button.

    Figure 5-22 also shows the result of using the Opposed option under the Solution heading.

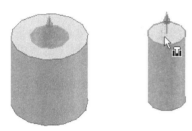

Identify the surfaces.

The Insert option

10 mm offset

The Opposed option

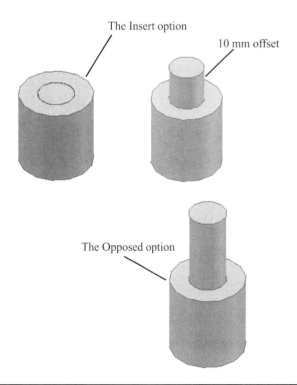

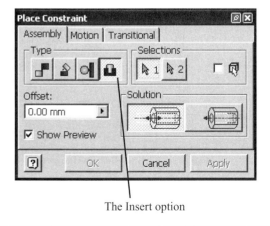

The Insert option

Figure 5-21

Figure 5-22

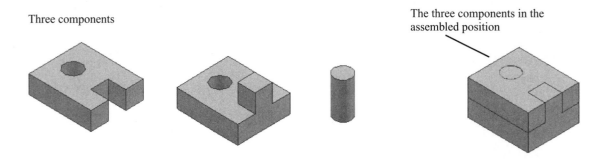

Three components

The three components in the assembled position

Figure 5-23

5-7 SAMPLE ASSEMBLY PROBLEM SP5-1

This section presents a sample assembly drawing. The assembly will then be used to create a presentation drawing including the animation command.

Figure 5-23 shows three components: Top Block, Bottom Block, and Peg. They were assembled using the techniques explained in Section 5-6.

To save an assembly

1. Create the assembly, then click the Save Copy As... heading under the File pull-down menu.

   The file will be saved as an .iam file.

5-8 PRESENTATION DRAWINGS

Presentation drawings are used to create exploded assembly drawings that can then be animated to show how the assembly is to be created from its components.

To create a presentation drawing

1. Click on the New Drawing tool.

   The Open dialog box will appear. See Figure 5-24.

2. Click the Standard.ipn tool, then OK.

Presentation drawings are created using the .ipn format.

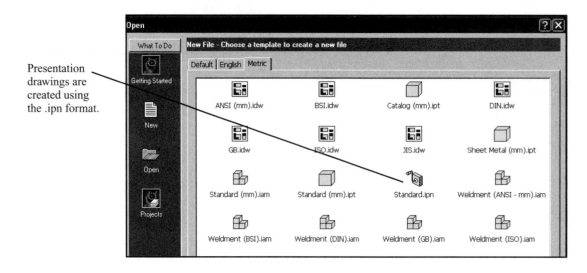

Figure 5-24

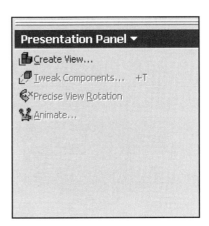

The Presentation Panel Bar

Figure 5-25

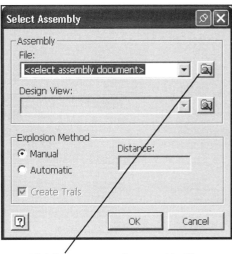

Click here to access the assembly files.

Figure 5-26

The Presentation Panel bar will appear. See Figure 5-25.

3. Click the Create View tool in the Presentation Panel bar.

The Select Assembly dialog box will appear. See Figure 5-26.

4. Click the Explore directories box.

The Open dialog box will appear, listing all the existing assembly drawings. See Figure 5-27.

5. Select the appropriate assembly drawing, then click Open.

The Select Assembly dialog box will reappear listing the selected assembly under the File heading. See Figure 5-28.

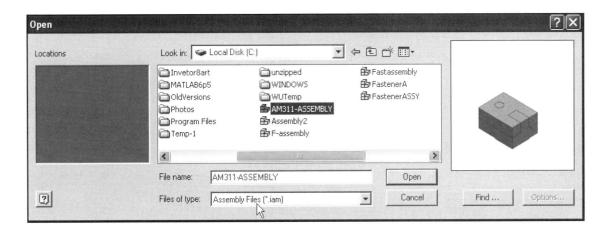

Figure 5-27

Figure 5-28

6. Click OK.

The assembly will appear. See Figure 5-29.

To create an exploded assembly drawing

1. Click the Tweak Components tool in the Presentation Panel bar.

The Tweak Component dialog box will appear. See Figure 5-30. The Direction option will automatically be selected.

A presentation drawing ready for tweaking.

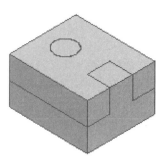

Figure 5-29

2. Select the direction of the tweak by selecting one of the assembly's vertical edge lines.

The Tweak Component dialog box will switch to the Components option.

3. Select the peg, then hold the left mouse button down and drag the peg to a position above the assembly.

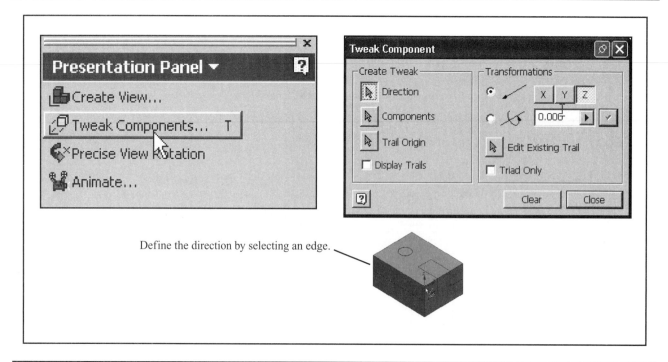

Define the direction by selecting an edge.

Figure 5-30

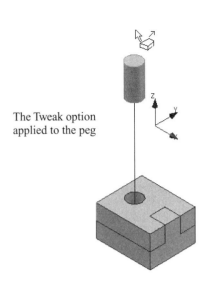

The Tweak option
applied to the peg

Figure 5-31

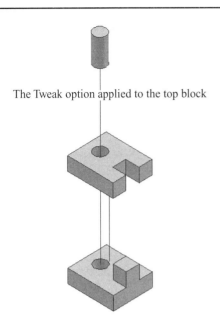

The Tweak option applied to the top block

Figure 5-32

See Figure 5-31.

4. Select the top block and drag it to a position above the bottom block.

See Figure 5-32.

5. Click the Clear box on the Tweak Component dialog box, then click the Close box.

To save the presentation drawing

1. Click the Save Copy As... heading on the File pull-down menu.

The Save Copy As dialog box will appear.

2. Enter the file name and click the Save box.

The drawing will be saved as an .ipn drawing.

5-9 ANIMATION

Presentation drawings can be animated using the Animation tool.

To animate a presentation drawing

1. Click on the Animate tool on the Presentation Panel bar.

The Animation dialog box will appear. See Figure 5-33. The control buttons on the Animation dialog box are similar to those found on CD players.

2. Click the Play forward button.

The assembly will be slowly reassembled in the reverse of the order used to tweak the components.

3. Click the Reset button to re-create the original presentation drawing.

Click here to play the animation.

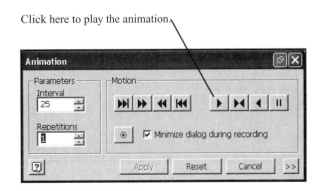

Figure 5-33

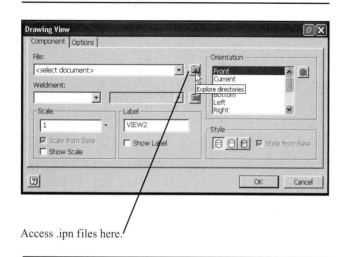

Access .ipn files here.

Figure 5-34

5-10 ISOMETRIC DRAWINGS

Isometric drawings can be created directly from presentation drawings. Assembly numbers (balloons) can be added to the isometric drawings to create exploded isometric assembly drawings.

To create an isometric drawing

1. Click on the New tool, then the Metric tab.

   The Open dialog box will appear.

2. Select the ANSI (mm).idw tool, then click OK.

The Drawing Management tools will appear in the panel bar. See Section 4-3 for a further explanation of the Drawing Management tools.

3. Click the Base View tool.

   The Drawing View dialog box will appear. See Figure 5-34.

4. Click the Explore directories button.

   The Open dialog box will appear. See Figure 5-35.

5. Select the appropriate presentation drawing (file type is .ipn), then click Open.

   The Drawing View dialog box will reappear. See Figure 5-36.

6. Select the Iso Top Right orientation and set the Scale as needed. Select the Hidden Lines Removed option under the Style heading.

Figure 5-37 shows the resulting isometric view. Figure 5-38 shows the isometric drawing created using the shaded option.

5-11 ASSEMBLY NUMBERS

Assembly numbers are added to an isometric drawing using the Balloon tool.

To add balloons

1. Locate the cursor in the panel bar area and right-click the mouse.
2. Click the Drawing Annotation option.

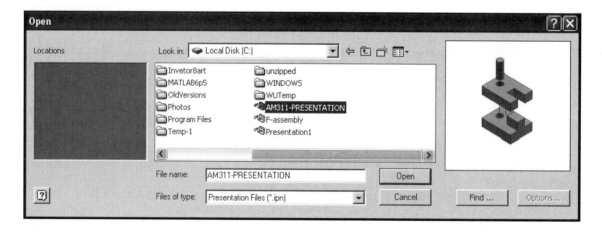

Figure 5-35

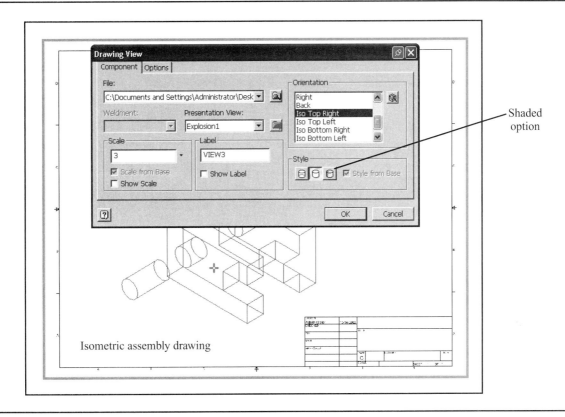

Shaded
option

Isometric assembly drawing

Figure 5-36

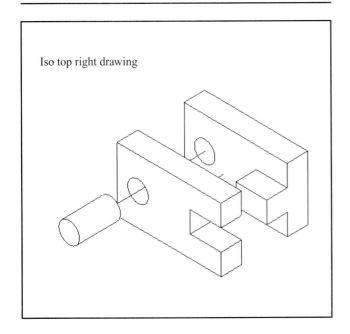

Iso top right drawing

Figure 5-37

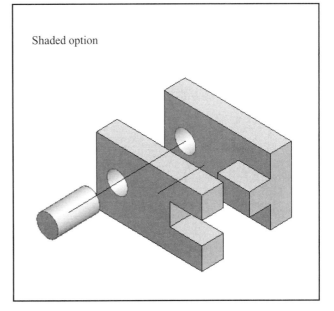

Shaded option

Figure 5-38

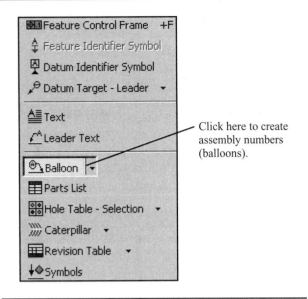

Click here to create assembly numbers (balloons).

Figure 5-39

Figure 5-40

The drawing annotation tools will be listed in the panel bar. See Figure 5-39.

3. Click the topmost edge line of the bottom block.

The Parts List – Item Numbering dialog box will appear. See Figure 5-40.

4. Click OK, then drag the cursor away from the selected edge line.
5. Locate a position away from the component and click the left mouse button. Move the cursor in a

horizontal direction and click the left mouse button again.

6. Right-click the mouse and select the Continue option.
7. Add balloons to the other components.
8. Move the cursor to the center of the screen and click the right mouse button, then select the Done option.

See Figure 5-41. Any excess tweak lines can be hidden by first clicking on them and clicking the right mouse button and selecting the Visibility option.

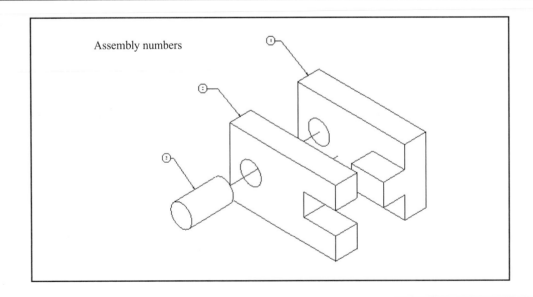

Assembly numbers

Figure 5-41

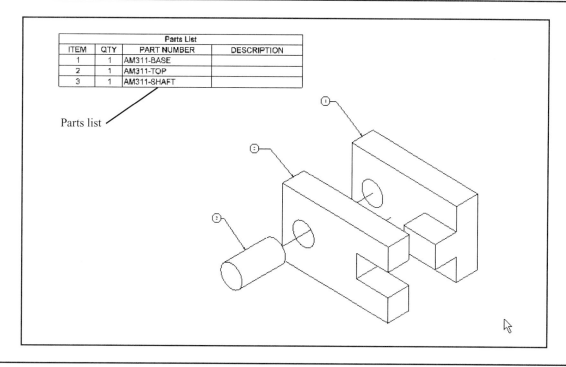

Figure 5-42

5-12 PARTS LIST

A parts list can be created from an isometric drawing after the balloons have been assigned using the Parts List tool on the Drawing Annotation Panel bar.

The Drawing Annotation Panel bar is accessed by moving the cursor into the panel bar area, then right-clicking the mouse and selecting the Drawing Annotation option.

To create a parts list

1. Click the Parts List tool on the Drawing Annotation Panel bar.
2. Move the cursor into the area around the isometric drawing.

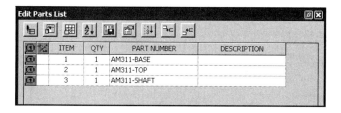

Figure 5-43

A broken red line will appear when the cursor is in the area.

3. Click the left mouse button and move the cursor away from the isometric drawing area.

The parts list will appear and move with the cursor.

4. Select a location for the parts list and left-click the mouse.

Figure 5-42 shows the resulting drawing. The parts list was generated using information from the original model drawings and the presentation drawings.

To edit a parts list

1. Move the cursor onto the parts list and right-click the mouse.
2. Click the Edit Parts List option.

The Edit Parts List dialog box will appear. See Figure 5-43. Any data presented in red numbers or letters are systems-generated output and cannot be edited. In the example shown the QTY numbers are in red. Additional components would have to be added to the drawing to change these numbers.

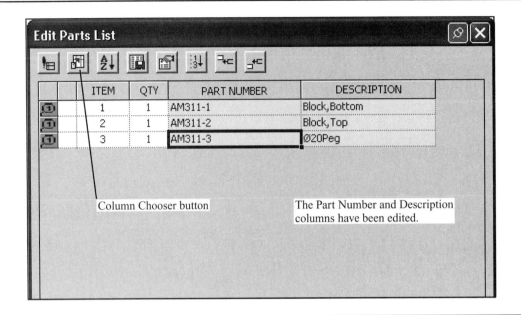

Figure 5-44

The Edit Parts List dialog box may be edited in a manner similar to that used with most spreadsheet programs. Click on a cell and either delete or add text. Figure 5-44 shows an edited parts list.

Naming parts

Each company or organization has its own system for naming parts. In the example given in this book the noun, modifier format was used.

To add a new column

Say two additional columns were required for the parts list shown in Figure 5-44: Material and Notes.

1. Click the Column Chooser button at the top of the Edit Parts List dialog box.

The Parts List Column Chooser dialog box will appear. See Figure 5-45.

Column names already defined

Enter new column names here.

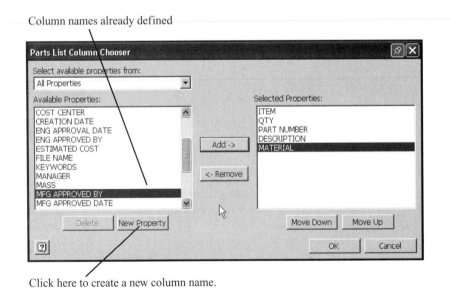

Click here to create a new column name.

Figure 5-45

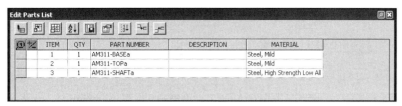

Red lettering indicates values that cannot be changed. Changes must be made to the original part file.

Figure 5-46

2. Scroll down the Available Properties listing to see if the new colum headings are listed.
3. Material is listed, so click on the listing, then click the Add box in the middle of the screen.

The heading Material will appear in the Selected Properties area.

The heading Notes is not listed, so it must be defined.

4. Click on the New Property box.

The Define New Property dialog box will appear. See Figure 5-45.

5. Type in the name of the new column, then click OK.

Note that only uppercase letters are used to define column headings.

6. Click OK on the Parts List Column Chooser dialog box.

Figure 5-46 shows the revised column in the parts list. The word Default appears under the Material heading because the material for a model will be assigned to the model drawing and brought forward into the parts list. The Material column can be changed only by going back and assigning a material to the model.

Figure 5-47 shows the edited parts list on the drawing.

5-13 TITLE BLOCK

All drawings include a title block, usually located in the lower left corner of the drawing sheet, as Figure 5-48 shows. Text may be added to a title block under existing headings, or new headings may be added.

To add text to a title block

1. Right-click the drawing name in the Model browser box, then click the Properties option.

Parts List				
ITEM	QTY	PART NUMBER	DESCRIPTION	MATERIAL
1	1	AM311-BASEa		Steel, Mild
2	1	AM311-TOPa		Steel, Mild
3	1	AM311-SHAFTa		Steel, High Strength Low Alloy

The material specifications were entered as the objects were created. See Section 3-32. If no material is defined, the word Default will appear.

Figure 5-47

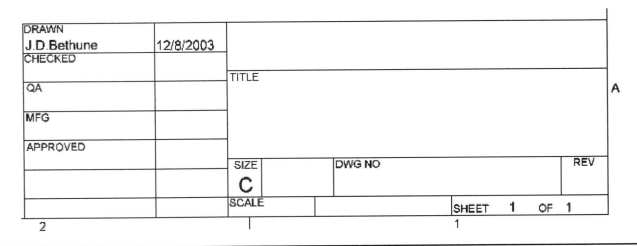

Figure 5-48

See Figure 5-49. The drawing's Properties dialog box will appear. Text can be typed into the Properties dialog box and will appear on the title block. Figure 5-50 shows the Summary input.

2. Click the Project tab on the Properties dialog box and enter the appropriate information.

See Figure 5-51. Figure 5-52 shows the completed title block.

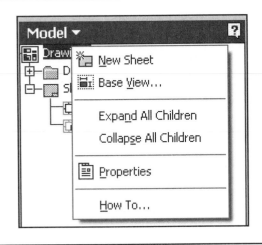

Figure 5-49

The title block included with Inventor is only one possible format. Each company and organization will have its own specifications.

5-14 DRAWING SHEETS

Drawings are prepared on predefined standard-size sheets of paper. Each standard size has been assigned a letter value. Figure 5-53 shows the letter values and the sheet size assigned to each. All these sizes and more are available within Inventor.

Figure 5-54 shows a drawing done on a C-size drawing sheet. Note the letter C in the title block.

To change a sheet size

1. Locate the cursor on the Sheet:1 heading in the browser box and right-click the mouse.
2. Select the Edit Sheet option.

The Edit Sheet dialog box will appear. See Figure 5-55.

3. Select the D option and click OK.

See Figure 5-56. Note that the letter C has been replaced with the letter D.

Figure 5-50

Figure 5-51

Figure 5-52

Standard Drawing Sheet Sizes
Inches

A = 8.5 ¥ 11
B = 11 ¥ 17
C = 17 ¥ 22
D = 22 ¥ 34
E = 34 ¥ 44

Standard Drawing Sheet Sizes
Millimeters

A4 = 210 ¥ 297
A3 = 297 ¥ 420
A2 = 420 ¥ 594
A1 = 594 ¥ 841
A0 = 841 ¥ 1189

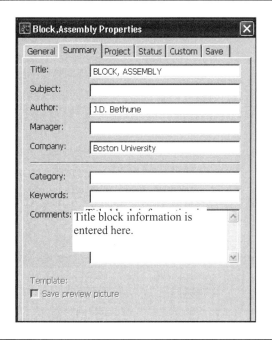

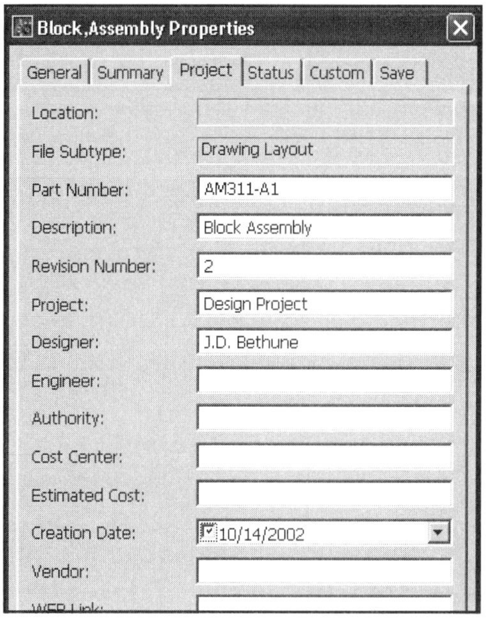

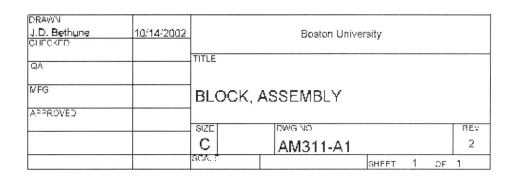

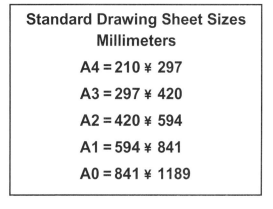

Figure 5-53

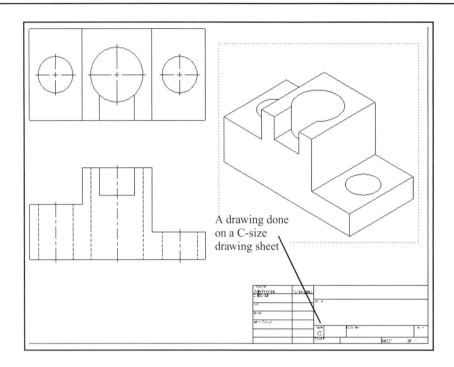

A drawing done on a C-size drawing sheet

Figure 5-54

5-15 OTHER TYPES OF DRAWING BLOCKS

Release blocks

Figure 5-57 shows an enlarged view of the title block. The area on the left side of the block is called a *release block.* After a drawing is completed it is first checked. If the drawing is acceptable, the checker will initial the drawing and forward it to the next approval person. Which person(s) and which department approve new drawings varies, but until a drawing is "signed off," that is, all required signatures have been entered, it is not considered a finished drawing.

Revision blocks

Figure 5-58 shows a sample revision block. It was created using the Revision Table tool located on the Drawing Annotation Panel. Drawings used in industry are constantly being changed. Products are improved or corrected, and drawings must reflect and document these changes.

Drawing changes are listed in the revision block by number. Revision blocks are usually located in the upper right corner of the drawing.

Each drawing revision is listed by number in the revision block. A brief description of the change is also included. It is important that the description be as accurate and complete as possible. The zone, on the drawing where the revision is located, is also specified.

The revision number is added to the field of the drawing in the area where the change was made. The revision letter is located within a "flag" to distinguish it from dimensions and drawing notes. The flag is created using the Revision Tag tool located on the Drawing Annotation Panel. See Figure 5-58. The Revision tag tool is a flyout from the Revision Table tool.

Select drawing sheet size here.

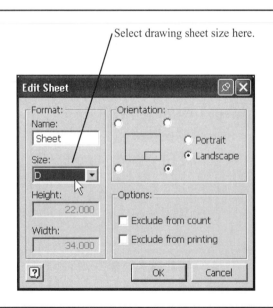

Figure 5-55

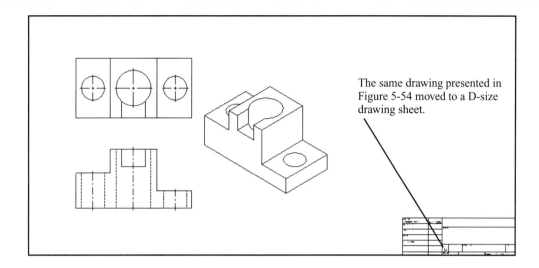

The same drawing presented in Figure 5-54 moved to a D-size drawing sheet.

Figure 5-56

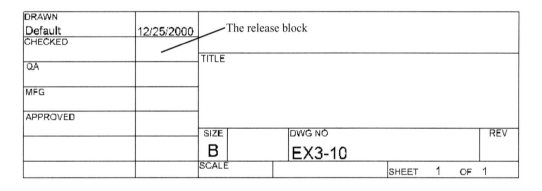

The release block

DRAWN		
Default	12/25/2000	
CHECKED		TITLE
QA		
MFG		
APPROVED		

		SIZE	DWG NO		REV
		B	EX3-10		
		SCALE		SHEET 1 OF 1	

Figure 5-57

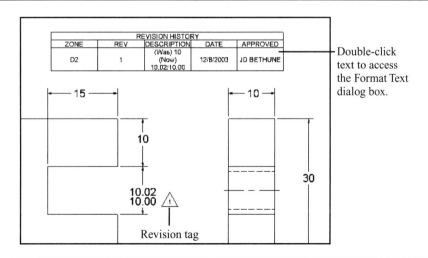

REVISION HISTORY				
ZONE	REV	DESCRIPTION	DATE	APPROVED
D2	1	(Was) 10 (Now) 10.02/10.00	12/8/2003	JD BETHUNE

Double-click text to access the Format Text dialog box.

Revision tag

Figure 5-58

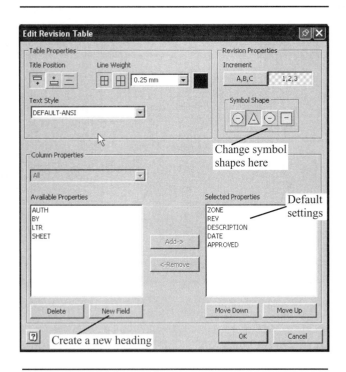

Figure 5-59

The text in the revision block can be edited by locating the cursor on the text, then double-clicking the left mouse button.

To edit the revision block

1. Move the cursor onto the revision block.

Filled green circles will appear around the revision block.

2. Right-click the mouse and select the Edit option.

The Edit Revision Table dialog box will appear. See Figure 5-59. Four different revision tag shapes are available. The block's headings may be edited or rearranged as needed.

ECOs

Most companies have systems in place that allow engineers and designers to make quick changes to drawings. These change orders are called *engineering change orders* (ECOs), *engineering orders* (EOs), or *change orders* (COs), depending on the company's preference. Change orders are documented on special drawing sheets that are usually stapled to a print of the drawing. Figure 5-60 shows a sample change order attached to a drawing.

After a number of change orders have accumulated, they are incorporated into the drawing. This process is called a *drawing revision,* which is different from a revision to the drawing. Drawing revisions are usually identified by a letter located somewhere in the title block. The revision letters may be included as part of the drawing number or in a separate box in the title block. Whenever you are working on a drawing make sure you have the latest revision and all appropriate change orders. Companies have recording and referencing systems for listing all drawing revisions and drawing changes.

Drawing notes

Drawing notes are used to provide manufacturing information that is not visual, for example, finishing instructions, torque requirements for bolts, and shipping instructions.

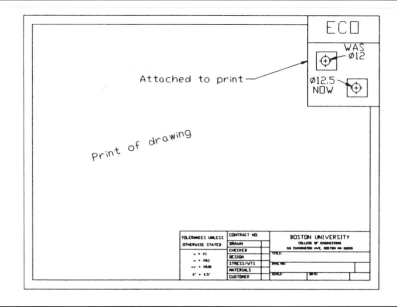

Figure 5-60

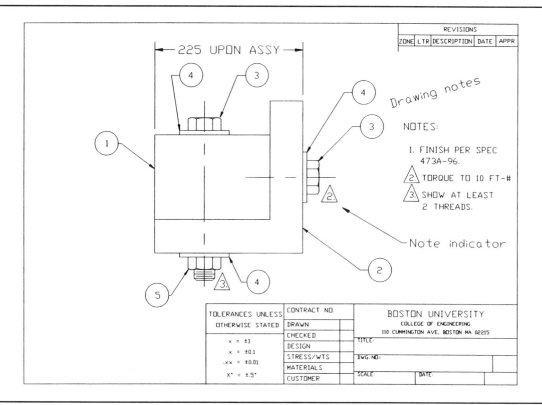

Figure 5-61

Drawing notes are usually listed on the right side of the drawing above the title block. Drawing notes are listed by number. If a note applies to a specific part of the drawing, the note number in enclosed in a triangle. The note numbers enclosed in triangles are also drawn next to the corresponding areas of the drawing. See Figure 5-61.

5-16 SAMPLE PROBLEM SP5-2

Figure 5-62 shows a group of parts that are to be assembled together. Each of the three parts was created using Standard (mm).ipt format.

1. Draw the required individual parts.
2. Start a new assembly drawing, and place the required parts on the drawing.
3. If necessary, use the Rotate Component tool to reposition a component.

   See Figure 5-63.

4. Assemble the components. The components may be grounded as they are assembled.

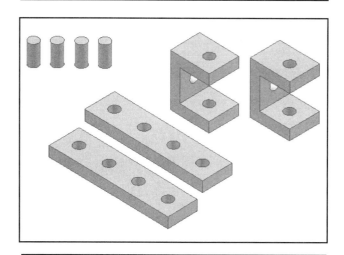

Figure 5-62

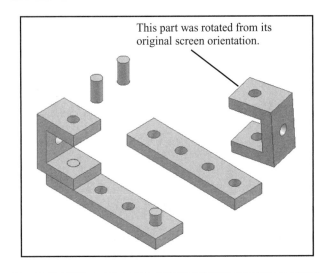

Figure 5-63

Figure 5-64 shows the final assembly.

5-17 TOP-DOWN ASSEMBLIES

A *top-down assembly* is an assembly that creates new parts as the assembly is created. Figure 5-65 shows a rotator assembly that was created using the top-down method. This section will explain how the assembly was created.

To start an assembly

1. Click on the New tool.
2. Click the Metric tab, then select Standard (mm).iam.

    The Assembly Panel will appear. See Figure 5-66.

To change the sketch plane

The parts created for this assembly were created on the XZ plane. This gives the assembly a more realistic appearance.

1. Click the Tools heading at the top of the screen and select Application Options.
2. Click the Part tab.
3. Select the Sketch on X-Z Plane button, then Apply and OK.

    See Figure 5-67.

To create a part

1. Click the Create Component tool on the Assembly Panel bar.

    The Create In-Place Component dialog box will appear.

2. Change the file name to PLATE.
3. Click the browse box located to the right of the Template box.

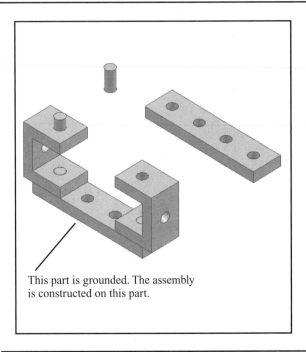

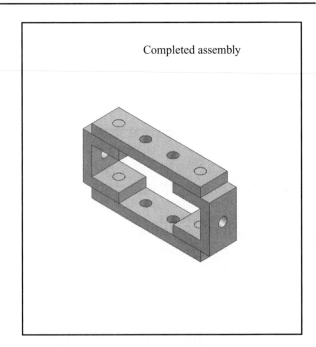

Figure 5-64

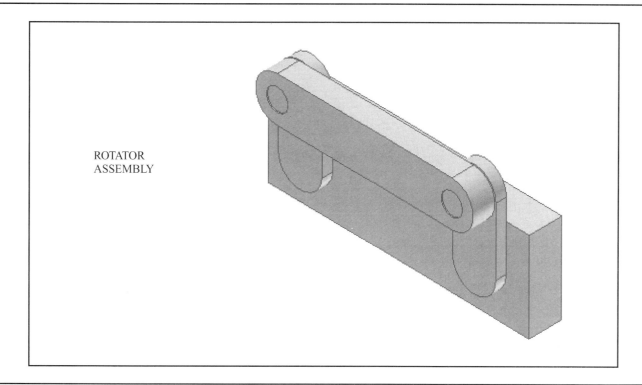

ROTATOR
ASSEMBLY

Figure 5-65

Figure 5-66

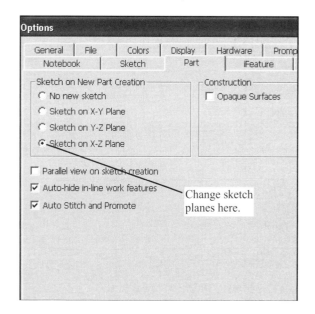

Change sketch
planes here.

Figure 5-67

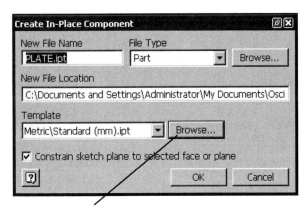

Click here to set metric units for component.

Figure 5-68

4. Click the Metric tab, then select the Standard (mm).ipt option, then OK.

See Figure 5-68. The 2D Sketch Panel will appear in the panel box area.

5. Right-click the mouse and select the Isometric View option.
6. Use the Two point rectangle, Line, and General Dimension tools to create the PLATE shown in Figure 5-69.

To add work points and work axis

1. Right-click the mouse and select the Finish Sketch option.

The Part Features panel will reappear.

2. Click the Work Point option and locate work points at both ends of the central line.
3. Click the Work Axis tool and add a work axis between the two work points.

The work points and work axis will be listed in the browser area.

4. Click on the PLATE origin in the browser area.
5. Select the Work Axis tool, then click the XZ Plane in the browser area and add a work axis through both work points.
6. Save the PLATE.

See Figure 5-70.

To create LINK-L

1. Click on the Look At tool on the Standard toolbar, then click one of the lines on the PLATE.

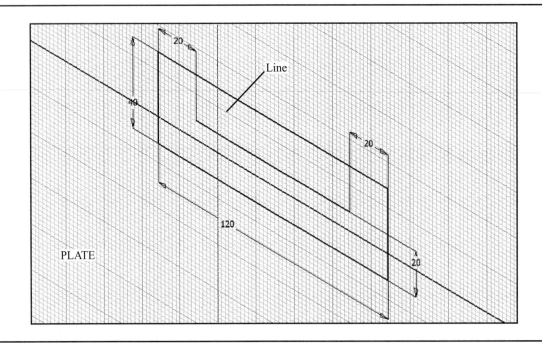

Figure 5-69

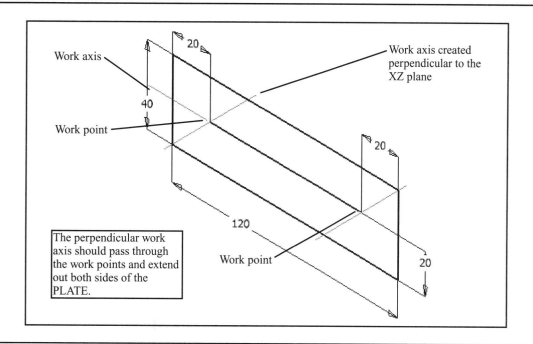

Figure 5-70

The drawing will return to a two-dimensional view.

2. Click on the Create Component tool and create file LINK-L with Metric units.
3. Click the drawing screen to access the 2D Sketch Panel.
4. Use the Circle and Line tools to create LINK-L as shown in Figure 5-71.
5. Change to the isometric view.

6. Right-click the mouse and select the Finish Sketch option.
7. Use the Work Point tool and create work points at the center of both circles. Use the Work Axis tool to create a work axis between the two hole centers.
8. Use the Work Axis tool and the XZ Plane option listed under the LINK-1 origin in the browser area to create two work axes through the two work points, perpendicular to the XZ plane.

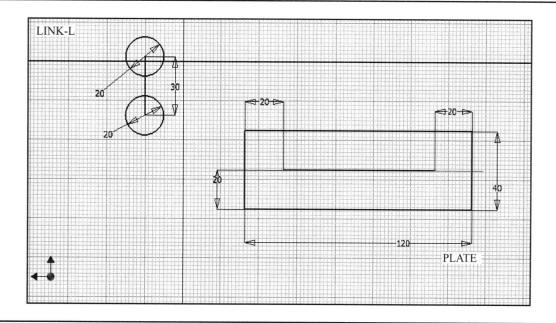

Figure 5-71

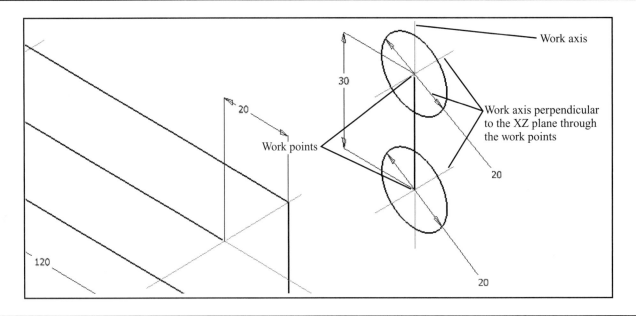

Figure 5-72

See Figure 5-72.

9. Save LINK-L.

To create LINK-R and the CROSSLINK

Two more links are required to complete the assembly.

1. Use the Create Component tool and create LINK-R using the dimensions presented in Figure 5-73.

2. Use the Create Component tool to create CROSSLINK using the dimensions presented in Figure 5-73.

3. Save LINK-R and CROSSLINK.

To save the assembly

1. Double-click the word Assembly in the browser area.

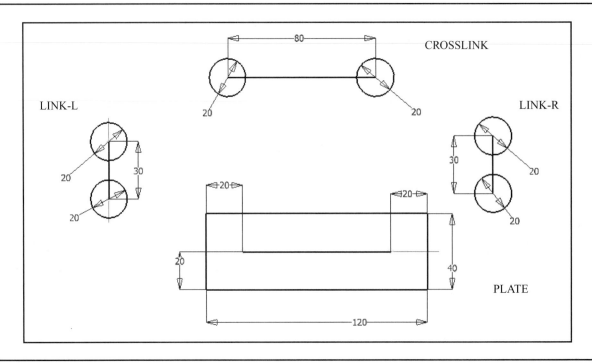

Figure 5-73

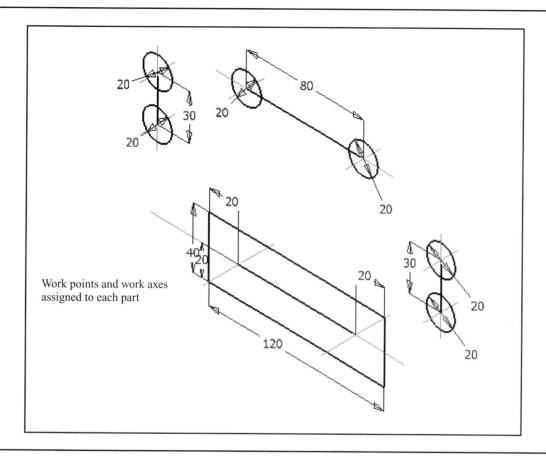

Work points and work axes assigned to each part

Figure 5-74

2. Click the Save As tool listed under the File heading at the top of the screen.
3. Save the file as ROTATOR.

The word Assembly will be changed to ROTATOR in the browser area.

To complete LIKNK-R and CROSSLINK

1. Change the drawing to the isometric view.
2. Click on LINK-R in the browser area to activate the LINK-R drawing and add two work points, a work axis between the hole centers, and two work axes perpendicular to the XZ plane.
3. Save LINK-R.
4. Click on CROSSLINK in the browser area to activate the CROSSLINK drawing and add two work points, a work axis between the hole centers, and two work axes perpendicular to the XZ plane.
5. Save CROSSLINK.

See Figure 5-74.

To assemble the parts

1. Double-click the word ROTATOR in the browser area to return to the Assembly Panel.
2. Click the Constraint tool.

The Place Constraint dialog box will appear.

3. Select the Mate option.

The Mate option may automatically be selected.

4. Mate the work axes as shown in Figure 5-75.
5. Click Apply.
6. Assure that the Mate tool is still active and mate the work axes shown in Figure 5-76.
7. Assure that the Mate tool is still active and mate the work axes shown in Figures 5-77 and 5-78.
8. Use the Look At tool to create a two-dimensional view of the assembly.

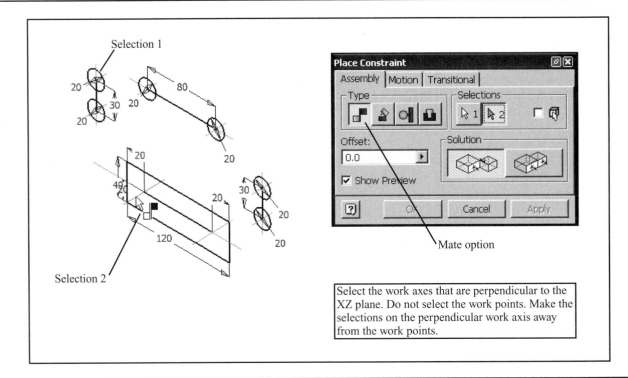

Figure 5-75

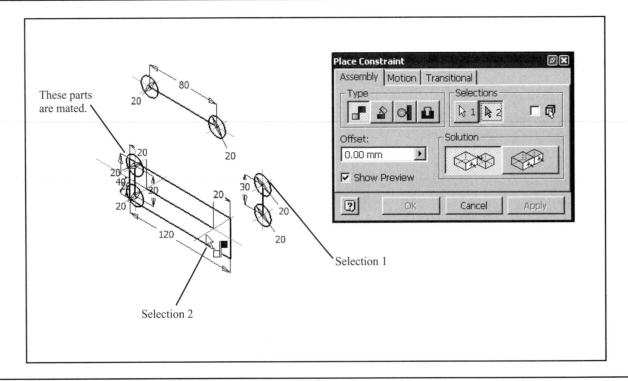

Figure 5-76

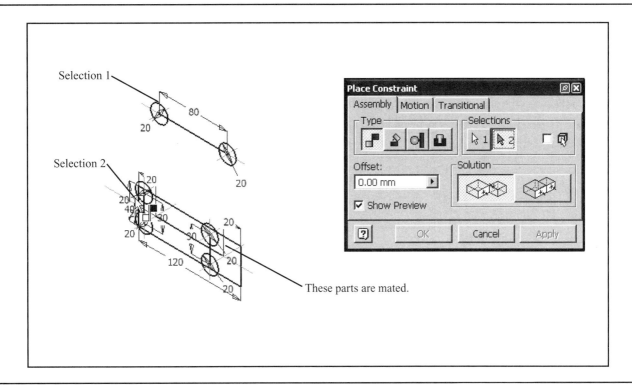

Figure 5-77

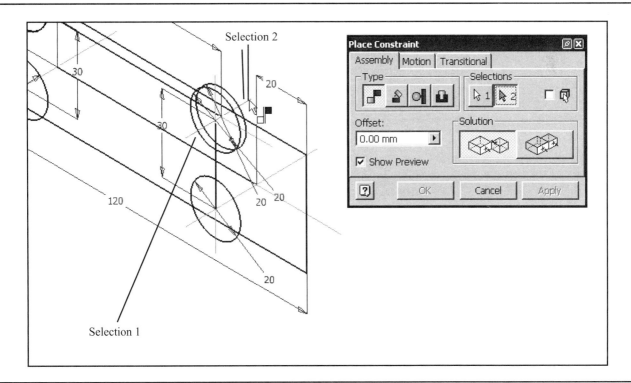

Figure 5-78

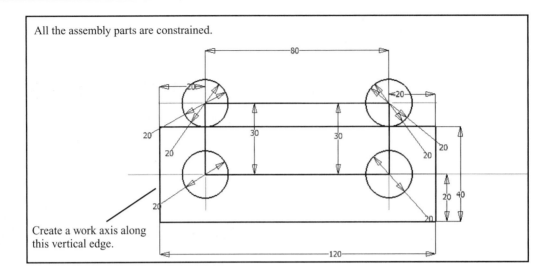

Figure 5-79

Figure 5-79 shows the resulting assembly drawing.

To animate the Links

1. Double-click the word ROTATOR in the browser area to access the Assembly Panel.
2. Create a work axis on the edge of the PLATE as shown in Figure 5-79.
3. Click the Constraint tool.

The Place Constraint dialog box will appear.

4. Select the Angle option.

The default angle setting is 0.00 deg.

5. Select the vertical work axes on the PLATE in step 2, then select the vertical axis of LINK-R as shown in Figure 5-80.
6. Apply the constraints.

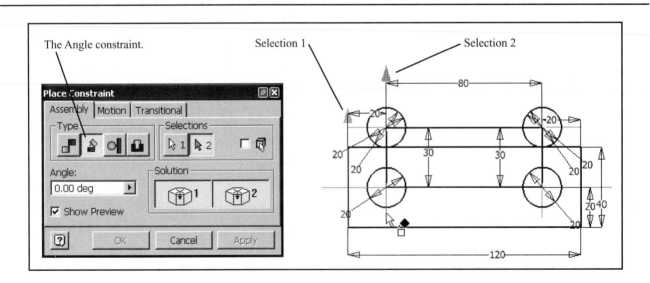

Figure 5-80

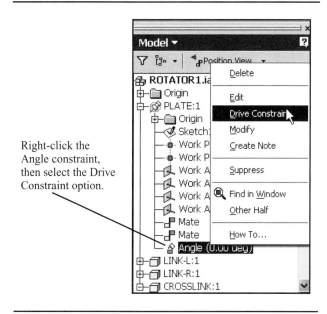

Right-click the Angle constraint, then select the Drive Constraint option.

Figure 5-81

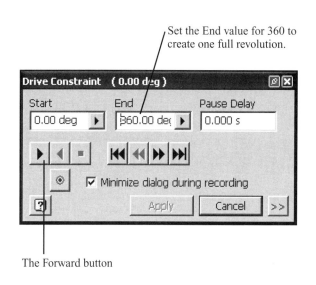

Set the End value for 360 to create one full revolution.

The Forward button

Figure 5-82

See Figure 5-81.

7. Save the assembly.

To set the assembly in motion

1. Right-click the Angle (0.00 deg) constraint in the browser area.

See Figure 5-81. The Drive Constraint dialog box will appear. See Figure 5-82.

2. Set the End angle for 360.
3. Click the Forward button.

The assembly should rotate freely. If the rotation is not correct, check the constraints.

To control the speed of the rotation

1. Click the ≪ button on the Drive Constraint dialog box.

The box will expand. See Figure 5-83.

2. Set the Increment value to 5 deg.
3. Click the Forward button.

The higher the Increment value the faster the assembly will rotate. The Repetitions setting is used to control the number of revolutions generated.

Click here to access the increment settings.

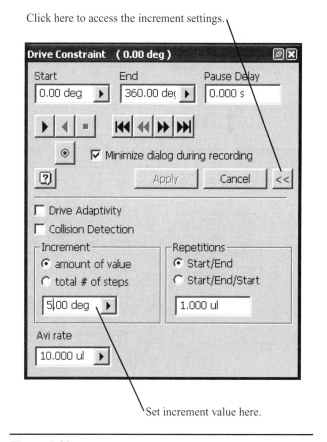

Set increment value here.

Figure 5-83

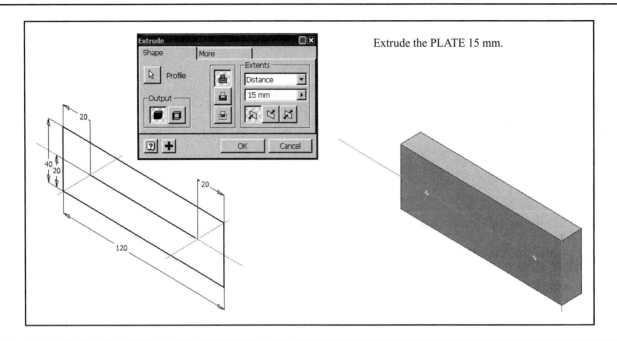

Extrude the PLATE 15 mm.

Figure 5-84

To complete the PLATE

1. Right-click the LINK-L, LINK-R, and CROSSLINK headings in the browser area and hide the parts by selecting the Visibility option.
2. Move the cursor into the drawing area, right-click the mouse, and select the Isometric View option.

See Figure 5-84.

3. Double-click PLATE in the browser area to change to the Part Features panel bar.
4. Select the Extrude tool and set the Extents distance to 15 mm.
5. Click OK.
6. Create a New Sketch on the front surface of the PLATE.
7. Create two Point, Hole Centers on the two work points, right-click the mouse, and select the Done option.
8. Click the Return tool, then create two Ø10 holes as shown in Figure 5-85.

To complete LINK-L and LINK-R

1. Right-click LINK-L in the browser area and select the Visibility option to make LINK-L visible.
2. Right-click the word Sketch under the LINK-L heading and select the Edit Sketch option.
3. Draw two tangent lines and two Ø10 circles as shown in Figure Figure 5-86.
4. Click the Return tool and extrude LINK-L 5 mm.
5. Extrude the top circle 10 forward, and extrude the lower circle 20 through and back as shown.
6. Edit LINK-R to the same dimensions and features.

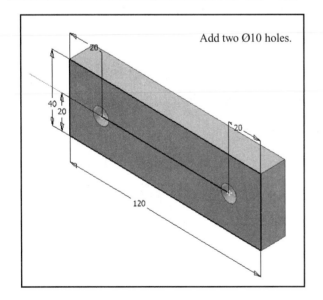

Add two Ø10 holes.

Figure 5-85

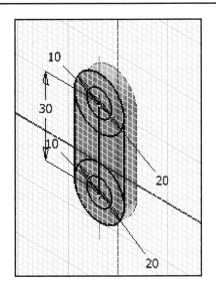
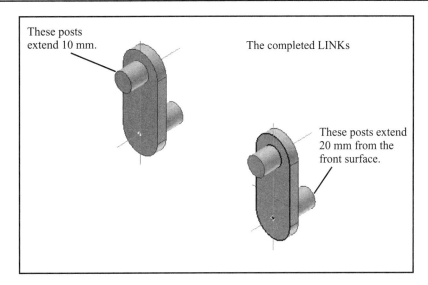

Figure 5-86

To complete the CROSSLINK

1. Hide LINK-L and LINK-R, then right-click CROSSLINK in the browser area and make the CROSSLINK visible.
2. Add two tangent lines and extrude the CROSSLINK 10 mm.

   See Figure 5-87.

3. Create a New Sketch and add two Ø10 holes as shown.

To align the assembly

1. Double-click the word ROTATOR in the browser area and make all parts visible.
2. Use the Move Component tool and position the parts so they are clearly visible.

   In this example the CROSSLINK was moved away from the LINKs. See Figure 5-88.

3. Click the Constraint tool and select the Mate option.

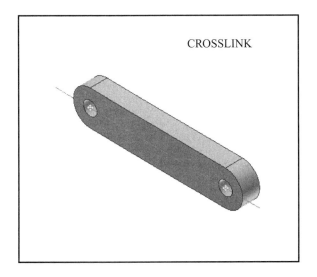

Figure 5-87

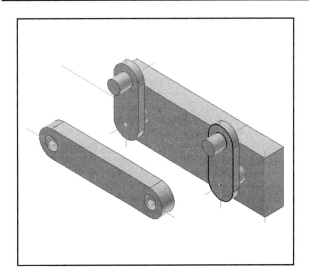

Figure 5-88

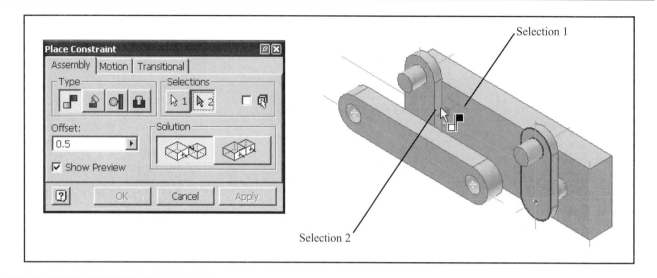

Figure 5-89

4. Set the Offset value for 0.5.
5. Select the front face of the PLATE and the back edge line of the LINK as shown.
6. Apply the constraint.

   See Figure 5-89.

7. Repeat the procedure for the other LINK.
8. Use the Look At tool and select the lower right surface of the PLATE to see the 0.5 offset between the PLATE and the LINKs.

See Figure 5-90.

9. Return the drawing to the isometric view.
10. Click the Constraint tool and mate the CROSSLINK with the LINKs using a 0.5 offset.

Presentations

Figure 5-91 shows a presentation drawing of the ROTATOR, and Figure 5-92 shows an exploded isometric drawing created using the .idw format.

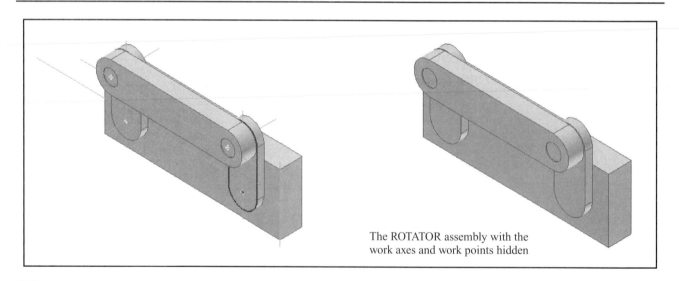

The ROTATOR assembly with the work axes and work points hidden

Figure 5-90

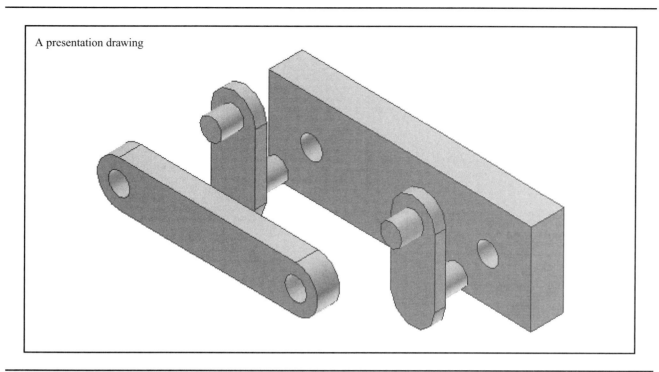

A presentation drawing

Figure 5-91

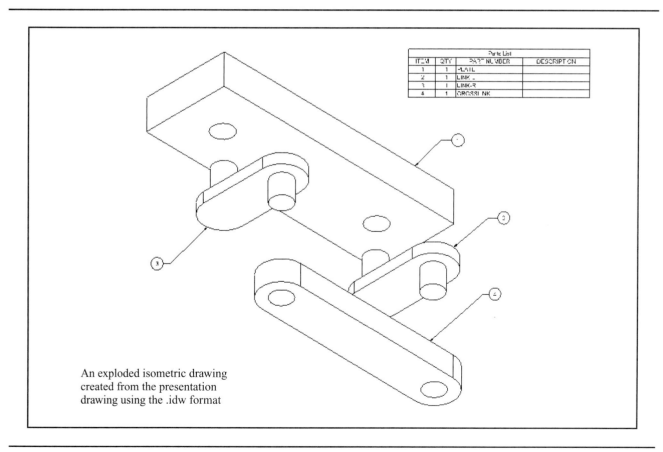

Parts List			
ITEM	QTY	PART NUMBER	DESCRIPTION
1	1	PLATE	
2	1	LINK-L	
3	1	LINK-R	
4	1	CROSSLINK	

An exploded isometric drawing created from the presentation drawing using the .idw format

Figure 5-92

5-18 EXERCISE PROBLEMS

EX5-1 MILLIMETERS

A dimensioned block is shown. Redraw this block and save it as SQBLOCK. See Section 5-3. Use the SQBLOCK to create assemblies as shown.

EX5-1A

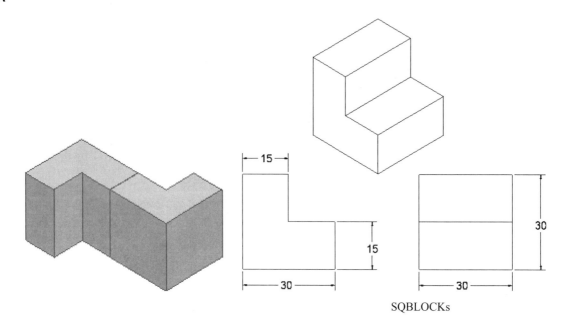

SQBLOCKs

EX5-1B

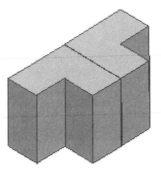

EX5-1C

20° angle between the two SQBLOCKs

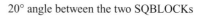
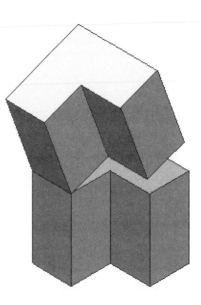

EX5-1D

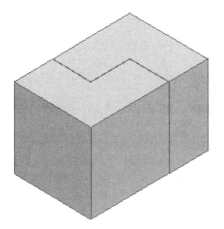

EX5-1F

EX5-1E

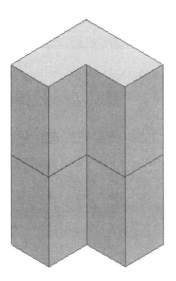

EX5-1G

10 mm offset

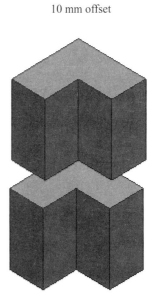

Pages 168 through 170 show a group of parts. These parts are used to create the assemblies presented as exercise problems in this section. Use the given descriptions, part numbers, and materials when creating BOMs for the assemblies.

EX5-2 MILLIMETERS

Redraw the following models and save them as Standard .ipn files.

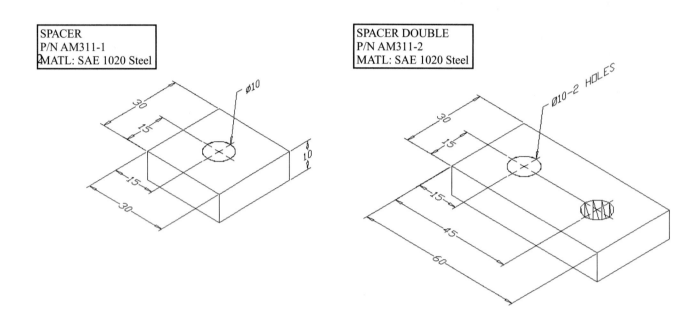

SPACER
P/N AM311-1
MATL: SAE 1020 Steel

SPACER DOUBLE
P/N AM311-2
MATL: SAE 1020 Steel

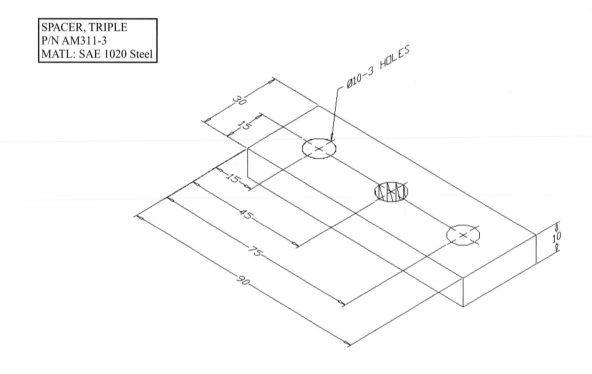

SPACER, TRIPLE
P/N AM311-3
MATL: SAE 1020 Steel

SPACER, QUAD
P/N AM311-4
MATL: SAE 1020 Steel

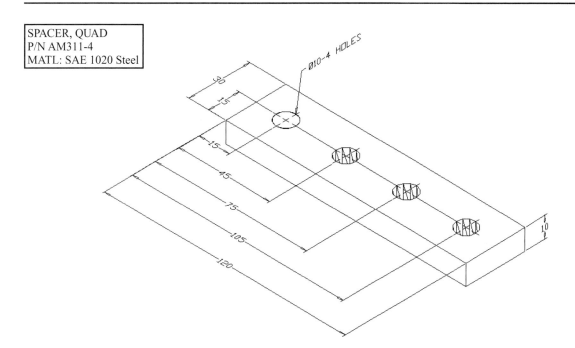

PEGS
MATL: Steel

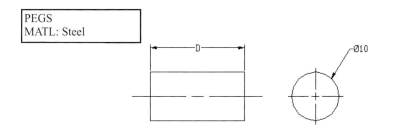

DESCRIPTION	PART NO.	D
PEG, SHORT	PG20-1	20
PEG	PG30-1	30
PEG, LONG	PG40-1	40

ALL DISTANCES IN MILLIMETERS

L-BRACKET
P/N BK20-1
MATL: SAE 1040 Steel

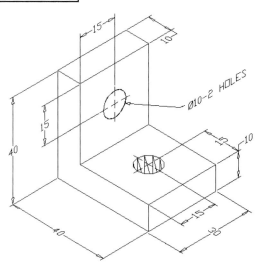

Z-BRACKET
P/N BK20-2
MATL: SAE 1040 Steel

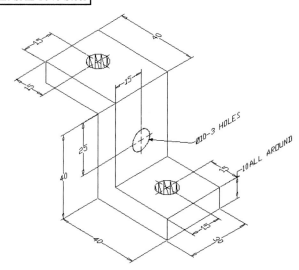

C-BRACKET
P/N BK20-3
SAE 1040 Steel

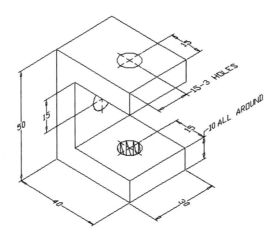

PLATE, BASE
SAE 1020 Steel

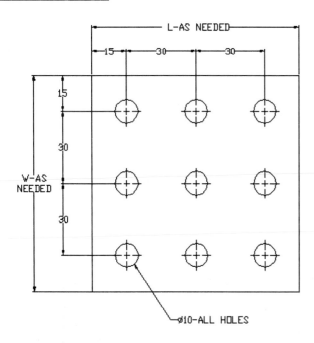

PART NO.	TOTAL NO OF HOLES	L	W	HOLE PATTERN
PL110-9	9	90	90	3x3
PL110-16	16	120	120	4x4
PL110-6	6	60	90	2x3
PL110-8	8	60	120	2x4
PL80-4	4	60	60	2x2

EX5-3 MILLIMETERS

Draw an exploded isometric assembly drawing of Assembly 1. Create a BOM.

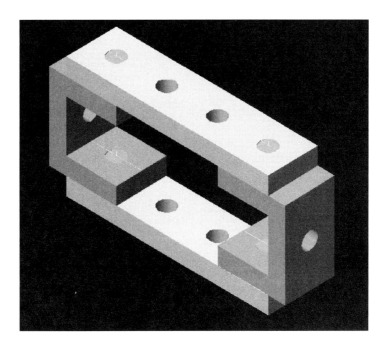

ASSEMBLY 1

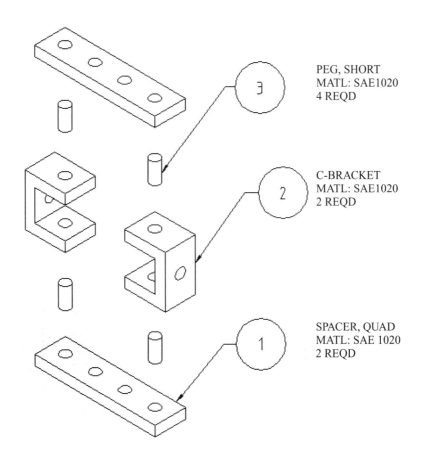

PEG, SHORT
MATL: SAE1020
4 REQD

C-BRACKET
MATL: SAE1020
2 REQD

SPACER, QUAD
MATL: SAE 1020
2 REQD

EX5-4 MILLIMETERS

Draw an exloded isometric assembly drawing of Assembly 2. Create a BOM.

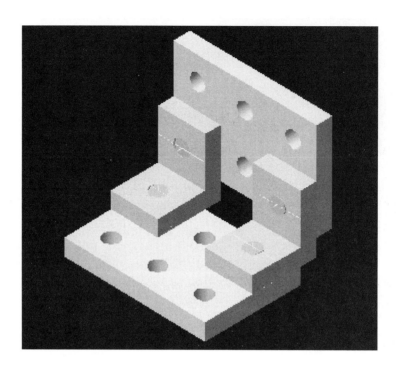

ASSEMBLY 2

PEG20
6064-T4 AL
4 REQD

L-BRACKET
6064-T4 AL
2 REQD

PL100-6
6064-T4 AL
2 REQD

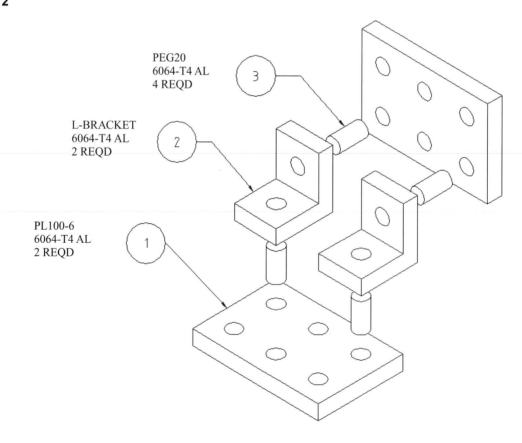

EX5-5 MILLIMETERS

Draw an exploded isometric assembly drawing of Assembly 3. Create a BOM.

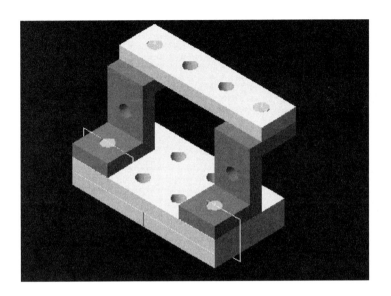

ASSEMBLY 3

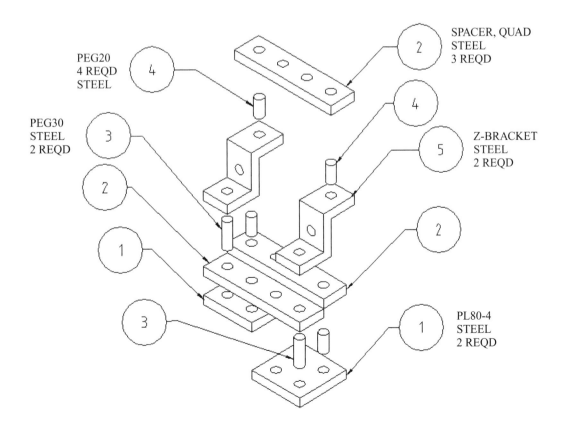

EX5-6 MILLIMETERS

Draw an exploded isometric assembly drawing of Assembly 4. Create a BOM.

ASSEMBLY 4

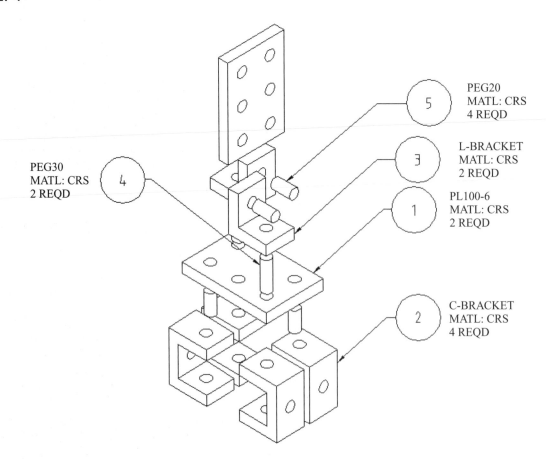

PEG30
MATL: CRS
2 REQD

PEG20
MATL: CRS
4 REQD

L-BRACKET
MATL: CRS
2 REQD

PL100-6
MATL: CRS
2 REQD

C-BRACKET
MATL: CRS
4 REQD

EX5-7 MILLIMETERS

Draw an exploded isometric assembly drawing of Assembly 5. Create a BOM.

ASSEMBLY 5

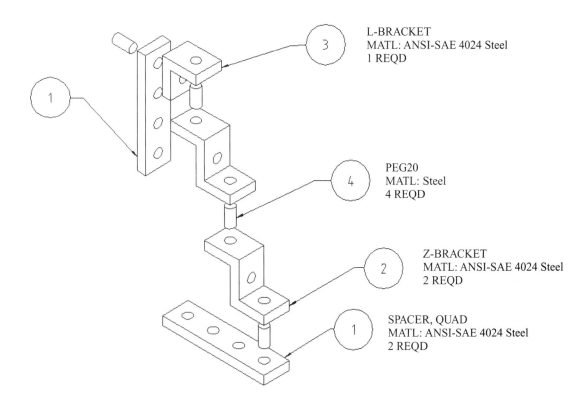

L-BRACKET
MATL: ANSI-SAE 4024 Steel
1 REQD

PEG20
MATL: Steel
4 REQD

Z-BRACKET
MATL: ANSI-SAE 4024 Steel
2 REQD

SPACER, QUAD
MATL: ANSI-SAE 4024 Steel
2 REQD

EX5-8

Draw an exploded isometric assembly drawing of Assembly 6. Create a BOM.

EX5-9

Create an original assembly based on the parts shown on pages 168–170. Include a scene, an exploded isometric drawing with assembly numbers, and a BOM. Use at least 12 parts.

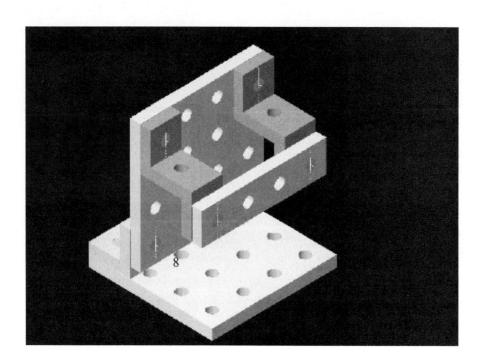

ASSEMBLY 6

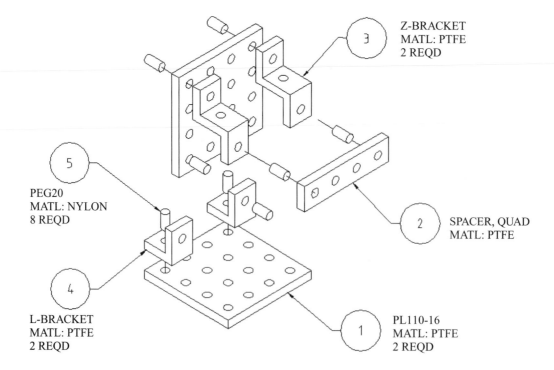

EX5-10

Draw the ROTATOR ASSEMBLY shown. Include the following:

A. An assembly drawing

B. An exploded presentation drawing

C. An isometric drawing with assembly numbers

D. A parts list

E. An animated assembly drawing; the LINKs should rotate relative to the PLATE. The LINKs should carry the CROSSLINK. The CROSSLINK should remain parallel during the rotation.

HINT: This problem is explained in Section 5-17.

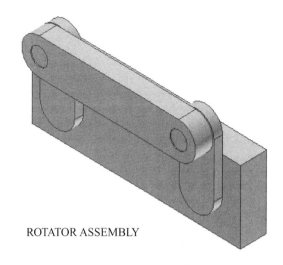

ROTATOR ASSEMBLY

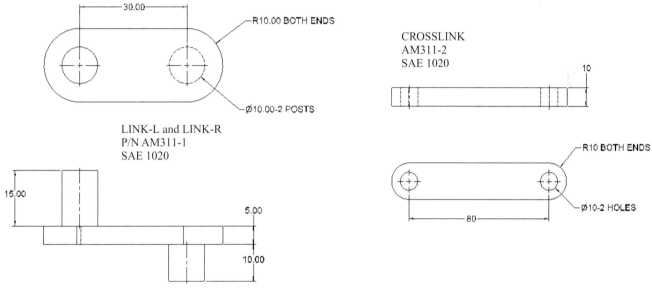

LINK-L and LINK-R
P/N AM311-1
SAE 1020

CROSSLINK
AM311-2
SAE 1020

PLATE
AM311-1
SAE 1020

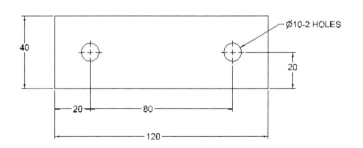

EX5-11

Draw the FLY ASSEMBLY shown. Include the following:

A. An assembly drawing
B. An exploded presentation drawing
C. An isometric drawing with assembly numbers
D. A parts list
E. An animated assembly drawing; the FLYLINK shold rotate around the SUPPORT base.

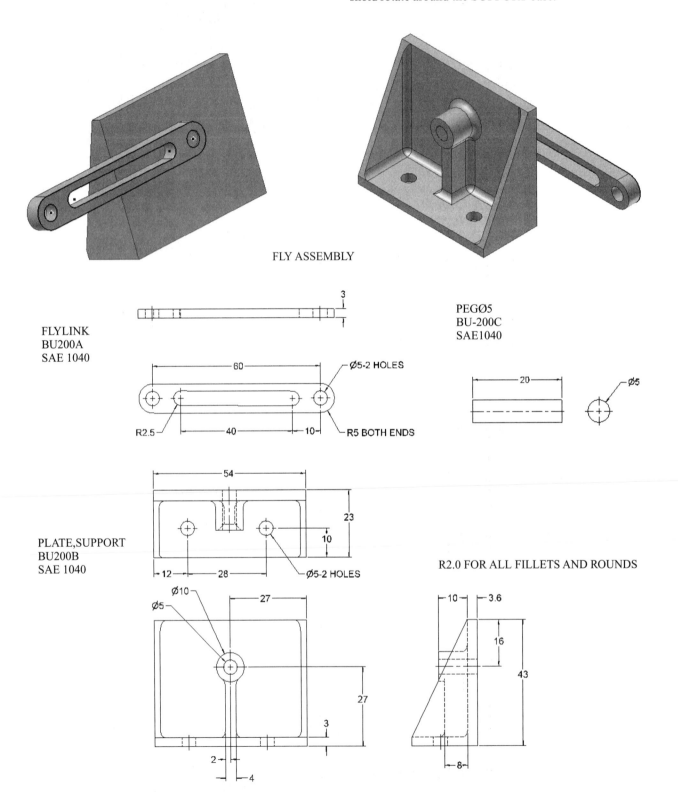

FLY ASSEMBLY

FLYLINK
BU200A
SAE 1040

PEGØ5
BU-200C
SAE1040

Ø5-2 HOLES
R2.5
R5 BOTH ENDS

PLATE,SUPPORT
BU200B
SAE 1040

R2.0 FOR ALL FILLETS AND ROUNDS

Ø5-2 HOLES
Ø10
Ø5

EX5-12

Draw the ROCKER ASSEMBLY shown. Include the following:

A. An assembly drawing
B. An exploded presentation drawing
C. An isometric drawing with assembly numbers
D. A parts list
E. An animated assembly drawing

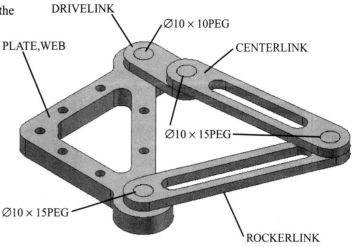

DRIVELINK
AM312-2
SAE 1040
5 mm THK

DRIVELINK
Ø10 × 10PEG
CENTERLINK
PLATE,WEB
Ø10 × 15PEG
Ø10 × 15PEG
ROCKERLINK

ROCKERLINK
AM312-4
SAE 1040
5 mm THK

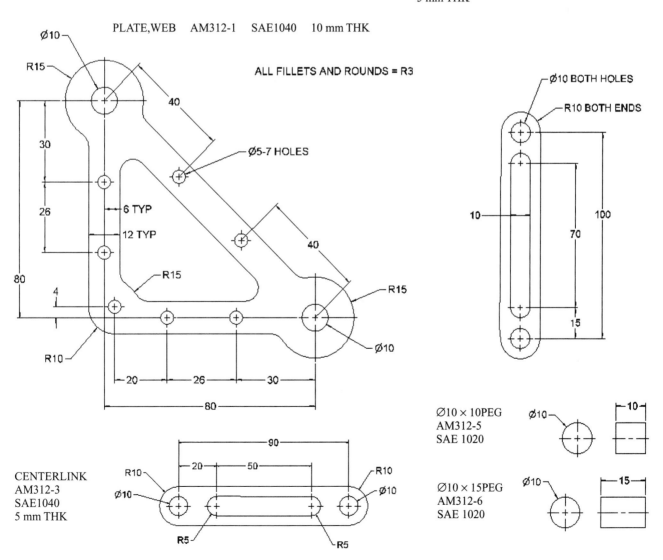

PLATE,WEB   AM312-1   SAE1040   10 mm THK

ALL FILLETS AND ROUNDS = R3

Ø10
R15
40
Ø5-7 HOLES
30
26
6 TYP
12 TYP
R15
80
40
4
R15
R10
Ø10
20
26
30
80

Ø10 BOTH HOLES
R10 BOTH ENDS
10
100
70
15

CENTERLINK
AM312-3
SAE1040
5 mm THK

90
20
50
R10
R10
Ø10
Ø10
R5
R5

Ø10 × 10PEG
AM312-5
SAE 1020
Ø10
10

Ø10 × 15PEG
AM312-6
SAE 1020
Ø10
15

EX5-13

Draw the LINK ASSEMBLY shown. Include the following:

A. An assembly drawing

B. An exploded presentation drawing

C. An isometric drawing with assembly numbers

D. A parts list

E. An animated assembly drawing; the HOLDER ARM should rotate between −30° and +30°.

LINK ASSEMBLY

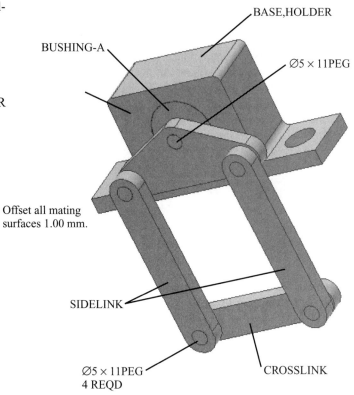

BASE,HOLDER

BUSHING-A

∅5 × 11PEG

Offset all mating surfaces 1.00 mm.

SIDELINK

∅5 × 11PEG
4 REQD

CROSSLINK

HOLDER ARM
AM-311-A3
7075-T6 AL
5 mm THK

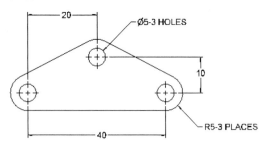

∅5-3 HOLES

R5-3 PLACES

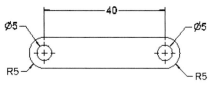

CROSSLINK
AM-311-A4
7075-T6 AL
5 mm THK

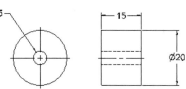

BUSHING
AM-311-A5
TEFLON

BASE,HOLDER
AM-311-A1
6061-T6 AL

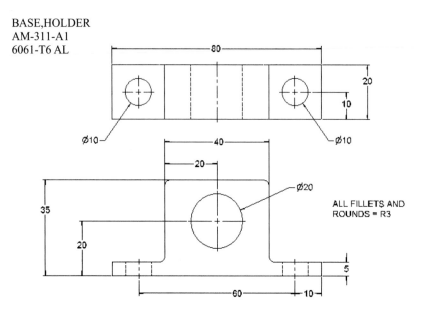

ALL FILLETS AND
ROUNDS = R3

LINK
AM-311-A2
7075-T6 AL
5 mm THK
2 REQD

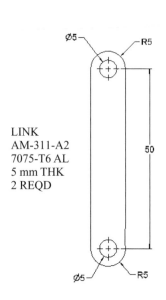

Threads and Fasteners

6-1 INTRODUCTION

This chapter explains how to draw threads and washers. It also explains how to select fasteners and how to design using fasteners, washers, and keys.

Threads are created in Inventor using either the Hole or the Thread tool located on the Part Features panel bar. See Figure 6-1. Predrawn fasteners may be accessed using the Standard Parts library. The Standard Parts library is explained starting with Section 6-9.

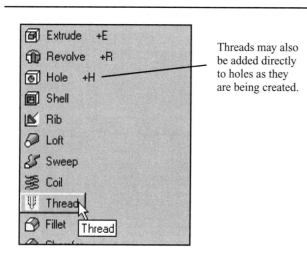

Threads may also be added directly to holes as they are being created.

Figure 6-1

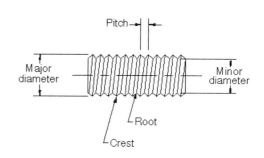

Figure 6-2

6-2 THREAD TERMINOLOGY

Figure 6-2 shows a thread. The peak of a thread is called the *crest,* and the valley portion is called the *root.* The *major diameter* of a thread is the distance across the thread from crest to crest. The *minor diameter* is the distance across the thread from root to root.

The *pitch* of a thread is the linear distance along the thread from crest to crest. Thread pitch is usually referred to in terms of a unit of length such as 20 threads per inch or 1.6 mm per thread.

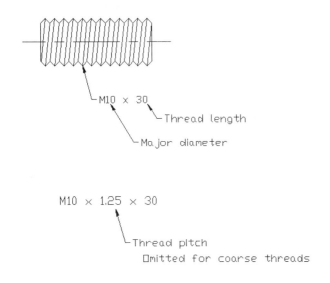

Figure 6-3

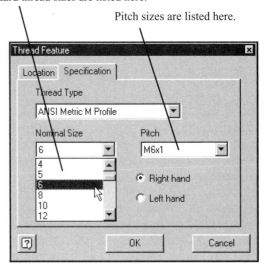

Figure 6-4

6-3 THREAD CALLOUTS–METRIC UNITS

Threads are specified on a drawing using drawing callouts. See Figure 6-3. The M at the beginning of a drawing callout specifies that the callout is for a metric thread. Holes that are not threaded use the Ø symbol.

The number following the M is the major diameter of the thread. An M10 thread has a major diameter of 10 mm. The pitch of a metric thread is assumed to be a coarse thread unless otherwise stated. The callout M10 × 30 assumes a coarse thread, or a thread length of 1.5 mm per thread. The number 30 is the thread length in millimeters. The "×" is read as "by," so the thread is called a "ten by thirty."

The callout M10 × 1.25 × 30 specifies a pitch of 1.25 mm per thread. This is not a standard coarse thread size, so the pitch must be specified.

Figure 6-4 shows a listing of standard metric thread sizes. Other sizes may be located by scrolling through the given Nominal Sizes. Inventor lists metric threads according to ANSI Metric M Profile standards.

Whenever possible use preferred thread sizes for designing. Preferred thread sizes are readily available and are usually cheaper than nonstandard sizes. In addition, tooling such as wrenches is also readily available for preferred sizes.

6-4 THREAD CALLOUTS–ANSI UNIFIED SCREW THREADS

ANSI Unified Screw Threads (English units) always include a thread form specification. Thread form specifications are designated by capital letters, as shown in Figure 6-6, and are defined as follows.

UNC — Unified National Coarse
UNF — Unified National Fine
UNEF — Unified National Extra Fine
UN — Unified National, or constant-pitch threads

An ANSI (English units) thread callout starts by defining the major diameter of the thread followed by the pitch specification. The callout .500-13 UNC means a thread whose major diameter is .500 in. with 13 threads per inch. The thread is manufactured to the Unified National Coarse (UNC) standards.

There are three possible classes of fit for a thread: 1, 2, and 3. The different class specifications specify a set of manufacturing tolerances. A class 1 thread is the loosest and a class 3 the most exact. A class 2 fit is the most common.

The letter A designates an external thread, B an internal thread. The symbol × means "by" as in 2 × 4, "two by four." The thread length (3.00) may be followed by the word

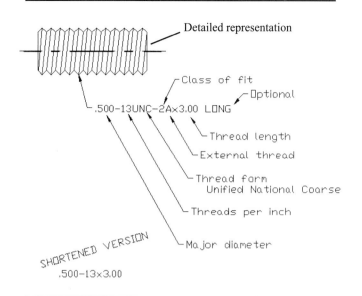

Detailed representation

Class of fit

Optional

.500-13UNC-2A×3.00 LONG

Thread length

External thread

Thread form
Unified National Coarse

Threads per inch

Major diameter

SHORTENED VERSION
.500-13×3.00

Figure 6-5

LONG to prevent confusion about which value represents the length.

Drawing callouts for ANSI (English unit) threads are sometimes shortened, such as in Figure 6-5. The callout .500-13 UNC-2A × 3.00 LONG is shortened to .500-13 × 3.00. Only a coarse thread has 13 threads per inch, and it should be obvious whether a thread is internal or external, so these specifications may be dropped. Most threads are class 2, so it is tacitly accepted that all threads are class 2 unless otherwise specified. The shortened callout form is not universally accepted. When in doubt, use a complete thread callout.

A listing of standard ANSI (English unit) threads, as presented in Inventor, is shown in Figure 6-6. Some of the drill sizes listed use numbers and letters. The decimal equivalents to the numbers are listed in Figure 6-6.

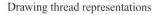

Drawing thread representations

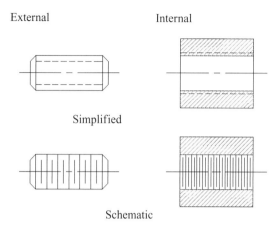

External

Internal

Simplified

Schematic

Figure 6-7

6-5 THREAD REPRESENTATIONS

There are three ways to graphically represent threads on a technical drawing: detailed, schematic, and simplified. Figure 6-5 shows an external detailed representation, and Figure 6-7 shows both the external and internal simplified and schematic representations.

Figure 6-8 shows an internal and an external thread created using Inventor. The threads will automatically be sized to the existing hole. Threads may be created only around existing holes and cylinders.

The decimal equivalents for threads specifed by numbers.

#1 - Ø.073
#2 - Ø.086
#3 - Ø.090
#4 - Ø.112
#6 - Ø.126
#6 - Ø.138
#8 - Ø.164
#10 - Ø.190
#12 - Ø.216

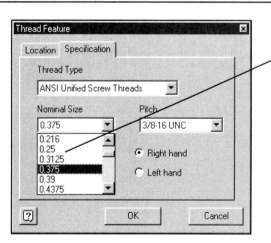

Select the thread size here.

Figure 6-6

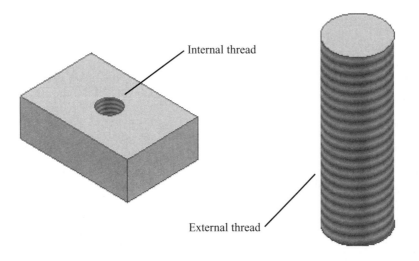

Internal thread

External thread

Figure 6-8

6-6 INTERNAL THREADS

Figure 6-9 shows an object with a Ø6.0 hole drilled through its center.

To add threads to an existing hole

1. Click on the Thread tool on the Part Features panel bar.

The Thread Feature dialog box will appear. See Figures 6-4 and 6-6.

2. Click on the existing hole.

The threads will automatically be created to match the hole's diameter.

3. Click OK.

Figure 6-10 shows the resulting threaded hole.

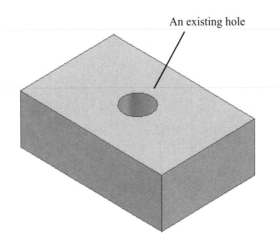

An existing hole

Figure 6-9

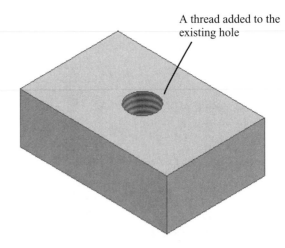

A thread added to the existing hole

Figure 6-10

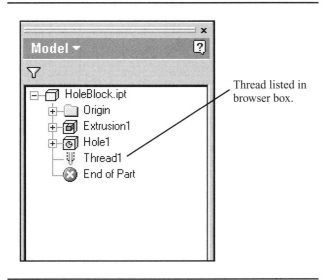

Thread listed in browser box.

Figure 6-11

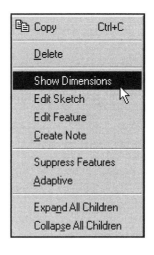

Right-click the mouse and select the Show Dimensions option.

Figure 6-12

When a thread is added to an existing hole, a thread listing will be included in the browser box. See Figure 6-11. The listing will confirm that a thread has been added, but it will not define the size or type of thread.

To determine the thread size

1. Right-click on the Hole listing in the browser box.

    A dialog box will appear. See Figure 6-12.

2. Select the Show Dimensions option.

    Figure 6-13 shows the resulting dimensions. The dimensions define the hole's diameter, and because Inventor will match the thread size to the existing hole diameter, the thread is an M6.

6-7 THREADED BLIND HOLES

The internally threaded holes presented in the last section passed completely through the material. This section shows how to draw holes that do not pass completely through the object but have a defined depth.

Figure 6-14 shows a tapped hole. It is drawn using the simplified representation. Note that there are three separate portions to the hole representation: the threaded portion,

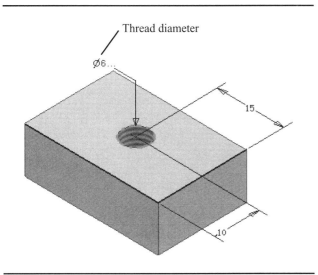

Thread diameter

Figure 6-13

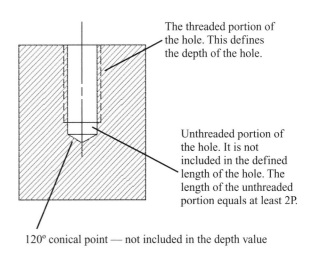

The threaded portion of the hole. This defines the depth of the hole.

Unthreaded portion of the hole. It is not included in the defined length of the hole. The length of the unthreaded portion equals at least 2P.

120° conical point — not included in the depth value

Figure 6-14

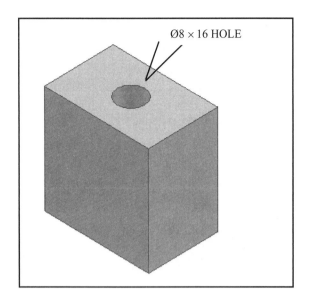

Ø8 × 16 HOLE

Figure 6-15

Click here to control where thread starts.

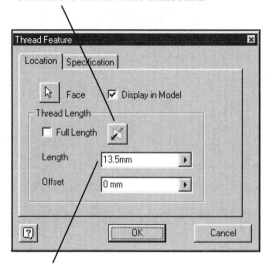

Define thread length here.

Figure 6-16

the unthreaded portion, and the 120° conical point. Only the threaded portion of the tapped hole is used to define the hole's depth. The unthreaded portion and the conical point are shown but are not included in the depth calculation.

A tapped hole is manufactured by first drilling a pilot hole that is slightly smaller than the major diameter of the threads. The threads are then cut into the side of the pilot hole. The tapping bit has no cutting edges on its bottom surface, so if it strikes the bottom of the hole, the bit can be damaged. Convention calls for the unthreaded portion of the pilot hole to extend about the equivalent of two thread pitches (2P) beyond the end of the threaded portion. The conical portion is added to the bottom surface of the pilot hole.

To draw a threaded blind hole-metric

Figure 6-15 shows an object with an existing Ø8 × 16 deep hole. Inventor will automatically apply an M8 thread to the Ø8 hole. From Figure 6-24 the pitch for an M8 thread is 1.25 mm per thread for a coarse thread.

A distance equal to two pitches (2P) is recommended between the bottom of the hole and the end of the threads. One pitch equals 1.25 mm, so 2P = 2.50 mm. The existing hole is 16 mm, so the thread depth is 13.5 mm.

1. Click on the Thread tool on the Part Features panel bar.

The Thread Feature dialog box will appear. See Figure 6-16.

2. Set the Length for 13.5 and the specifications for an M8 × 1.25.
3. Click the directional button located in the middle of the dialog box to ensure that the thread starts at the top surface plane.

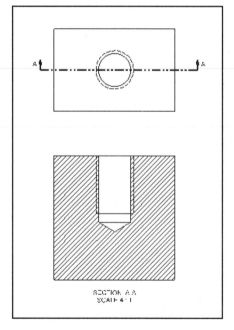

SECTION A-A
SCALE 4 : 1

A sectional view of a blind hole created using Inventor.

Figure 6-17

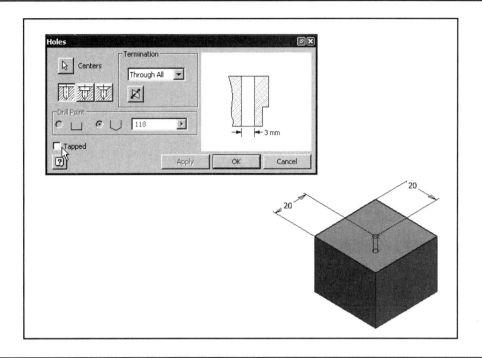

Figure 6-18

Figure 6-17 shows a sectional view of the threaded blind hole. Note how hidden lines are used to represent threads in both the top and the section views.

To draw a blind hole–ANSI threads

Pitch for ANSI threads is defined as threads per inch. A ¼-20 UNC thread has a pitch of 20 threads per inch. The length in inches of one thread is ¹⁄₂₀ or 0.05 in. Therefore, $2P = 2(0.05) = 0.10$ in.

If a block includes a hole 1.50 deep, then the appropriate thread length is $1.50 - 0.10 = 1.40$ in.

6-8 CREATING THREADED HOLES USING THE HOLE COMMAND

Threaded holes may be created directly using the Hole command. Figure 6-18 shows a $40 \times 40 \times 30$ mm block with a hole center defined.

To create a threaded through hole

1. Click on the Hole command.

   The Holes dialog box will appear. See Figure 6-18.

2. Click the Termination box and select the through All option.

3. Click the Tapped box. A check mark will appear.
4. Click on Thread and Types select the ANSI Metric M Profile option.

   See Figure 6-19.

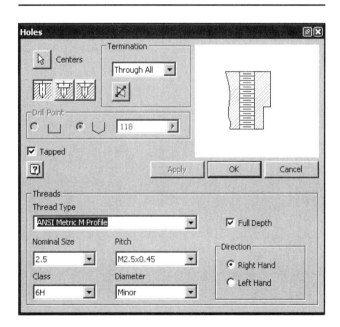

Figure 6-19

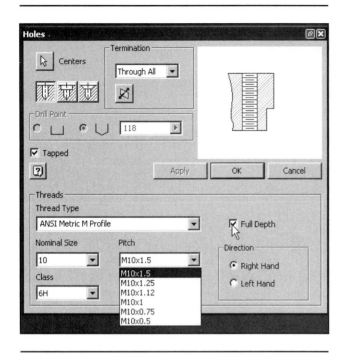

Figure 6-20

5. Set the Nominal Size for 10 and 2.5 × 0.45 click the Pitch box and select the M10 × 1.5 pitch.

More than one pitch size is available. In addition to the 1.50 pitch a 1.25 or 0.75 option is also available. These are Fine and Extra Fine designations. See Figure 6-20.

6. Click OK.

To create a blind threaded hole

1. Click on the Hole command.

   The Holes dialog box will appear. See Figure 6-21.

2. Set the Termination for Distance, and click the Tapped box.

3. Set the Nominal Size for 10 and the Pitch for M10 × 1.5.

4. Click the Full Depth box so that no check appears.

5. Set the thread depth for 8 and the hole depth for 10.

   The hole depth should, with rare exceptions, be greater than the thread depth.

6. Click OK.

6-9 EXTERNAL THREADS

External threads can be added to existing cylinders using the same methods described for internal threaded holes; however, most design use standards parts, that is, screws, nuts, and washers purchased from a vendor. Inventor includes an extensive library of standard parts that will be explained and used in this chapter.

The length of an external thread is an important design consideration. This section explains how to size screws that fit into tapped holes, and screws that pass though an object and are secured by a nut.

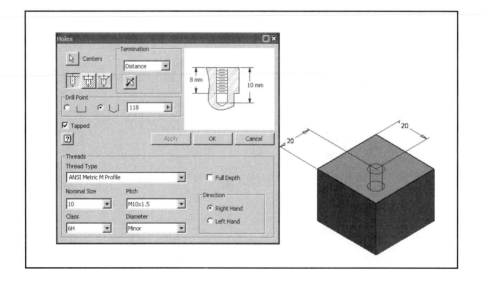

Figure 6-21

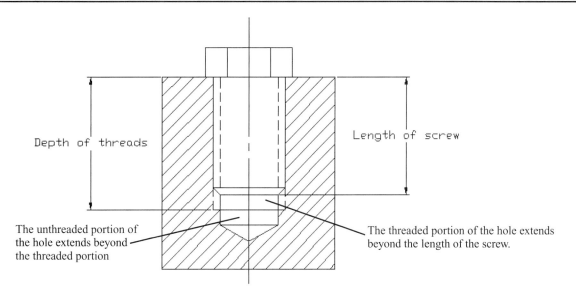

Figure 6-22

To detemine the pitch length of a thread

The depth of a blind threaded hole must always be greater than the length of a screw inserted into it. If the internal threads are too short, then the screw will not seat completely. A general rule of thumb is to allow at least two unused threads beyond the end of the screw. This helps assure a good fit between the screw and the threaded hole. See Figure 6-22.

Figures 6-23 and 6-24 show partial listings of standard screw lengths available, from the Standard Parts library. Whenever possible, always select screws of stardard size and length.

The thread callout for both UNC (Unified National Coarse) and UNF (Unified National Fine) includes the thread's pitch. For example, a ½"-20 UNF thread has 20 threads per inch, or a pitch length of 0.06". A ½."-13 UNC has 13 threads per inch, or a pitch length of 0.08.

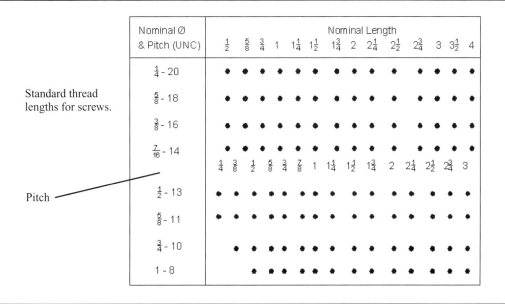

Standard thread lengths for screws.

Pitch

Figure 6-23

Nominal Ø & Pitch Length	Nominal Length												
	8	10	14	16	20	25	30	35	40	45	50	55	60
M5×0.8	•	•	•	•	•	•	•	•	•	•	•		
M6×1		•	•	•	•	•	•	•	•	•	•	•	•
M8×1.25			•	•	•	•	•	•	•	•	•	•	•
M10×1.5				•	•	•	•	•	•	•	•	•	•
M12×1.75					•	•	•	•	•	•	•	•	•
M14×2					•	•	•	•	•	•	•	•	•
M16×2						•	•	•	•	•	•	•	•
M20×2.5							•	•	•	•	•	•	•
M24×3									•	•	•	•	•

Figure 6-24

Metric thread callouts do not include a pitch specification for coarse threads. Fine threads require a pitch definition in the thread callout. Figure 6-24 lists an M10 thread as having a pitch of 1.50. This means that one thread is 1.50 mm long. The pitch lengths specified in Figure 6-24 are for coarse threads

6-10 SIZING A THREADED HOLE TO ACCEPT A SCREW

Say we wish to determine the length of thread and the depth of hole needed to accept an M10 × 25 hex head screw and to create a drawing of the screw in the threaded hole. The 25 mm length was derived from the listings in Figure 6-24. The length of the threaded hole must extended two pitches (2P) beyond the end of the screw, and the untapped portion of the hole must extend two pitches (2P) beyond the threaded portion of the hole. In this example the thread pitch equals 1.50 mm. Therefore, 2P = 2(1.50) = 3.0 mm. This is the minimum distance and can be increased but never decreased.

The two-pitch length requirement for the distance between the end of the screw and the end of the threaded portion of the hole determines that the thread depth should be 25.0 + 3.0 = 28.0 mm. The two-pitch length requirement between the end of the threaded portion of the hole and the bottom of the hole requires that the hole must have a depth of 28.0 + 3.0 or 31.0 mm.

It is important that complete hole depths be specified, as they will serve to show any interference with other holes or surfaces.

To draw a blind threaded hole

1. Create a new Standard (mm).ipt drawing and create a 40 × 40 × 60 block.

2. Use the Point, Hole Center and the Hole command to create an M10 × 28 deep thread and a 31 deep hole.

   See Figure 6-25.

3. Save the threaded block as ThreadedBlock:1.

   Figure 6-26 shows the finished hole.

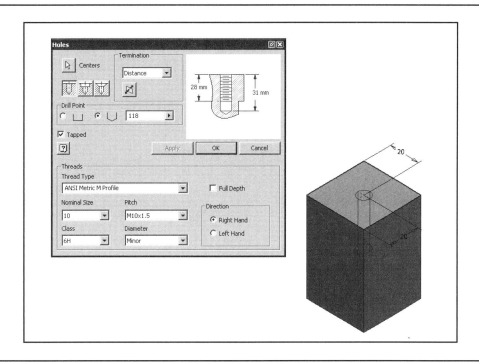

Figure 6-25

To access an M10 × 25 hex head bolt

1. Create a new Standard (mm).iam drawing called M10ASSEMBLY.
2. Use the Place Component command and locate one copy of the ThreadedBlock:1 block on the drawing screen.
3. Click the word Model at the head of the browser box and select the Library option, then click the Standard Parts option.

See Figure 6-27. The Standard Parts library title page will appear. See Figure 6-28.

4. Click the Fasteners option.

   The browser box will change.

5. Click the Screws and Threaded Bolts option, then click the Hex Head Types option.
6. Scroll through the options and select the Hex Bolt (Regular Thread–Metric) option.

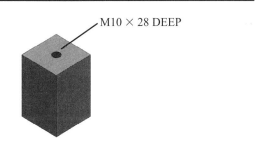

Figure 6-26

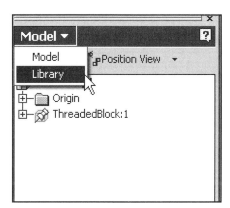

Figure 6-27

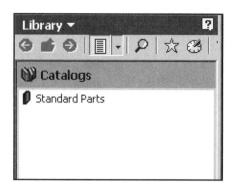
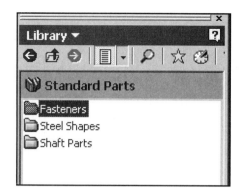

Figure 6-28

7. Click the model heading in the browser box, click the assembly name, then click the file pull-down menu and select Save All. Assign the assembly name and click OK. Return to the Standard Parts library as described in step 3.

See Figure 6-29.

8. Set the nominal diameter for 10, the pitch for 1.5, and the nominal length for 25.

9. Move the cursor into the picture area of the screw. An eyedropper will appear. Hold down the left mouse button. The eyedropper will fill up, that is, become solid black. Drag the eyedropper into the drawing area. The screw will appear on the drawing.

See Figure 6-30.

The M10 bolt will appear on the drawing screen with the ThreadedBlock:1. See Figure 6-31. If the bolt interferes with the ThreadedBlock:1, use the Move Component option to position the bolt away from the block.

To insert the M10 bolt

1. Click the Place Constraint tool on the panel bar.

The Place Constraint dialog box will appear. See Figure 6-31.

2. Click the Insert option
3. Click the bottom surface of the bolt's hex head, then click the threaded hole in the ThreadedBlock:1.

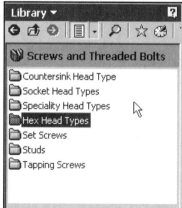

Figure 6-29

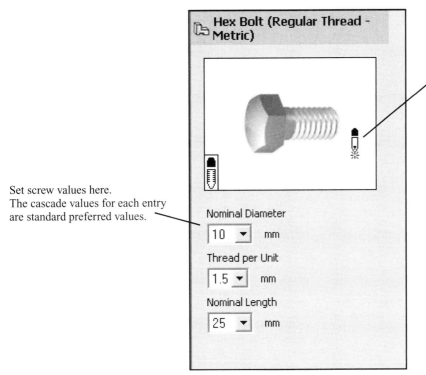

Move the cursor into the picture area and left-click and hold the mouse button. The eyedropper will fill up, becoming solid black. Drag the part into the drawing area.

Set screw values here.
The cascade values for each entry are standard preferred values.

Figure 6-30

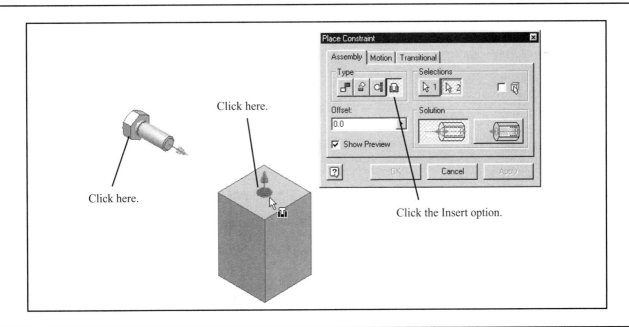

Click here.

Click here.

Click the Insert option.

Figure 6-31

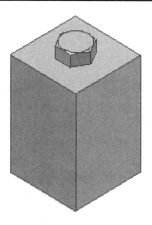

Figure 6-32

Figure 6-32 shows the resulting assembly.

4. Save the assembly.

Figure 6-33 shows a top view and a sectional view of the M10ASSEMBLY created using the ISO.idw option. Note the open area between the bottom of the M10 bolt and the bottom of the hole.

6-11 SCREWS AND NUTS

Screws often pass through an object or group of objects and are secured using nuts. The threads of a nut must match the threads of the screw.

Nuts are manufactured at a variety of heights depending on their intended application. Nuts made for heavy loads are thicker than those intended for light loads. In general, nut thickness can be estimated as 0.88 of the major diameter of the thread. For example, if the nut has an M10 thread, the thickness will be about .88(10) = 8.8 mm.

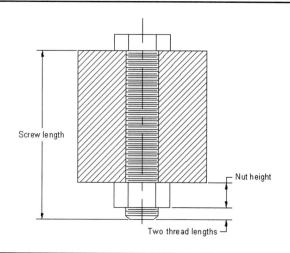

Figure 6-34

It is good practice to specify a screw length that allows for at least two threads beyond the nut. This will ensure that all threads of the nut are in contact with the screw threads. See Figure 6-34.

To calculate the screw thread length

In the following example an M10 bolt will pass through a box that has a height of 30 mm and be held in place using an M10 nut. The required length of the screw is calculated by adding the height of the box to the height of the nut, then adding at least two thread lengths (2P).

For an M10 coarse thread 2P = 1.60 mm. The height of the nut = .88(10) = 8.80 mm, and the height of the box = 30 mm:

$$30.00 + 8.80 + 1.60 = 40.4 \text{ mm}$$

Refer to Figure 6-24 and find the nearest M10 standard thread length that is greater than 40.4. The table shows the

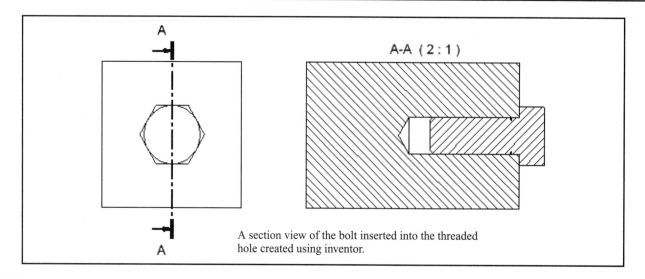

A section view of the bolt inserted into the threaded hole created using inventor.

Figure 6-33

next available standard length that is greater than 40.4 is 45 mm.

To add an M10 × 45 hex head screw to a drawing

1. Draw a 40 × 40 × 30 box.
2. Locate a Ø11 hole in the center of the top surface of the box.

The hole does not have threads. It is a clearance hole and so should be slightly larger than the M10 thread. See Figure 6-35.

3. Save the block as Ø11BLOCK.

To add a bolt and nut to the drawing

1. Start a new assembly drawing called M10NUT. Use the Standard (mm).iam format.
2. Use the Place Component tool and place a copy of the Ø11BLOCK on the drawing screen.
3. Use the Save All command on the File pull-down menu to save and name the new assembly.
4. Click the Model heading in the browser box and access the Standard Parts library.
5. Click Standard Parts, then Fasteners, Screws and Threaded Bolts, Hex Head Types, Hex Bolt (Regular Thread–Metric).

The Hex Bolt dialog box will appear.

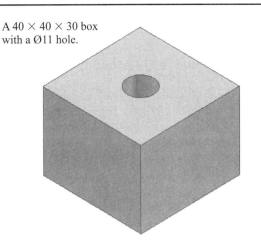

A 40 × 40 × 30 box with a Ø11 hole.

Figure 6-35

6. Set the Nominal Diameter to 10, the pitch to 1.5, and the Nominal Length to 45.
7. Move the cursor into the picture area of the bolt in the browser area. Hold down the left mouse button. An eyedropper will appear and fill up. When the eyedropper is full, drag the dropper into the drawing area and release the left button.
8. Locate the bolt, press the left mouse button, then press the right mouse button. Select the Done option.

See Figure 6-36.

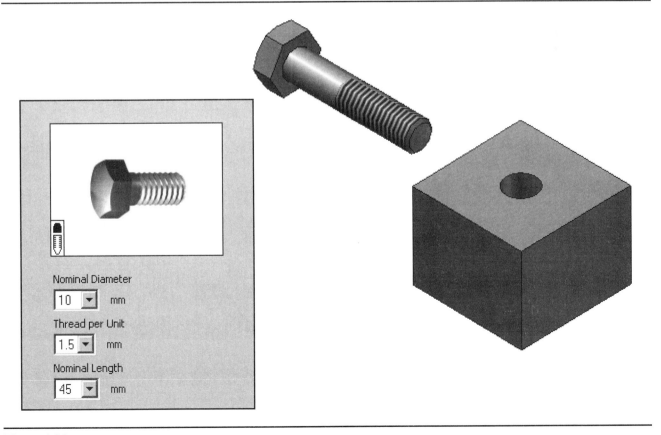

Figure 6-36

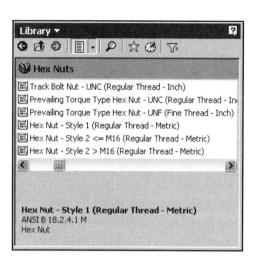

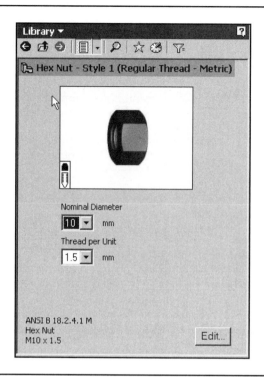

Figure 6-37

9. Use the Return button on the Library menu to return to the Fastener menu. Select Nuts, Hex Nuts, then Hex Nut–style 1 (Regular Thread–Metric). Set the Nominal Diameter for 10 and the Thread per Unit for 1.5.

See Figure 6-37.

10. Drag a copy of the nut onto the drawing screen.

See Figure 6-38.

To assemble the components

1. Click the Place Constraint tool.
2. Insert the bolt into the hole as described in Section 6-9.

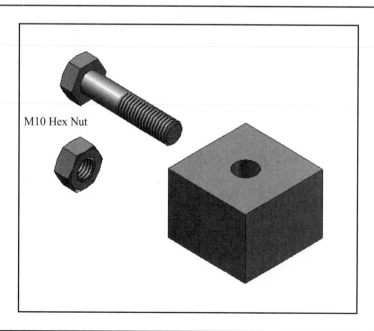

M10 Hex Nut

Figure 6-38

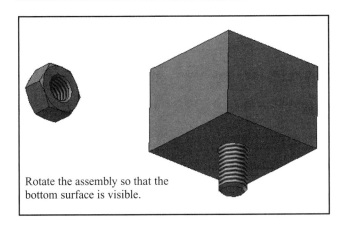

Rotate the assembly so that the bottom surface is visible.

Figure 6-39

The nut is to be located on the bottom surface of the block, which is presently not visible.

3. Click the Rotate command on the Standard toolbar and rotate the block so that the bottom surface is visible.

See Figure 6-39.

4. Click the Place Constraint tool and select the Insert option.

5. Click the bottom of the nut and the edge line of the ∅11 hole.

See Figure 6-40.

6. Apply the constraints.

Figure 6-41 shows the resulting assembly.

6-12 TYPES OF THREADED FASTENERS

There are many different types of screws generally classified by their head types. Figure 6-42 shows six of the most commonly used types.

The choice of head type depends on the screw's application and design function. A product design for home use would probably use screws that had slotted heads, as most homes possess a blade screwdriver. Hex head screws can be torqued to higher levels than slotted pan heads but require socket wrenches. Flat head screws are used when the screw is located in a surface that must be flat and flush.

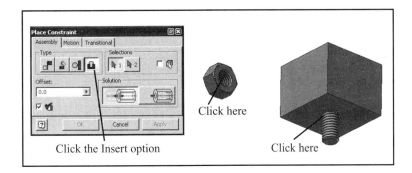

Click the Insert option   Click here   Click here

Figure 6-40

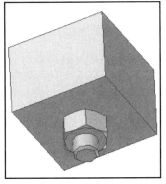

Resulting assembly

Figure 6-41

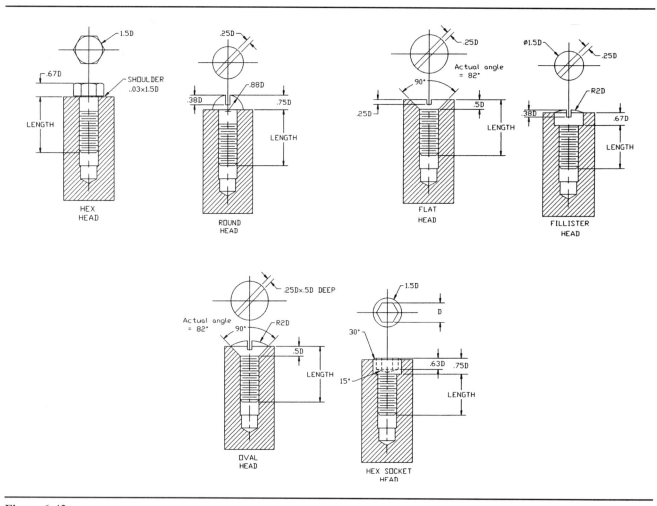

Figure 6-42

Sometimes a screw's head shape is selected to prevent access. For example, the head of the screw used to open most fire hydrants is pentagon-shaped and requires a special wrench to open it. This is to prevent unauthorized access that could affect a district's water pressure.

Screw connections for oxygen lines in hospitals have left-handed threads. They are the only lines that have left-handed threads, to ensure that no patient needing oxygen is connected to anything but oxygen.

Inventor's Standard Parts library lists many different types of fasteners. See Figure 6-43. There are many subfiles to each of the fastener headings.

6-13 FLAT HEAD SCREWS– COUNTERSUNK HOLES

Flat head screws are inserted into countersunk holes. The procedure is first to create a countersunk hole on a component, then to create an assembly using the component along with the appropriate screw listed in the Standard Parts library.

The following example uses an M8 × 50 hexagon socket flat head screw.

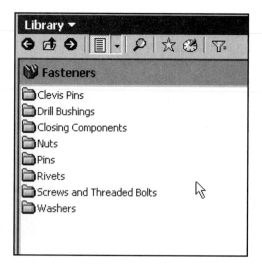

Figure 6-43

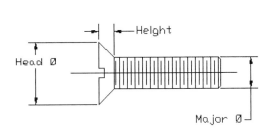

Nominal Size	Major Ø-Max	Head Ø Theoretical Max.	Head Ø Actual-Min	Head Height
M2×0.4	2.00	4.4	3.5	1.2
M3×0.5	3.00	6.3	5.2	1.7
M4×0.7	4.00	9.4	8.0	1.6
M5×0.8	5.00	10.4	8.9	2.0
M6×1.0	6.00	12.6	10.9	3.3
M8×1.25	8.00	17.3	15.4	4.6
M10×1.5	10.00	20.0	17.8	5.0

Figure 6-44

Figure 6-44 shows a limited listing of slotted flat head screws. The chart includes the head diameter and height. These values will be needed to size the countersunk hole.

To create a countersunk hole

1. Create a 40 × 40 × 80 block.
2. Locate a hole's center point in the center (20 × 20) of the top surface of the block using the Point, Hole Center command.
3. Go to the Part Features panel bar and click the Hole tool.
4. Click the Countersink and Tapped boxes.

See Figure 6-45.

The pitch length of an M8 thread is 1.25, so two thread lengths (2P) equals 2.50.

The hole's threads must be at least 50.00 + 2.50 = 52.50, and the pilot hole must be at least 52.50 + 2.50 = 55.00 deep.

5. Click the Full Depth box (remove check mark) and set the Threads option for ANSI Metric M Profile, the thread depth for 52.5, the hole depth for 55, and the head diameter for 15.4.

Figure 6-46 shows the countersunk hole located in the 40 × 40 × 80 block.

6. Save the block.

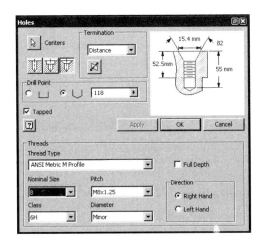

Figure 6-45

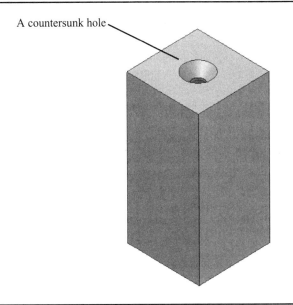

A countersunk hole

Figure 6-46

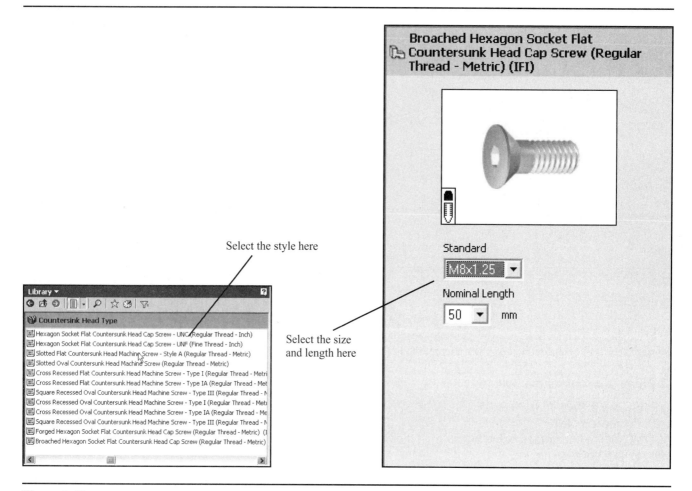

Select the style here

Select the size
and length here

Figure 6-47

To insert a countersunk screw

1. Create an assembly drawing using the Standard.iam format.
2. Use the Place Component tool and locate one copy of the block on the drawing screen.
3. Use the Save All command to save and name the assembly.
4. Access the Standard Parts library located under the Model heading in the browser box.
5. Select the Fastener option, then Screws and Threaded bolts, then Countersink Head Types.
6. Select the Broached Hexagon Socket and set the diameter for M8 × 1.25 and the Nominal Length for 50.

   See Figure 6-47.

7. Locate the cursor in the picture area of the nut and click and hold the left mouse button. Drag the nut into the drawing area.
8. Click the Place Constraint tool, then select the Insert option on the Place Constraint dialog box.

9. Insert the screw into the block.

Figure 6-48 shows the resulting assembly. Figure 6-49 shows a top and a section view of the countersunk screw inserted into the block. Note that the portion of the hole below the bottom of the M8 × 50 screw is clear; that is, it does not show the unused threads. This is a drawing convention that is intended to add clarity to the drawing. Inventor will automatically omit the unused threads.

6-14 COUNTERBORES

A counterbored hole is created by first drilling a hole, then drilling a second larger hole aligned with the first. Counterbored holes are often used to recess the heads of fasteners.

Say we wish to fit a 3/8-16 UNC × 1.50 LONG hex head screw into a block that includes a counterbored hole, and that after assembly the head of the screw is to be below the surface of the block.

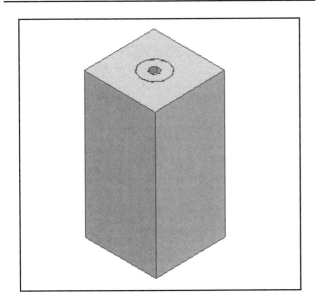

Figure 6-48

To determine the counterbore depth

Figure 6-50 shows a partial listing of standard hex head heights and distances across the flats. These values can be obtained from manufacturers' catalogs, many of which are posted on the web, or by using the approximations presented in Figure 6-42.

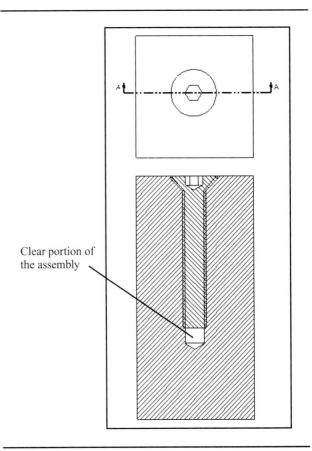

Clear portion of the assembly

Figure 6-49

Nominal Ø - in.		Width across the Flats	Width across the Corners	Head Height
$\frac{1}{4}$	.2500	.438	.505	.188
$\frac{5}{16}$	.3125	.500	.577	.235
$\frac{3}{8}$	.3750	.562	.650	.268
$\frac{7}{16}$	.4375	.625	.722	.316
$\frac{1}{2}$	.5000	.750	.866	.364
$\frac{5}{8}$	.6250	.938	1.083	.444
$\frac{3}{4}$	.7500	1.125	1.299	.524

Nominal Ø-mm	Width across the Flats	Width across the Corners	Head Height
M5×0.8	8.00	9.24	3.58
M6×1.0	10.00	11.54	4.38
M8×1.25	13.00	15.01	5.68
M10×1.5	16.00	18.48	6.85
M12×1.75	18.00	20.78	7.95
M14×2.00	21.00	24.25	9.25

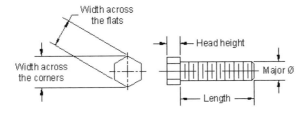

Figure 6-50

In this example the head height is .268 in. The counterbore must have a depth greater than the head height. A distance of .313 (⁵⁄₁₆) was selected.

To determine the thread length

The screw is 1.50 in. long and has a pitch of 16 threads per inch. Each thread is therefore ¹⁄₁₆ or .0625 in. It is recommended that there be at least two threads beyond the end of the screw. Two thread pitches would be 2(.0625) = .125 in.

The thread depth is 1.50 + .125 or 1.625 in. minimum; however, the thread is created below the counterbore, so for Inventor the value must include the depth of the counterbore. The thread depth is 1.625 + .313 = 1.938 in.

To determine the depth of the hole

The hole should extend at least two pitch lengths beyond the threaded portion of the hole, plus the depth of the counterbore, so 1.938 + .125 = 2.063 in.

To determine the counterbore's diameter

The distance across the corners of the screw is listed in Figure 6-50 as .650 in. The counterbored hole must be at least this large plus an allowance for the tool (socket wrench) needed to assemble and disassemble the screw. In general, the diameter is increased 0.125 in. or ¹⁄₁₆ in. [1.6 mm] all around to allow for tooling needs.

If space is a concern, then designers will change the head type of a screw so that the tooling will not add to the required diameter. For example, a socket head type screw may be used if design requirements permit.

The minimum counterbore diameter is .650 + .125 = .775. For this example 0.8125 or ¹³⁄₁₆ in. was selected.

To draw a counterbored hole

1. Draw a 3.00 × 3.00 × 5.00 block.
2. Locate a hole's center point in the center of the 3.00 × 3.00 surface.
3. Access the Part Features menu and click the Hole tool.

   The Holes dialog box will appear.

4. Click the Counterbore, Tapped and Full Depth (turned off) bolt.

Define the size requirements here.

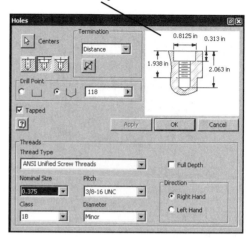

Figure 6-51

5. Set the hole depth for 2.063, the thread depth for 1.938, the counterbore diameter for .8125, and the counterbore depth for .313.
6. Select 3/8-16 UNC threads.

   See Figure 6-51.

7. Save the block.

To assemble the screw

1. Create an assembly drawing using the Standard (in).iam format.
2. Use the Place Component tool to place a copy of the counterbored block on the screen.
3. Use the Save All command to save and name the assembly.
4. Access the Standard Parts library and select a 3/8-16 UNC × 1.50 hex head bolt and drag it onto the drawing.

   See Figure 6-52.

5. Use the Place Constraint tool to insert the bolt into the hole.

   Figure 6-53 shows the bolt inserted into the counterbored hole. Note the tooling clearance around the hex head and the clearance between the top of the bolt and the top surface of the block.

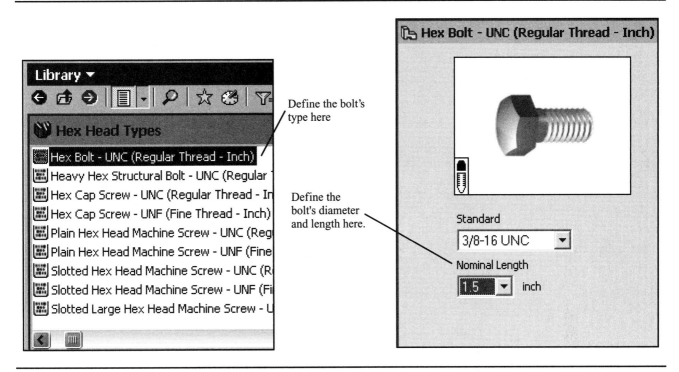

Define the bolt's type here

Define the bolt's diameter and length here.

Figure 6-52

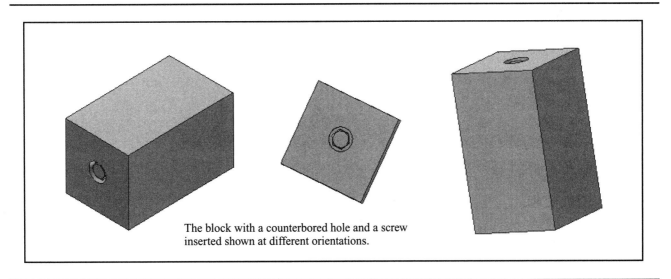

The block with a counterbored hole and a screw inserted shown at different orientations.

Figure 6-53

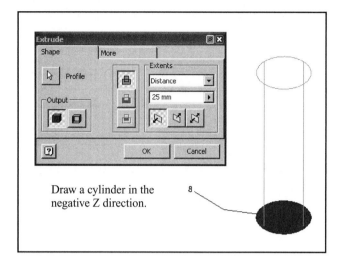

Draw a cylinder in the negative Z direction.

8

Figure 6-54

Specify the dimensioning system to be used to define the chamfer here.

Define the chamfer's size here.

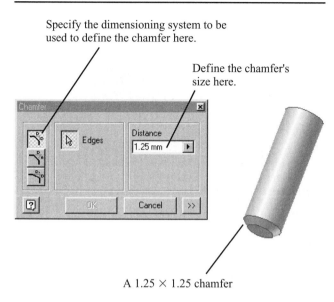

A 1.25 × 1.25 chamfer

Figure 6-55

6-15 TO DRAW FASTENERS NOT INCLUDED IN THE STANDARD PARTS LIBRARY

The Standard Parts library is a partial listing of fasteners. There are many other sizes and styles available. Inventor can be used to draw specific fasteners that can then be saved and used on assemblies.

Say we wish to draw an M8 × 25 hex head screw and that this size is not available in the Standard Parts library.

To draw an M8 × 25 hex head screw

1. Create a Standard.ipt drawing.
2. Draw a Ø8 × 25 cylinder. Draw the cylinder with its top surface on the XY plane so that it extends in the negative Z direction.

   See Figure 6-54.

3. Add a 1.25 × 1.25 chamfer to the bottom of the cylinder.

   See Figure 6-55. In general, the chamfer will approximately equal by one pitch length.

4. Create a new sketch plane on the top surface of the cylinder.
5. Use the Polygon command and draw a hexagon centered on the top surface of the cylinder. Make

the hexagon 12 mm across the flats (see Section 2-10) and extrude it to a height of 5.4 mm above the XY plane.

See Figure 6-56.

The head height and distance across the flats were determined using the general values defined in Figure 6-42. Specific values may be obtained from manufacturers, many of whom list their products on the web, or from reference books such as *Machinery's Handbook.*

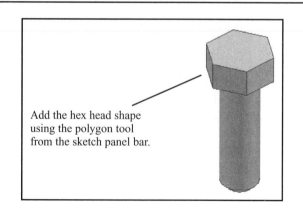

Add the hex head shape using the polygon tool from the sketch panel bar.

Figure 6-56

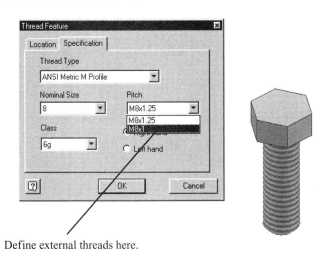

Define external threads here.

Figure 6-57

6. Add threads to the cylinder, using the Thread command. Click the cylinder, and Inventor will automatically create threads to match the cylinder's diameter.

See Figure 6-57. Check the thread specification to assure that correct threads were created. Note that the coarse pitch of 1.25 was automatically selected, but the fine pitch of 1.00 is also available.

7. Save the drawing as M8 × 25 Hex Head Screw.

6-16 FASTENERS FROM THE WEB

Many manufactures post their catalogs or production specifications on the web. For example, Figure 6-58 shows a page from the W.M. Berg Catalog available on line at www.wmberg.com. It is typical of the type of information available.

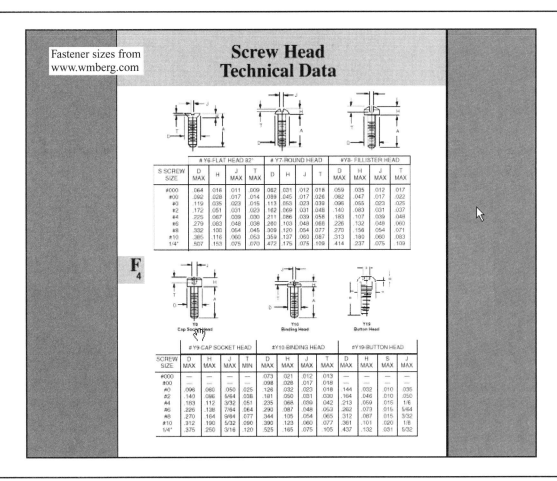

Figure 6-58   (Courtesy of W.M. Berg.)

Nominal Ø Pitch	Width across the Flats	Width across the Corners	Thickness
M5×0.8	8.00	9.24	4.70
M6×1.0	10.00	11.55	5.20
M8×1.25	13.00	15.01	6.80
M10×1.5	16.00	18.48	8.40
M12×1.75	18.00	20.78	10.80
M14×2.00	21.00	24.25	12.80

Nominal Ø	Width across the Flats	Width across the Corners	Thickness
1/4 .2500	.438	.505	.226
5/16 .3125	.500	.577	.273
3/8 .3750	.562	.650	.337
7/16 .4375	.625	.722	.385
1/2 .5000	.750	.866	.448
5/8 .6250	.938	1.083	.496
3/4 .7500	1.125	1.299	.559

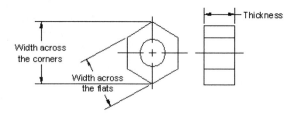

Figure 6-59

6-17 SAMPLE PROBLEM SP6-1

Nuts are used with externally threaded objects to hold parts together. There are many different styles of nuts. The Standard Parts library includes listings for hex, slotted hex, and wing nuts. Figure 6-59 shows a partial listing of both ANSI and metric standard hex nuts.

The threads of a nut must be exactly the same as the external threads inserted into them. For example, if a screw with an M6 × 1.0 thread is selected, an M6 × 1.0 nut thread must be selected.

The head height of the nut must be considered when determining the length of a bolt. It is good practice to have at least two threads extend beyond the nut to help assure that the nut is fully secured. Strength calculations are based on all nut threads' being engaged, so having threads extend beyond a nut is critical.

Many web sites include CAD drawings of their products that can be downloaded and used on Inventor drawings. In general, Inventor will read files using the SAT or STEP format. Files using a .dwg format can be downloaded to Mechanical Desktop, then into Inventor.

Figure 6-60 shows two blocks, each 25 mm thick with a center hole of Ø9.00 mm. The holes are clearance holes and do not include threads. The blocks are to be held together using an M8 hex head screw and a compatible nut.

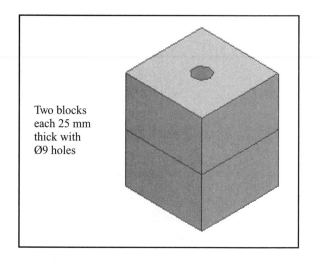

Two blocks each 25 mm thick with Ø9 holes

Figure 6-60

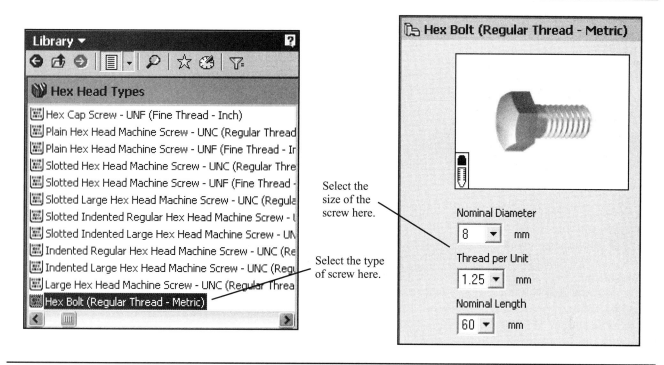

Select the size of the screw here.

Select the type of screw here.

Figure 6-61

To determine the minimum thread length required

Each block is 25 mm thick, for a total of 50 mm. The nut height, from Figure 6-59, for an M8 hex nut is 6.80, so the minimum thread length that will pass through both parts and the nut is 56.80 mm. Two threads must extend beyond

the nut to assure that is fully secured. From Figure 6-59 the length of an M8 thread is given as 1.25 mm, so two threads equal 2.50 mm. Therefore, the minimum thread length must be $50.00 + 6.80 + 2.50 = 59.3$ mm.

Bolts are manufactured in standard lengths, some of which are listed in the Standard Parts library. If the required thread length was not available from the library, manufacturers' catalogs would have to be searched and a new screw drawing created.

To select a screw

1. Access the Standard Parts library.
2. Select the Fasteners option, then Screws and Threaded Bolts, then Hex Head Types, then **Hex Bolt (Regular Thread-Metric.)**

See Figure 6-61. The standard thread length that is closest to, but still greater than, 59.3 is 60 mm. The 60 mm length is selected and applied to the drawing.

3. Define the values for an M8 × 1.25 × 60 hex head screw and drag it into the drawing.

See Figure 6-62.

4. Insert the screw into the two assembled parts.

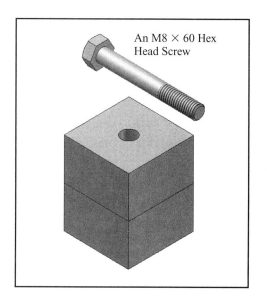

An M8 × 60 Hex Head Screw

Figure 6-62

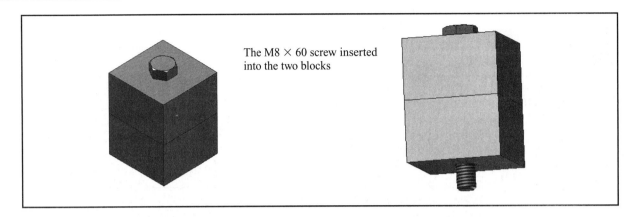

The M8 × 60 screw inserted into the two blocks

Figure 6-63

See Figure 6-63. Note that the screw extends beyond the bottom of the two assembled parts.

To select a nut

1. Access the Standard Parts library and select the Fasteners option, then Nuts, then Hex Nuts, then the Hex Nut - Style 1 (Regular Thread - Metric) listing.

2. Define the nut as an M8 × 1.25 Hex Nut.

   See Figure 6-64.

3. Drag a copy of the nut into the drawing area.

   Figure 6-65 shows the nut added to the drawing screen.

4. Use the Place Constraint tool and insert the nut onto the screw so that it is flush with the bottom surface of the blocks.

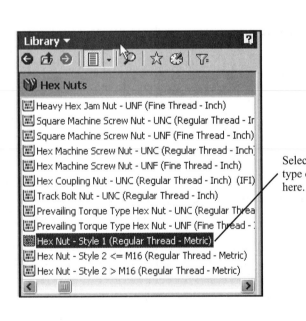

Select the type of nut here.

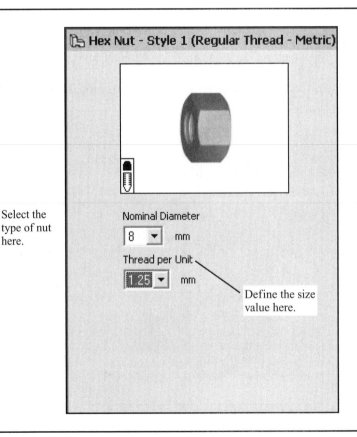

Define the size value here.

Figure 6-64

Figure 6-65

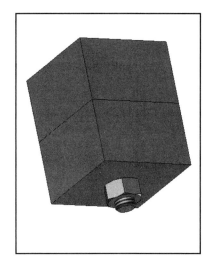

Figure 6-66

Figure 6-66 shows the nut inserted onto the screw.

The nuts listed in the Standard Parts library represent only a partial listing of the sizes and styles of nuts available. If a design calls for a nut size or type not listed in Standard Parts library, refer to manufacturers' specifications, then draw the nut and save it as an individual drawing. It can then be added to the design drawings as needed.

6-18 WASHERS

Washers are used to increase the bearing area under fasteners or as spacers. Washers are identified by their inside diameter, outside diameter, and thickness. In addition, washers can be designated N, R, or W for narrow, regular, and wide, respectively. These designations apply only to the outside diameters; the inside diameter is the same.

Inventor lists washers by their nominal diameter. The nominal diameters differ from the actual inside diameter by a predetermined clearance allowance. For example, a washer with a nominal diameter of 8 has an actual inside diameter of 8.90, or 0.90 mm greater than the 8 nominal diameter. This means that washer sizes can easily be matched to thread sizes using nominal sizes. A washer with a nominal diameter designation of 8 will fit over a thread designated M8.

There are different types of washers including, among others, plain and tapered. The Standard Parts library includes a listing of both plain and tapered washers. Figure 6-67 shows a partial listing of washer sizes. Note the differences in the outside diameters of N, R, and W designations.

To insert washers onto a fastener

We again start with the two blocks shown in Figure 6-60. Each has a height of 25 mm. We known from the previous section that the nut height is 6.80 mm and that the requirement that at least two threads extend beyond the end of the nut adds 2.50 mm, yielding a total thread length requirement of 59.3. We now have to add the thickness of the two washers and recalculate the minimum required bolt length.

Say we selected a plain regular washer with a nominal size of 8. From Figure 6-67 the thickness is found to be 2.30 mm or a total of 4.60 mm for the two washers. This extends the minimum bolt length requirement to 59.3 + 4.60 = 63.9. The nearest standard thread bolt length listed in the Standard Parts library that is greater than the 63.9 requirement is 65.

To add washers to an assembly

1. Use the Place Component tool and locate two copies of the block on the drawing, then align the blocks.
2. Use the Save All command to save and name the assembly.

Nominal Ø mm	Series	Inside Ø	Outside Ø	Thickness
5	N	5.50	11.00	1.40
	R	5.50	15.00	1.75
	W	5.50	20.00	2.30
6	N	6.65	13.00	1.75
	R	6.65	18.80	1.75
	W	6.65	25.40	2.30
8	N	8.90	18.37	2.30
	R	8.90	24.48	2.30
	W	8.90	31.38	2.80
10	N	10.85	20.00	2.30
	R	10.85	28.00	2.80
	W	10.85	39.00	3.50
12	N	13.57	25.40	2.80
	R	13.57	34.00	3.50
	W	13.57	44.00	3.50
14	N	15.25	28.00	2.80
	R	15.25	39.00	3.50
	W	15.25	50.00	4.00

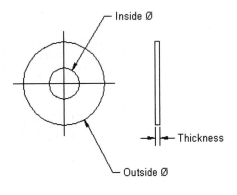

The designations N, R, and W indicate normal, regular, and wide sizes.

Nominal Ø in.	Series	Inside Ø	Outside Ø	Thickness
¼ .250	N	.281	.500	.063
	R	.281	.734	.063
	W	.281	1.000	.063
5/16 .312	N	.344	.625	.063
	R	.344	.875	.063
	W	.344	1.125	.063
3/8 .375	N	.406	.734	.063
	R	.406	1.000	.063
	W	.406	1.250	.100
7/16 .438	N	.469	.875	.063
	R	.469	1.125	.063
	W	.469	1.469	.100
½ .500	N	.531	1.000	.063
	R	.531	1.250	.100
	W	.531	1.750	.100
5/8 .625	N	.656	1.250	.100
	R	.656	1.750	.100
	W	.656	2.250	.100
¾ .750	N	.812	1.375	.100
	R	.812	2.000	.160
	W	.812	2.500	.160

Figure 6-67

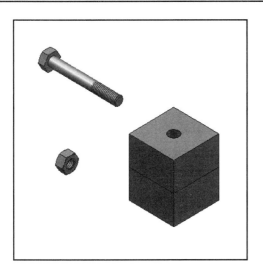

Figure 6-68

3. Access the Standard Parts library and select an M8 × 65 Hex Head screw and an M8 Hex Nut and drag them into the drawing.

See Figure 6-68.

4. Use the Standard Parts library and select the Fasteners option, then Washers, then Plain, then Plain Washer (Metric).

See Figure 6-69. A new dialog box will appear, allowing the selection of N, R, or W washers with an 8 nominal diameter. See Figure 6-70.

5. Select the 25.4 (Regular) washer and add two copies to the drawing.

See Figure 6-71.

6. Use the Place Constraint Insert option tool and align the washers with the holes in the blocks.

See Figure 6-72.

7. Use the Place Constraint tool and insert the M8 × 65 screw and the nut.

Figure 6-73 shows the resulting assembly.

The washers listed in the Standard Parts library represent only a partial listing of the washers available. If a design calls for a washer size or type not listed in the library, refer to manufacturers' specifications, then draw the washer and save it as an individual drawing. It can then be added to the design drawings as needed.

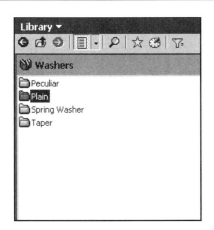

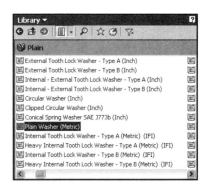

Figure 6-69

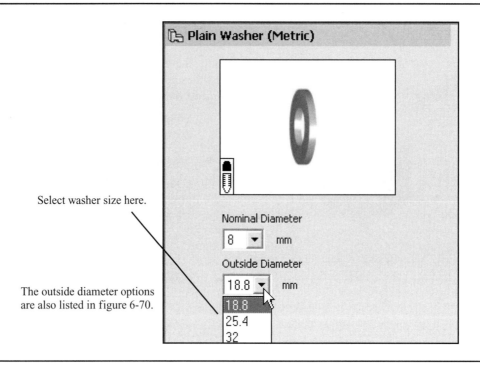

Select washer size here.

The outside diameter options are also listed in figure 6-70.

Figure 6-70

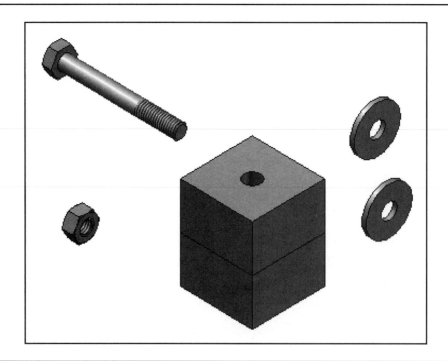

Figure 6-71

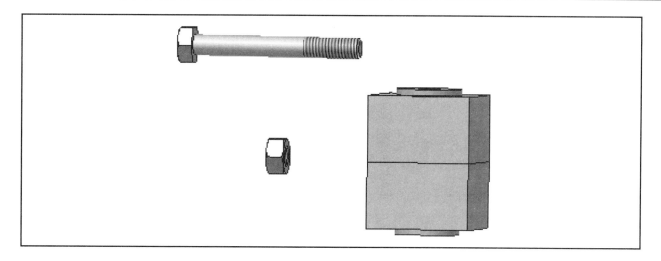

Figure 6-72

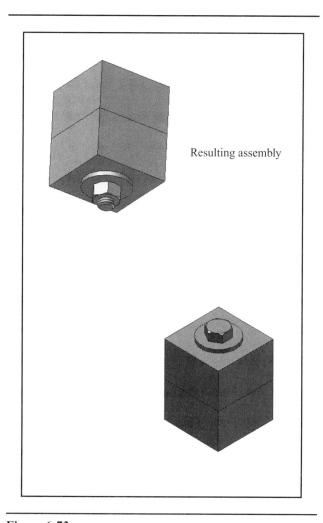

Resulting assembly

Figure 6-73

6-19 SETSCREWS

Setscrews are fasteners used to hold parts like gears and pulleys to rotating shafts or other moving objects to prevent slippage between the two objects. See Figure 6-74.

Most setscrews have recessed heads to help prevent interference with other parts.

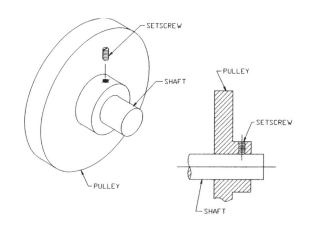

Figure 6-74

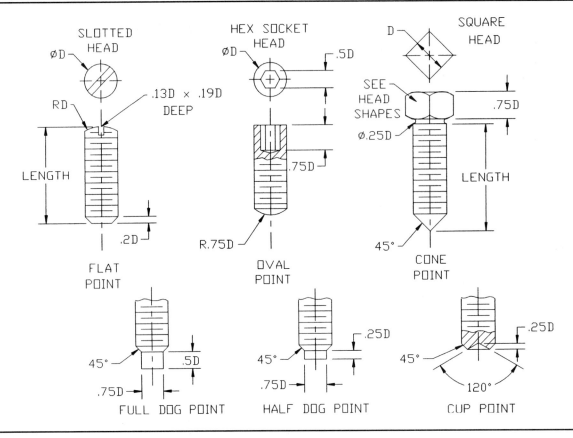

Figure 6-75

Many different head styles and point styles are available. See Figure 6-75. The dimensions shown in Figure 6-75 are general sizes for use in this book. For actual sizes, see the manufacturer's specifications.

A collar with two #10(.19)-32UNF threaded holes

Collar I.D. = .750
D.O. = 1.000
LENGTH = .750

Figure 6-76

Figure 6-76 shows a collar with two 10(.19)-32UNF threaded holes.

To add a setscrew

1. Create an assembly drawing using the Standard (mm).iam format.
2. Use the Place Component tool and place one copy of the collar on the drawing.
3. Use the Save All tool to save and name the assembly.
4. Access the Standard Parts library and select the Fasteners option, then Screws and Threaded Bolts, then Socket Head Types.
5. Select the Hexagon Socket Set Screw–Half Dog Point–UNF option and set the nominal diameter for 10-32 and the length for .38.

See Figure 6-77.

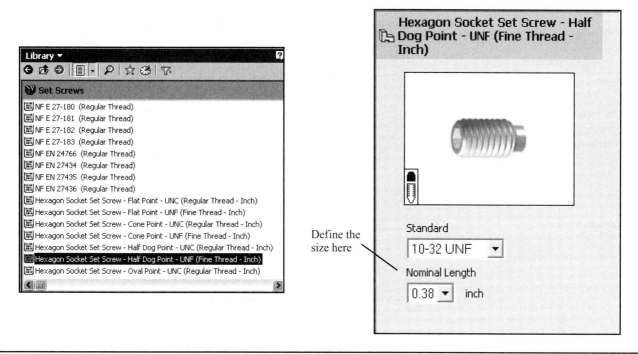

Figure 6-77

6. Click the CAD box and place a copy of the setscrew on the drawing.

   See Figure 6-78.

7. Use the Place Constraint tool and insert the setscrew into one of the threaded holes.

   Figure 6-79 shows the resulting assembly.

Figure 6-78

Figure 6-79

6-20 RIVETS

Rivets are fasteners that hold together adjoining or overlapping objects. A rivet starts with a head at one end and a straight shaft at the other end. The rivet is then inserted into the object, and the headless end is "bucked" or otherwise forced into place. A force is applied to the headless end that changes its shape so that another head is formed holding the objects together.

There are many different shapes and styles of rivets. Figure 6-80 shows five common head shapes for rivets. Hollow rivets are used on aircraft because they are very lightweight. A design advantage of rivets is that they can be drilled out and removed without damage to the objects they hold together.

Rivet types are represented on technical drawings using a coding system. See Figure 6-81. Since rivets are sometimes so small and the material they hold together so thin that it is difficult to clearly draw the rivets, some companies draw only the rivet's centerline in the side view and identify the rivets using a drawing callout.

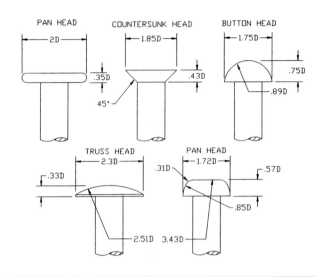

Figure 6-80

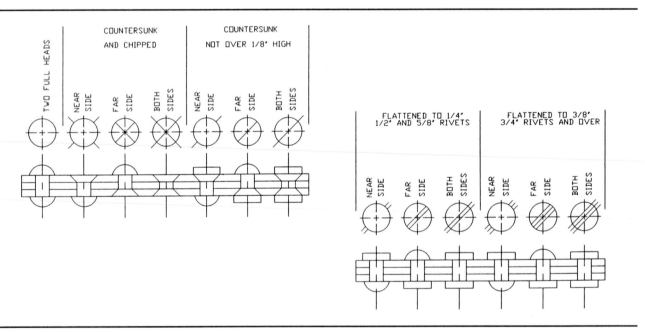

Figure 6-81

6-21 EXERCISE PROBLEMS

EX6-1 MILLIMETERS

Figure 6-82 shows three blocks. Assume that the block is 30 × 30 × 10 and that the hole is Ø9. Assemble the three blocks so that their holes are aligned and they are held together by a hex head bolt secured by an appropriate hex nut. Locate a washer between the bolt head and the top block and between the nut and the bottom block. Create all drawings using either an A4 or A3 drawing sheet, as needed. Include a title block on all drawing sheets.

A. Define the bolt.
B. Define the nut.
C. Define the washers.
D. Draw an assembly drawing including all components.
E. Create a BOM for the assembly.
F. Create a presentation drawing of the assembly.
G. Create an isometric exploded drawing of the assembly.
H. Create an animation drawing of the assembly.

EX6-2 MILLIMETERS

Figure 6-83 shows three blocks, one 30 × 30 × 50 with a centered M8 threaded hole, and two 30 × 30 × 10 blocks with centered Ø9 holes. Join the two 30 × 30 × 10 blocks to the 30 × 30 × 50 block using an M8 hex head bolt. Locate a regular plain washer under the bolt head.

A. Define the bolt.
B. Define the thread depth
C. Define the hole depth.
D. Define the washer.
E. Draw an assembly drawing including all components.
F. Create a BOM for the assembly.
G. Create a presentation drawing of the assembly.
H. Create an isometric exploded drawing of the assembly.
I. Create an animation drawing of the assembly.

Three blocks, each 40 × 40 × 10 with a centered Ø9 hole.
P/N AM311-10M

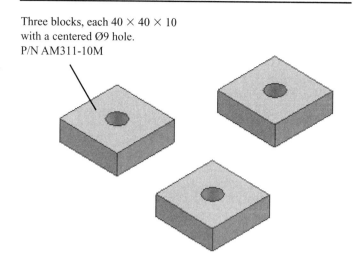

Assemble the three blocks using a hex head nut, a hex nut, and two plain, narrow washers

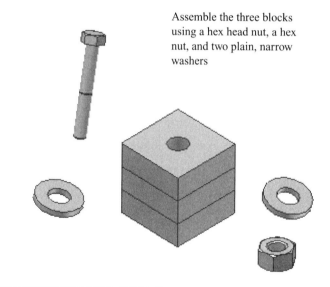

Figure 6-82

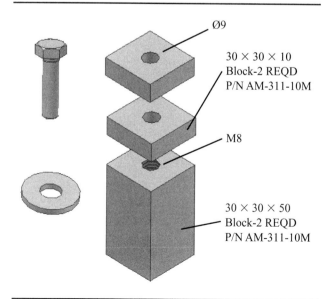

Ø9

30 × 30 × 10
Block-2 REQD
P/N AM-311-10M

M8

30 × 30 × 50
Block-2 REQD
P/N AM-311-10M

Figure 6-83

EX6-3 INCHES

Figure 6-84 shows three blocks. Assume that each block is $1.00 \times 1.00 \times 0.375$ and that the hole is $\varnothing.375$. Assemble the three blocks so that their holes are aligned and that they are held together by a $\frac{5}{16}$-18 UNC indented regular hex head bolt secured by an appropriate hex nut. Locate a washer between the bolt head and the top block and between the nut and the bottom block. Create all drawings using either an A4 or A3 drawing sheet, as needed. Include a title block on all drawing sheets.

A. Define the bolt.
B. Define the nut.
C. Define the washers
D. Draw an assembly drawing including all components.
E. Create a BOM for the assembly.
F. Create a presentation drawing of the assembly.
G. Create an isometric exploded drawing of the assembly.
H. Create an animation drawing of the assembly.

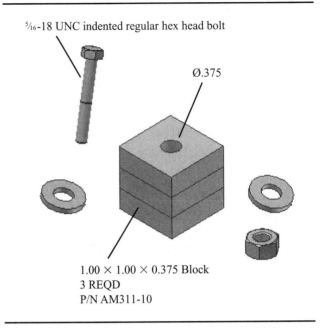

$\frac{5}{16}$-18 UNC indented regular hex head bolt

$\varnothing.375$

$1.00 \times 1.00 \times 0.375$ Block
3 REQD
P/N AM311-10

Figure 6-84

EX6-4 INCHES

Figure 6-85 shows three blocks, one $1.00 \times 1.00 \times 2.00$ with a centered threaded hole, and two $1.00 \times 1.00 \times 0.375$ blocks with centered $\varnothing.375$ holes. Join the two $1.00 \times 1.00 \times 0.375$ blocks to the $1.00 \times 1.00 \times 2.00$ block using a $\frac{5}{16}$-18 UNC hex head bolt. Locate a regular plain washer under the bolt head.

A. Define the bolt.
B. Define the thread depth
C. Define the hole depth.
D. Define the washer.
E. Draw an assembly drawing including all components.
F. Create a BOM for the assembly.
G. Create a presentation drawing of the assembly.
H. Create an isometric exploded drawing of the assembly.
I. Create an animation drawing of the assembly.

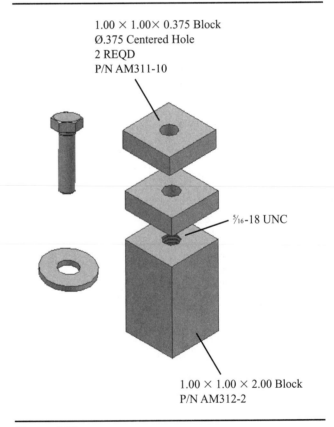

$1.00 \times 1.00 \times 0.375$ Block
$\varnothing.375$ Centered Hole
2 REQD
P/N AM311-10

$\frac{5}{16}$-18 UNC

$1.00 \times 1.00 \times 2.00$ Block
P/N AM312-2

Figure 6-85

EX6-5 INCHES OR MILLIMETERS

Figure 6-86 shows a centerng block. Create an assembly drawing of the block and insert three setscrews into the three threaded holes so that they extend at least 0.25 in. or 6 mm into the center hole.

A. Use the inch dimensions.
B. Use the millimeter dimensions.
C. Define the setscrews.
D. Draw an assembly drawing including all components.
E. Create a BOM for the assembly.
F. Create a presentation drawing of the assembly.
G. Create an isometric exploded drawing of the assembly.
H. Create an animation drawing of the assembly.

Centering Block
P/N BU2004-5
SAE 1020 Steel

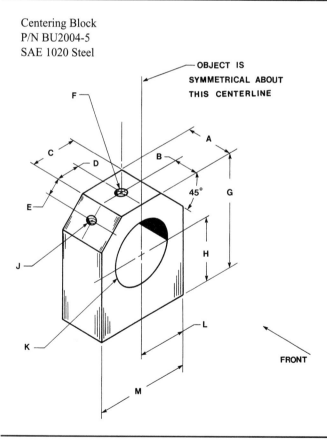

DIMENSION	INCHES	mm
A	1.00	26
B	.50	13
C	1.00	26
D	.50	13
E	.38	10
F	.190-32 UNF	M8X1
G	2.38	60
H	1.38	34
J	.164-36 UNF	M6
K	Ø1.25	Ø30
L	1.00	26
M	2.00	52

Figure 6-86

EX6-6 MILLIMETERS

Figure 6-87 shows two parts; a head cylinder and a base cylinder. The head cylinder has outside dimensions of $\varnothing40 \times 20$, and the base cylinder has outside dimensions of $\varnothing40 \times 50$. The holes in both parts are located on a $\varnothing24$ bolt circle. Assemble the two parts using hex head bolts.

A. Define the bolt.
B. Define the holes in the head cylinder, the counter-bore diameter and depth, and the clearance hole diameter.
C. Define the thread depth in the base cylinder
D. Define the hole depth in the base cylinder.
E. Draw an assembly drawing including all components.
F. Create a BOM for the assembly.
G. Create a presentation drawing of the assembly.
H. Create an isometric exploded drawing of the assembly.
I. Create an animation drawing of the assembly.

Cylinder Head
P/N EK130-1
SAE 1040 Steel

Counterbored holes on a $\varnothing24$ bolt circle.

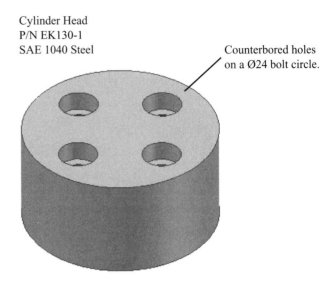

Cylinder Base
P/N EK130-2
SAE 1040 Steel

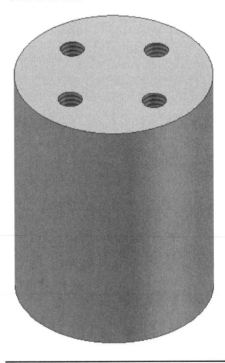

Figure 6-87

EX6-7 MILLIMETERS

Figure 6-88 shows a pressure cylinder assembly.

A. Draw an assembly drawing including all components.
B. Create a BOM for the assembly.
C. Create a presentation drawing of the assembly.
D. Create an isometric exploded drawing of the assembly.
E. Create an animation drawing of the assembly.

EX6-8 MILLIMETERS

Figure 6-88 shows a pressure cylinder assembly.

A. Revise the assembly so that it uses M10 × 35 hex head bolts.
B. Draw an assembly drawing including all components.
C. Create a BOM for the assembly.
D. Create a presentation drawing of the assembly.
E. Create an isometric exploded drawing of the assembly.
F. Create an animation drawing of the assembly.

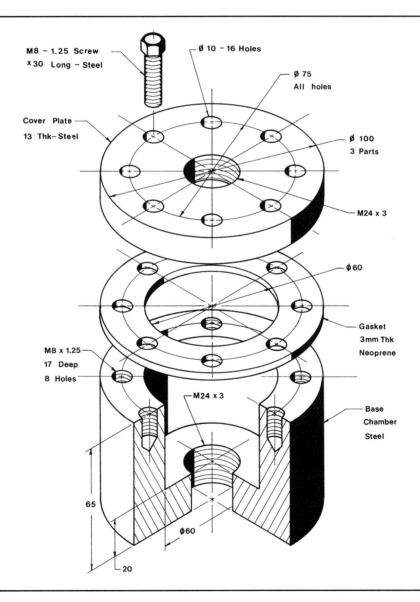

Figure 6-88

EX6-9 INCHES AND MILLIMETERS

Figure 6-89 shows a C-block assembly.

Use one of the following fasteners assigned by your instructor.

1. M12 hex head
2. M10 square head
3. ¼-20 UNC hex head
4. ⅜-16 UNC square head
5. M10 socket head
6. M8 slotted head
7. ¼-20 UNC slotted head
8. ⅜-16 UNC socket head

    A. Define the bolts.
    B. Define the nuts.
    C. Define the washers.
    D. Draw an assembly drawing including all components.
    E. Create a BOM for the assembly.
    F. Create a presentation drawing of the assembly.
    G. Create an isometric exploded drawing of the assembly.
    H. Create an animation drawing of the assembly.

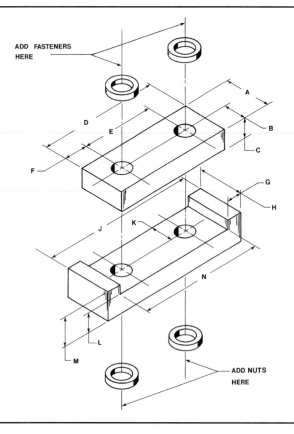

DIMENSION	INCHES	mm
A	1.25	32
B	.63	16
C	.50	13
D	3.25	82
E	2.00	50
F	.63	16
G	.38	10
H	1.25	32
J	4.13	106
K	.63	16
L	.50	13
M	.75	10
N	3.38	86

Figure 6-89

EX6-10 MILLIMETERS

Figure 6-90 shows an exploded assembly drawing. There are no standard parts, so each part must be drawn individually.

A. Draw an assembly drawing including all components.
B. Create a BOM for the assembly.
C. Create a presentation drawing of the assembly.
D. Create an isometric exploded drawing of the assembly.
E. Create an animation drawing of the assembly.

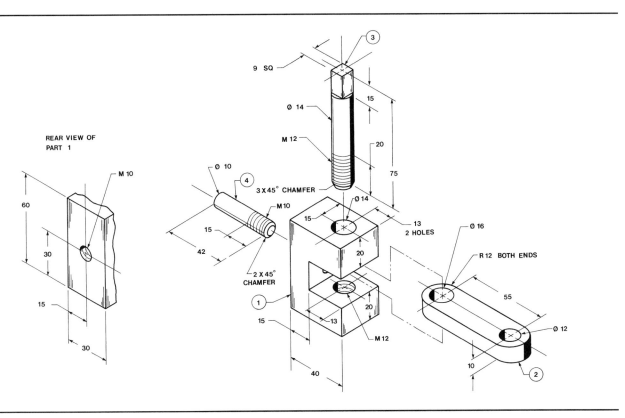

Figure 6-90

EX6-11 MILLIMETERS

Figure 6-91 shows an exploded assembly drawing.

A. Draw an assembly drawing including all components.
B. Create a BOM for the assembly.
C. Create a presentation drawing of the assembly.
D. Create an isometric exploded drawing of the assembly.
E. Create an animation drawing of the assembly.

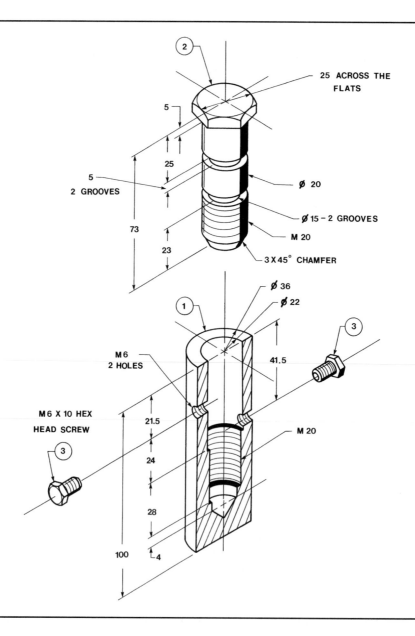

Figure 6-91

EX6-12 INCHES OR MILLIMETERS

Figure 6-92 shows an exploded assembly drawing. No dimensions are given. If Parts 3 and 5 have either M10 or ⅜-16 UNC threads, size parts 1 and 2. Based on these values estimate and create the remaining sizes and dimensions.

A. Draw an assembly drawing including all components.
B. Create a BOM for the assembly.
C. Create a presentation drawing of the assembly.
D. Create an isometric exploded drawing of the assembly.
E. Create an animation drawing of the assembly.

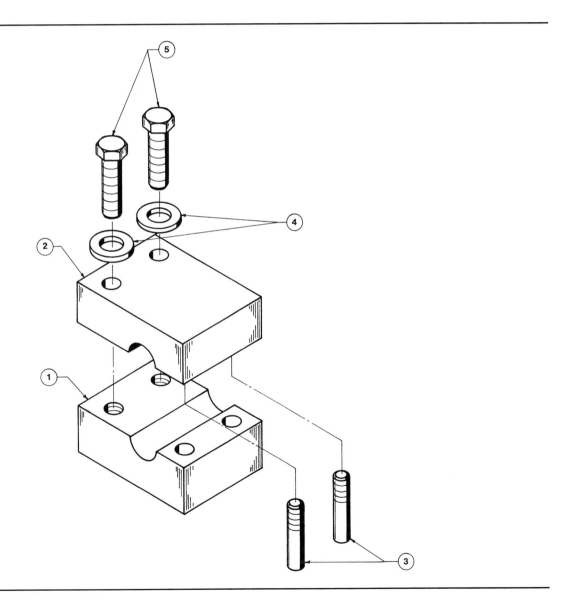

Figure 6-92

EX6-13 INCHES

Figure 6-93 shows an assembly drawing and detail drawings of a surface gauge.

A. Draw an assembly drawing including all components.
B. Create a BOM for the assembly.
C. Create a presentation drawing of the assembly.
D. Create an isometric exploded drawing of the assembly.
E. Create an animation drawing of the assembly.

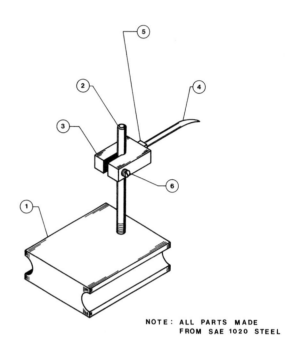

SIMPLIFIED SURFACE GAGE

NOTE: ALL PARTS MADE
FROM SAE 1020 STEEL

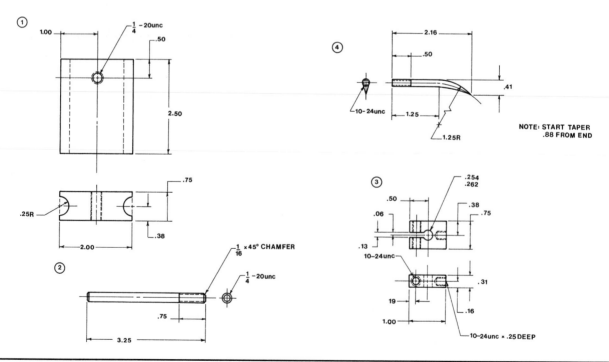

Figure 6-93

EX6-14 MILLIMETERS

Figure 6-94 shows an assembly made from parts defined on pages 168 through 170. Assemble the parts using M10 threaded fasteners.

A. Define the bolt.
B. Define the nut.
C. Draw an assembly drawing including all components.
D. Create a BOM for the assembly.
E. Create a presentation drawing of the assembly.
F. Create an isometric exploded drawing of the assembly.
G. Create an animation drawing of the assembly.
H. Consider possible interference between the nuts and ends of the fasteners both during and after assembly. Recommend an assembly sequence.

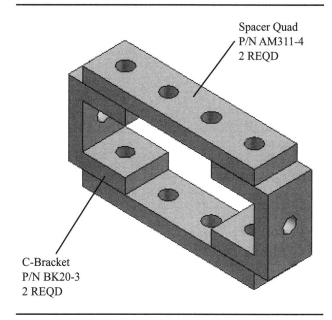

Spacer Quad
P/N AM311-4
2 REQD

C-Bracket
P/N BK20-3
2 REQD

Figure 6-94

EX6-15 MILLIMETERS

Figure 6-95 shows an assembly made from parts defined on pages 168 through 170. Assemble the parts using M10 threaded fasteners.

A. Define the bolt.
B. Define the nut.
C. Draw an assembly drawing including all components.
D. Create a BOM for the assembly.
E. Create a presentation drawing of the assembly.
F. Create an isometric exploded drawing of the assembly.
G. Create an animation drawing of the assembly.
H. Consider possible interference between the nuts and ends of the fasteners both during and after assembly. Recommend an assembly sequence.

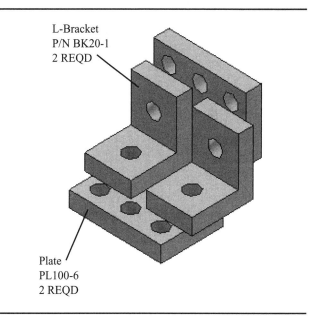

L-Bracket
P/N BK20-1
2 REQD

Plate
PL100-6
2 REQD

Figure 6-95

EX6-16 MILLIMETERS

Figure 6-96 shows an assembly made from parts defined on pages 168 through 170. Assemble the parts using M10 threaded fasteners.

A. Define the bolt.
B. Define the nut.
C. Draw an assembly drawing including all components.
D. Create a BOM for the assembly.
E. Create a presentation drawing of the assembly.
F. Create an isometric exploded drawing of the assembly.
G. Create an animation drawing of the assembly.
H. Consider possible interference between the nuts and ends of the fasteners both during and after assembly. Recommend an assembly sequence.

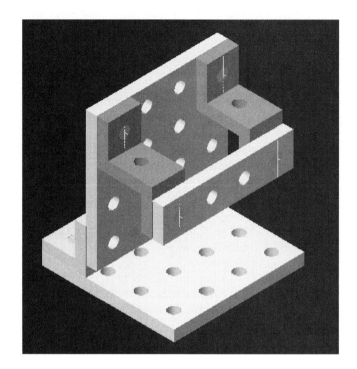

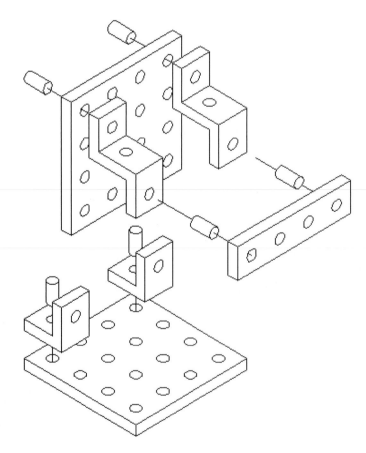

Figure 6-96

EX6-17

Design an Access Controller based on the information given. The controller works by moving an internal cylinder up and down within the base so the cylinder aligns with output holes A and B. Liquids will enter the internal cylinder from the top, then exit the base through holes A and B. Include as many holes in the internal cylinder as necessary to create the following liquid exit combinations.

1. A open, B closed
2. A open, B open
3. A closed, B open

The internal cylinder is to be held in place by an alignment key and a stop button. The stop button is to be spring-loaded so that it will always be held in place. The internal cylinder will be moved by pulling out the stop button, repositioning the cylinder, then reinserting the stop button.

Prepare the following drawings.

A. Draw an assembly drawing.
B. Draw detail drawings of each nonstandard part. Include positional tolerances for all holes.
C. Prepare a parts list.

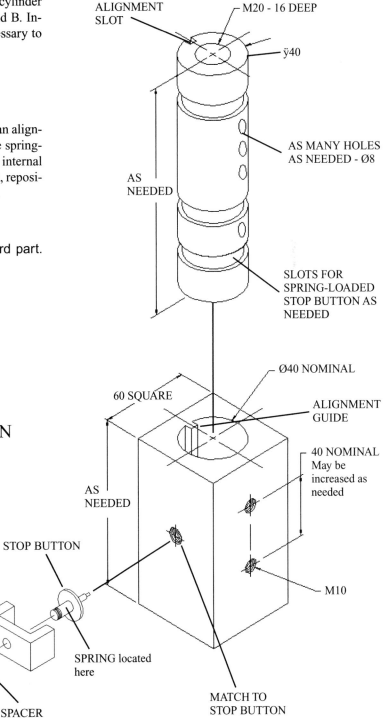

EX6-18

Design a hand-operated grinding wheel specifically for sharpening a chisel. The chisel is to be located on an adjustable rest while it is being sharpened. The mechanism should be able to be clamped to a table during operation using two thumbscrews.

A standard grinding wheel is Ø6.00″ and 1/2″ thick, and has an internal mounting hole with a 50.00±.03 bore.

Prepare the following drawings.

A. Draw an assembly drawing.
B. Draw detail drawings of each nonstandard part. Include positional tolerances for all holes.
C. Prepare a parts list.

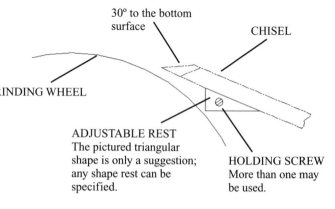

30° to the bottom surface

CHISEL

GRINDING WHEEL

ADJUSTABLE REST
The pictured triangular shape is only a suggestion; any shape rest can be specified.

HOLDING SCREW
More than one may be used.

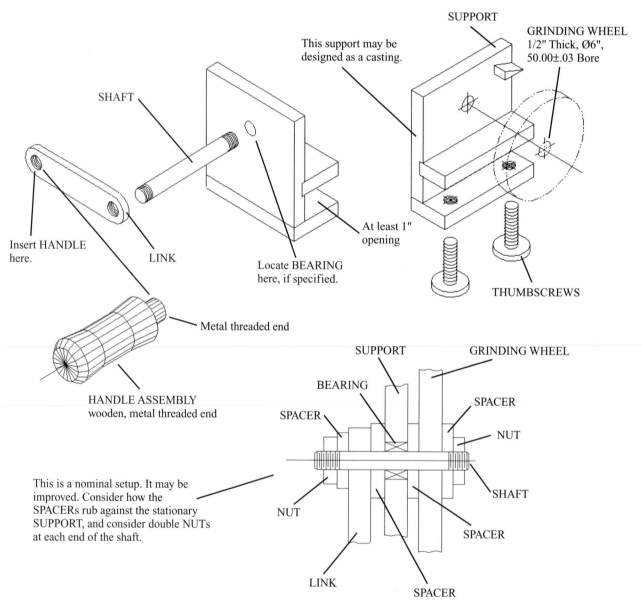

This support may be designed as a casting.

SUPPORT

GRINDING WHEEL
1/2″ Thick, Ø6″, 50.00±.03 Bore

SHAFT

Insert HANDLE here.

LINK

Locate BEARING here, if specified.

At least 1″ opening

THUMBSCREWS

Metal threaded end

HANDLE ASSEMBLY
wooden, metal threaded end

This is a nominal setup. It may be improved. Consider how the SPACERs rub against the stationary SUPPORT, and consider double NUTs at each end of the shaft.

SUPPORT   GRINDING WHEEL

BEARING

SPACER

SPACER

NUT

SPACER

NUT

SHAFT

LINK

SPACER

SPACER

EX6-19

Given the assembly shown, add the following fasteners.

1. Create an assembly drawing.
2. Create a parts list including assembly numbers.
3. Create a dimensioned drawing of the Support Block and specify a dimension for each hole including the thread size and the depth required.

Fasteners:

A.

1. M10 × 35 HEX HEAD BOLT
2. M10 × 35 HEX HEAD BOLT
3. M10 × 30 HEX HEAD BOLT
4. M10 × 30 HEX HEAD BOLT
5. M10 × 30 HEX HEAD BOLT
6. M10 × 25 HEX HEAD BOLT

B.

1. M10 × 35 HEX HEAD BOLT
2. M8 × 35 HEX HEAD BOLT
3. M10 × 30 HEX HEAD BOLT
4. M8 × 30 HEX HEAD BOLT
5. M6 × 30 HEX HEAD BOLT
6. M6 × 25 HEX HEAD BOLT

EX6-20

Redesign the ROTATOR ASSEMBLY so that LINK-L and LINK-R are the same part and they contain Ø10 holes rather than posts. Complete the assembly with suitable fasteners. Assure that is no interference as the assembly rotates.

The hole pattern is the same for all three parts.

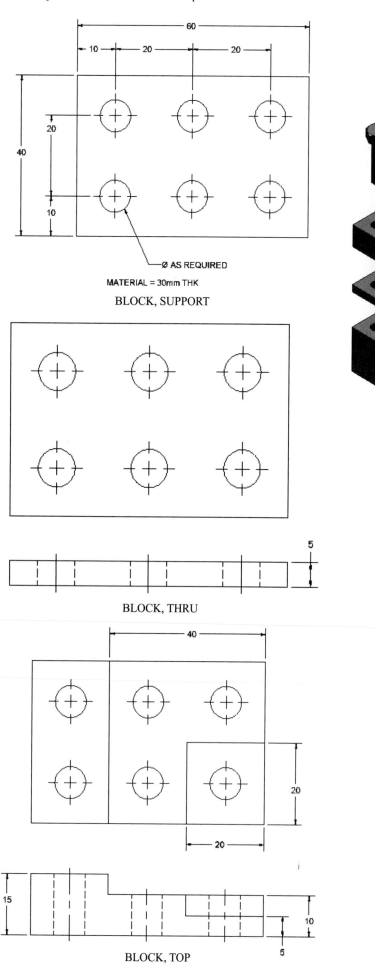

60

10 20 20

40

20

10

Ø AS REQUIRED

MATERIAL = 30mm THK

BLOCK, SUPPORT

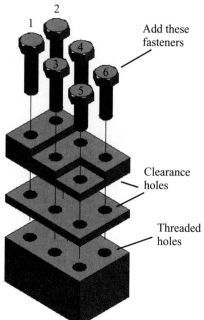

2
1
4
3
6
5

Add these fasteners

Clearance holes

Threaded holes

BLOCK, THRU

5

BLOCK, TOP

40

20

20

15

10

5

Dimensioning

7-1 INTRODUCTION

Inventor uses two types of dimensions: model and drawing. Model dimensions are created as the model is being constructed and may be edited to change the shape of a model. Drawing dimensions can be edited, but the changes will not change the shape of the model. If the shape of a model is changed, the drawing dimensions associated with the revised surfaces will change to reflect the new values.

Dimensions are usually applied to a drawing using either American National Standards Institute (ANSI) or International Organization for Standardization (ISO) standards. If English units are selected when a new drawing is started, the ANSI inch standards (ANSI (in).idw) will be invoked. If metric units are selected, the ISO standards (ISO.idw) may be invoked. This book uses ANSI standards for both inch and metric units.

Figure 7-1 shows the layout of a model that includes only the parametric dimensions. The Metric option was selected before the model was drawn. The model dimensions were created automatically as the model was created. The dimensions are numerically accurate, but they use the ISO form and do not conform to the dimensioning standards as defined in ANSI Y14.5. For example, the unit values are written above the dimension lines, and there are two 40 dimensions in the top view when only one is needed. There are also dimensions located on the surface of the views, and

Model dimensions were created as the model was created.

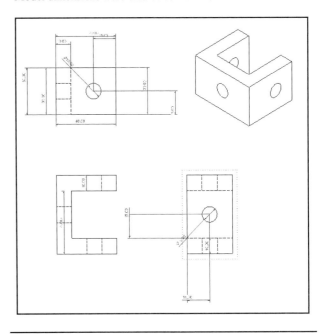

Figure 7-1

there are improperly crossed extension lines. It is possible to locate the parametric dimensions correctly by selecting their locations during the creation of the model, but it is much easier to locate dimensions on orthographic views.

7-2 TERMINOLOGY AND CONVENTIONS – ANSI

Some common terms

Figure 7-2 shows both ANSI and ISO style dimensions. The terms apply to both styles.

Dimension lines: In mechanical drawings, lines between extension lines that end with an arrowhead and include a numerical dimensional value located within the line.

Extension lines: Lines that extend away from an object and allow dimensions to be located off the surface of an object.

Leader lines: Lines drawn at an angle, not horizontal or vertical, that are used to dimension specific shapes such as holes. The start point of a leader line includes an arrowhead. Numerical values are drawn at the end opposite the arrowhead.

Linear dimensions: Dimensions that define the straight line distance between two points.

Angular dimensions: Dimensions that define the angular value, measured in degrees, between two straight lines.

Some dimensioning conventions

See Figure 7-3.

1. Dimension lines should be drawn evenly spaced, that is, the distance between dimension lines should be uniform. A general rule of thumb is to locate dimension lines about ½ in. or 15 mm apart.

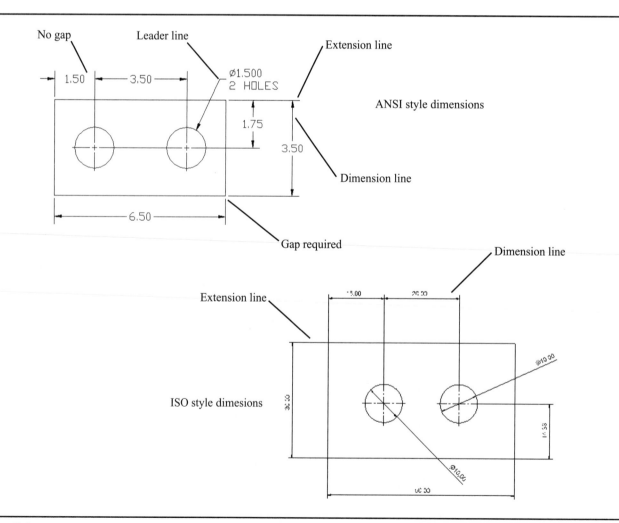

Figure 7-2

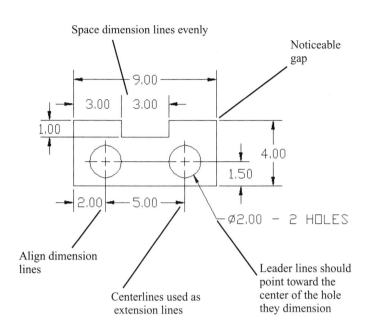

Figure 7-3

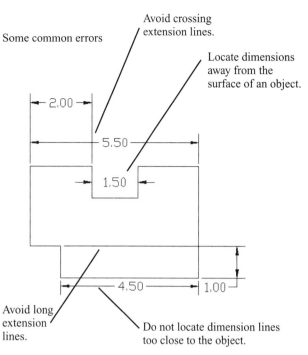

Figure 7-4

2. There should a noticeable gap between the edge of a part and the beginning of an extension line. This serves as a visual break between the object and the extension line. The visual difference between the line types can be enhanced by using different colors for the two types of lines.
3. Leader lines are used to define the size of holes and should be positioned so that the arrowhead points toward the center of the hole.
4. Centerlines may be used as extension lines. No gap is used when a centerline is extended beyond the edge lines of an object.
5. Align dimension lines whenever possible to give the drawing a neat, organized appearance.

Some common errors to avoid

See Figure 7-4.

1. Avoid crossing extension lines. Place longer dimensions farther away from the object than shorter dimensions.
2. Do not locate dimensions within cutouts; always use extension lines.
3. Do not locate any dimension close to the object. Dimension lines should be at least ¹/₂ in. or 15 mm from the edge of the object.
4. Avoid long extension lines. Locate dimensions in the same general area as the feature being defined.

7-3 CREATING DRAWING DIMENSIONS

Drawing dimensions are added to a drawing using the General Dimension tool. The General Dimension tool is located on the Drawing Annotation Panel bar.

This section will add drawing dimensions to the model view shown in Figure 7-5. The dimensions will be in compliance with ANSI standards.

Drawing dimensions are different from model dimensions. *Model dimensions* are created as the model is created and can be used to edit (change the shape) of the model. *Drawing dimensions* are attached to a specified distance. Changing a drawing dimension will not change the shape of the model.

To access the Drawing Annotation Panel bar

1. Move the cursor into the panel bar area, and right-click the mouse.

A small dialog box will appear.

2. Select the Drawing Annotation option.

The panel bar will change to the Drawing Annotation Panel bar shown in Figure 7-6. The General Dimension tool is the first tool listed.

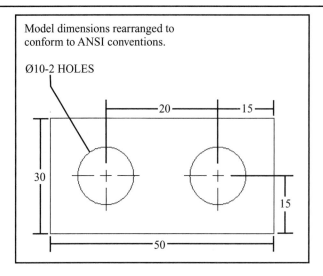

Model dimensions rearranged to conform to ANSI conventions.

Ø10-2 HOLES

Figure 7-5

To add centerlines to holes

Figure 7-7 shows a model whose dimensions have been hidden. The dimensions were not deleted, as they may be needed at some time to edit the model.

Centerlines were added to the model using the Center Mark tool located on the Drawing Annotation toolbar and as explained in Section 4-3.

To add horizontal dimensions

1. Click the General Dimension tool.
2. Move the cursor into the drawing area and first click the upper left corner of the model.

A green circle will appear on the corner, indicating that it has been selected.

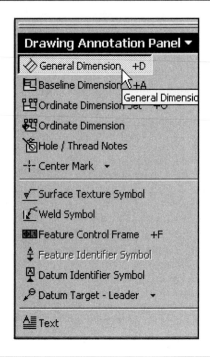

Figure 7-6

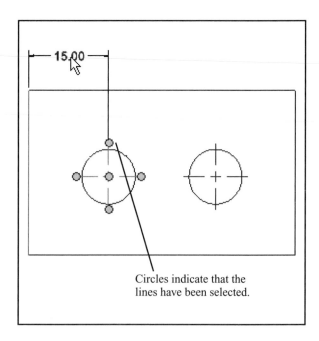

Circles indicate that the lines have been selected.

Figure 7-7

3. Click the top end of the left hole's vertical center-line.
4. Move the cursor, locating the dimension, then press the left mouse button.

Locate the dimension away from the edge of the model and position the text in the approximate center between the two extension lines.

5. Locate a second horizontal dimension between the two vertical hole centerlines.

Overall dimensions

Overall dimensions define the outside sizes of a model, the maximum length, width, and height. It is important that overall dimensions be easy to find and read, as they are often used to determine the stock sizes needed to produce the model.

Convention calls for overall dimensions to be located farther away from the model than any other dimensions. In Figure 7-8 the 50 overall dimension was located above the other two horizontal dimensions, that is, farther away from the model. The dimension could also have been located below the model.

Note that the spacing between the model's edge and the two horizontal dimensions is approximately equal to the distance between the overall dimension and the two horizontal dimensions.

Vertical dimensions

ANSI standards call for the text of vertical dimensions to be written *unidirectionally.* This means that the text should

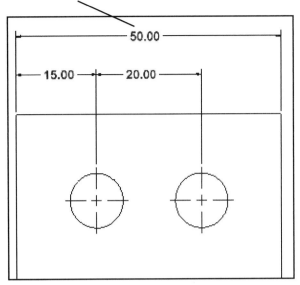

Overall horizontal dimension

Figure 7-8

be written horizontally and be read from left to right. Figure 7-9 shows two vertical dimensions added to the model. Both use unidirectional text. Note also that the overall height dimension is located the farthest away from the model's edge.

To create unidirectional text

1. Click the Format heading at the top of the screen, then select the Dimension Styles option.

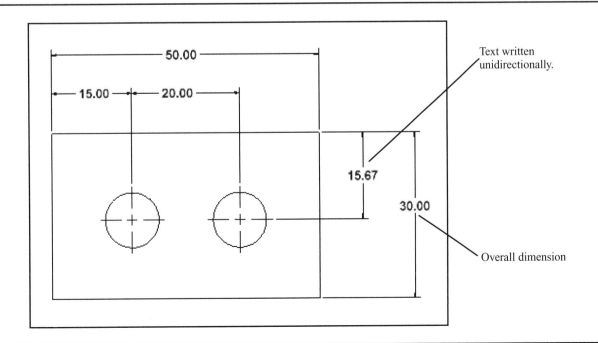

Figure 7-9

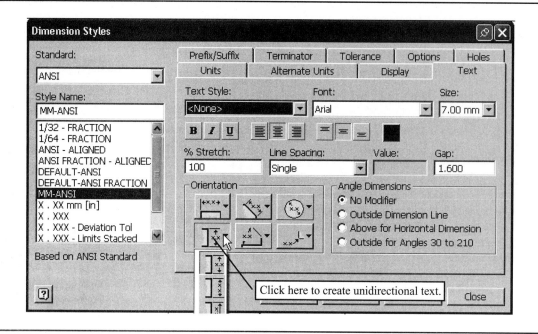

Figure 7-10

The Dimension Styles dialog box will appear. See Figure 7-10.

2. Click the Text tab.
3. Click the vertical dimension box in the Orientation box, then click the Save and Close boxes.

All vertical dimensions will now be written using unidirectional text.

To dimension holes

There are two Ø10 holes in the model. Two hole dimensions could be applied, or one dimension could be used with the additional note Ø10 - 2 HOLES. In general, it is desirable to use as few dimensions as possible to clearly and completely define the model's size. This helps prevent a cluttered and confusing drawing.

1. Click the General Dimension tool, then the edge of one of the holes.
2. Hold the left mouse button down and drag the dimension away from the model.
3. Locate the hole dimension and left-click the mouse.
4. Right-click the mouse and select the Done option.

See Figure 7-11.

To add text to the hole dimension

1. Move the cursor onto the hole dimension and right-click the mouse.

   A dialog box will appear.

2. Click the Text option.

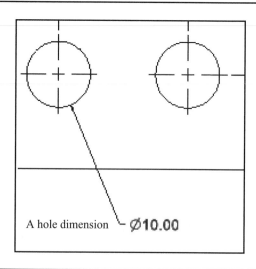

Figure 7-11

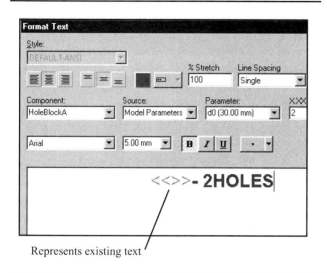

Represents existing text

Figure 7-12

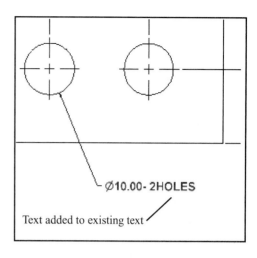

Ø10.00- 2HOLES

Text added to existing text

Figure 7-13

The Format Text dialog box will appear. See Figure 7-12. The ≪ ≫ symbol represents the existing text.

3. Locate the cursor to the right of the ≪ ≫ symbol and type - 2 HOLES.
4. Click the OK box.

Figure 7-13 shows the resulting dimension.

7-4 DRAWING SCALE

Drawings are often drawn "to scale" because the actual part is either too big to fit on a sheet of drawing paper or too small to be seen. For example, a microchip circuit must be drawn at several thousand times its actual size to be seen.

Drawing scales are written using the following formats.

SCALE: 1=1
SCALE: FULL
SCALE: 1000=1
SCALE: .25=1

In each example the value on the left indicates the scale factor. A value greater than 1 indicates that the drawing is larger than actual size. A value smaller than 1 indicates that the drawing is larger than actual size.

Regardless of the drawing scale selected the dimension values must be true size. Figure 7-14 shows the same rectangle drawn at two different scales. The top rectangle is drawn at a scale of 1 = 1, or its true size. The bottom rectangle is drawn at a scale of 2 = 1, or twice its true size. In both examples the 3.00 dimension remains the same.

7-5 UNITS

It is important to understand that dimensional values are not the same as mathematical units. Dimensional values are manufacturing instructions and always include a tolerance, even if the tolerance value is not stated. Manufacturers use a predefined set of standard dimensions that are applied to any dimensional value that does not include a written tolerance. Standard tolerance values differ from organization to organization. Figure 7-15 shows a chart of standard tolerances.

In Figure 7-16 a distance is dimensioned twice: once as 5.50 and a second time as 5.5000. Mathematically these two values are equal, but they are not the same manufacturing instruction. The 5.50 value could, for example, have a

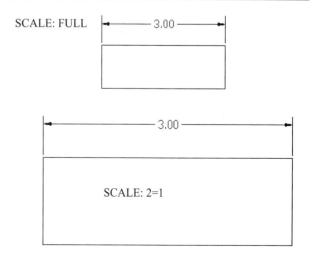

SCALE: FULL 3.00

3.00

SCALE: 2=1

Figure 7-14

TOLERANCES UNLESS
OTHERWISE STATED

X ± 1

.X ± .1

.xx ± .01

.XXX ± .005

X° ± 1°

.X° ± .1°

Figure 7-15

standard tolerance of ±.01, whereas the 5.5000 value could have a standard tolerance of ±.0005. A tolerance of ± .0005 is more difficult and, therefore, more expensive to manufacture than a tolerance of ±.01.

Figure 7-17 shows examples of units expressed in millimeters and in decimal inches. A zero is not required to the left of the decimal point for decimal inch values less than one. Millimeter values do not require zeros to the right of the decimal point. Millimeter and decimal inch values never include symbols; the units will be defined in the title block of the drawing.

To prevent a 0 from appearing to the left of the decimal point

1. Click the Format heading at the top of the screen, then select the Dimension Styles option.

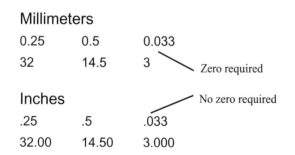

Millimeters

| 0.25 | 0.5 | 0.033 |
| 32 | 14.5 | 3 |

Zero required

Inches

No zero required

| .25 | .5 | .033 |
| 32.00 | 14.50 | 3.000 |

Figure 7-17

These dimensions are not the same. They have different tolerance requirements.

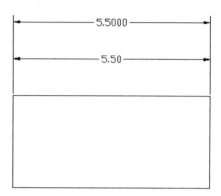

Figure 7-16

The Dimension Styles dialog box will appear. See Figure 7-18.

2. Click the New box.

This procedure creates a new dimension style that suppresses all leading zeros.

To change the number of decimal places in a dimension value

1. Click on the Format heading at the top of the screen, then select the Dimension Styles option.

The Dimension Styles dialog box will appear. See Figure 7-19.

2. Click the scroll arrow on the right side of the Precision box.

A listing of available precision settings will cascade down.

3. Click the desired precision value.

Save the changes if desired. You can now dimension using any of the dimension commands, and the resulting values will be expressed using the selected precision.

7-6 ALIGNED DIMENSIONS

Aligned dimensions are dimensions that are parallel to a slanted edge or surface. They are not horizontal or vertical. The unit values for aligned dimensions should be horizontal or unidirectional.

To create an aligned dimension

Figure 7-20 shows a wedge-shaped solid model presented in the Drawing Layout mode. The General Dimension

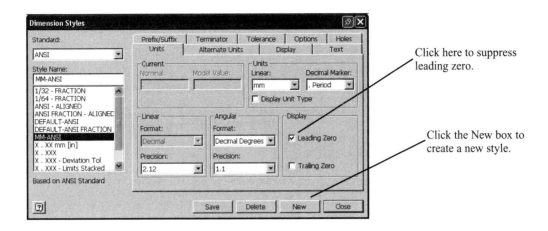

Click here to suppress leading zero.

Click the New box to create a new style.

Figure 7-18

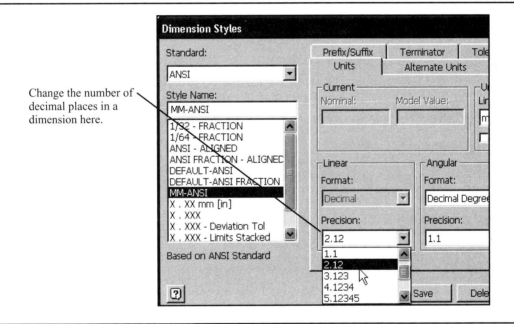

Change the number of decimal places in a dimension here.

Figure 7-19

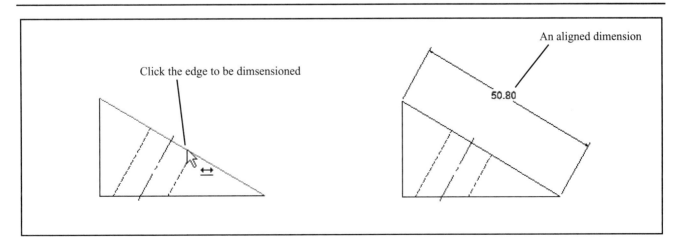

Click the edge to be dimsensioned

An aligned dimension

50.80

Figure 7-20

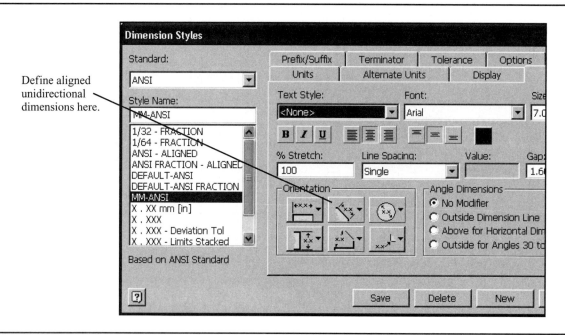

Define aligned unidirectional dimensions here.

Figure 7-21

tool was used to add the aligned dimension. General dimensions can have tolerances and fits added to them as they are created. See Figure 7-20. Power dimensions will be discussed in more detail in Chapter 9.

1. Access the Drawing Annotation panel bar by moving the cursor into the panel bar area and right-clicking the mouse, then select the Drawing Annotation option.
2. Click the General Dimension tool, then move the cursor onto the slanted edge to be dimensioned.

The edge will change color, indicating that it has been selected.

3. Move the cursor away from the edge and left-click the mouse.
4. Move the mouse as necessary to locate the aligned dimension.

See Figure 7-20.

To create a unidirectional aligned dimension

Aligned dimensions can be made unidirectional by using the Orientation option on the Dimension Styles dialog box.

1. Click the Format heading at the top of the screen, then select the Dimension Styles tool.

The Dimension Styles dialog box will appear. See Figure 7-21.

2. Click the Text tab, then the Aligned Dimension box in the Orientation area.
3. Select the unidirectional option.

7-7 RADIUS AND DIAMETER DIMENSIONS

Figure 7-22 shows an object that includes both arcs and circles. The general rule is to dimension arcs using a radius dimension, and circles using diameter dimensions. This convention is consistent with the tooling required to produce the feature shape. Any arc greater than 170° is considered a circle and is dimensioned using a diameter.

To create a radius dimension

1. Access the Drawing Annotation panel bar and click the General Dimension tool.
2. Move the cursor into the drawing area and click one of the filleted corners in the front view.
3. Move the cursor away from the fillet and locate the dimension by clicking the mouse.
4. Right-click the mouse and select the Done option.

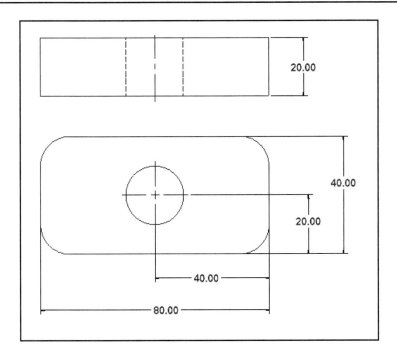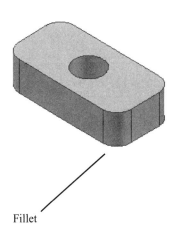

Fillet

Figure 7-22

Figure 7-23 shows the resulting dimension. There are four equal arcs on the object, and they all must be dimensioned. It would be better to add the words 4 CORNERS to the radius dimension than to include four radius dimensions.

To add text to an existing drawing dimension

1. Move the cursor to the fillet dimension.

Filled colored circles will appear, indicating that the dimension has been selected.

2. Right-click the mouse and select the Text option.

The Format Text dialog box will appear. See Figure 7-24. The symbol ≪ ≫ represents the existing text.

3. Type 4 CORNERS under the existing text and click the OK box.

Radius dimension for fillet

Figure 7-23

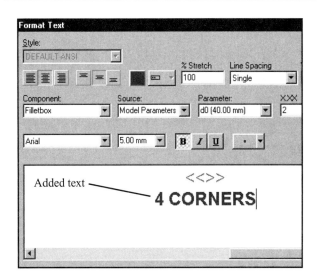

Figure 7-24

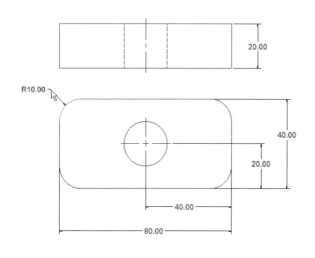

Figure 7-25

Dimensions for holes with depth

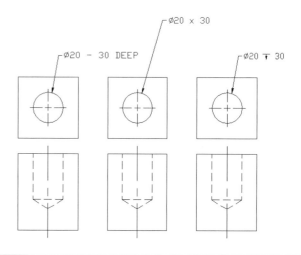

Figure 7-26

Figure 7-25 shows the resulting dimension.

7-8 DIMENSIONING HOLES

Holes are dimensioned by stating their diameter and depth, if any. Holes that go completely through an object are defined using only a diameter dimension. See Figure 7-25.

To dimension a through hole

1. Click on the Hole/Thread Note tool.
2. Click the edge of the hole and drag the dimension.
3. Locate and enter the dimension.

To dimension individual holes

Figure 7-26 shows three different methods that can be used to dimension a hole that does not go completely through an object. Depth values may be added using the Power Dimensioning dialog box, or the Edit Text or Edit Format options.

Figure 7-27 shows two methods of dimensioning holes in sectional views. The single line note version is the preferred method.

To dimension hole patterns

Figure 7-28 shows two different hole patterns dimensioned. The circular pattern includes the note Ø10 - 4 HOLES. This note serves to define all four holes within the object.

Figure 7-28 also shows a rectangular object that contains five holes of equal diameter, equally spaced from one another. The notation 5 × Ø10 specifies 5 holes of 10 diameter. The notation 4 × 20 (=80) means 4 equal spaces of 20. The notation (=80) is a reference dimension and is included for convenience. Reference dimensions are explained in Chapter 9.

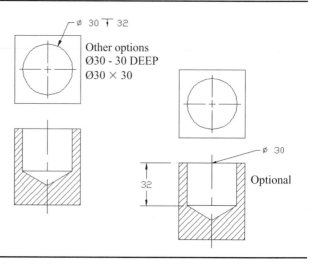

Figure 7-27

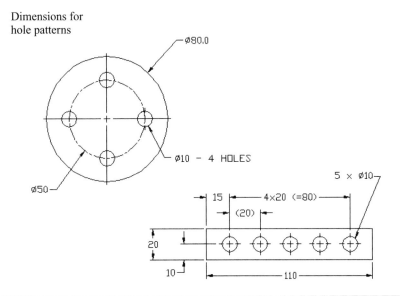

Figure 7-28

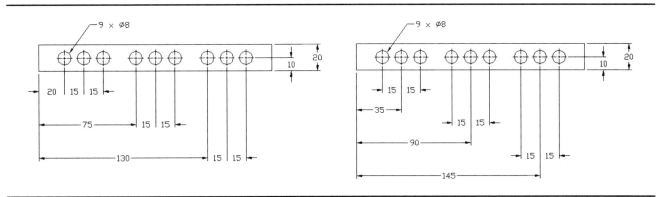

Figure 7-29

Figure 7-29 shows two additional methods for dimensioning repeating hole patterns. Figure 7-30 shows a circular hole pattern that includes two different hole diameters. The hole diameters are not noticeably different and could be confused. One group is defined by indicating letter (A); the other is dimensioned in a normal manner.

To create symbolic dimensions

In an attempt to eliminate language restrictions from drawings, ANSI standards permit the use of certain symbols for dimensions. Like musical notation that can be universally read by people who speak different languages, symbolic dimensions can be read by different people regardless of which language they speak. For example, the symbol Ø replaces the notation DIA.

Figure 7-31 shows a Ø10 hole that has a depth of 15. Inventor will automatically apply symbolic dimensions. The shown dimension is created as follows.

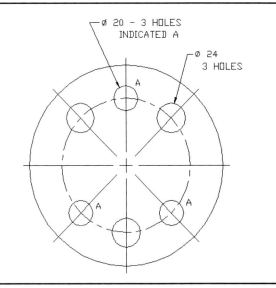

Figure 7-30

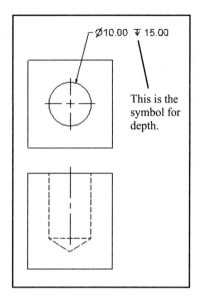

Figure 7-31

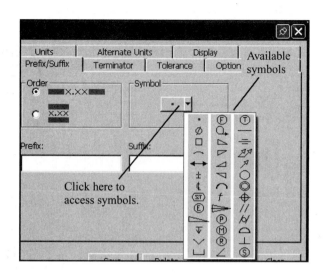

Figure 7-32

1. Access the Drawing Annotation panel bar and click the Hole/Thread Notes tool.
2. Click the edge of the circular view of the hole.
3. Drag the cursor away from the hole's edge and locate the dimension.
4. Right-click the mouse and select the Done option.

To add symbols to dimension text

1. Click on the Format heading at the top of the screen, then click the Dimension Styles option.

The Dimension Styles dialog box will appear. See Figure 7-32.

2. Click the Prefix/Suffix tab.

A prefix is placed before the existing dimension and a suffix is placed after the existing dimension.

3. Click the Symbol box.

A listing of symbols will appear.

4. Select the appropriate prefix and/or suffix symbol.

7-9 DIMENSIONING COUNTERBORED, COUNTERSUNK HOLES

Sections 6-13 and 6-14 explain how to draw counterbored and countersunk holes. This section shows how to dimension countersunk and counterbored holes

Figure 7-33 shows a 70 × 30 × 20 block that includes both a counterbored and a countersunk hole. The sizes for the holes were determined based on the fastener information given in Sections 6-13 and 6-14. The Holes dialog boxes used to create the holes are also shown.

The information used to create the holes will automatically be used to create the hole's dimension. Prefixes and suffixes may be added to the dimension, but the dimension may not be changed. To change a dimension, the original model must be changed. Once the model is changed, the dimension will automatically reflect the changes.

To dimension a counterbored hole

The counterbored hole has a clear hole that goes completely through the block.

1. Access the Drawing Annotation Panel bar and click the Hole/Thread Notes tool.
2. Click the edge of the circular view of the hole.
3. Drag the cursor away from the hole's edge and locate the dimension.
4. Right-click the mouse and select the Done option.

Note how the dimensions match exactly the values used to create the hole. See Figures 7-33 and 7-34.

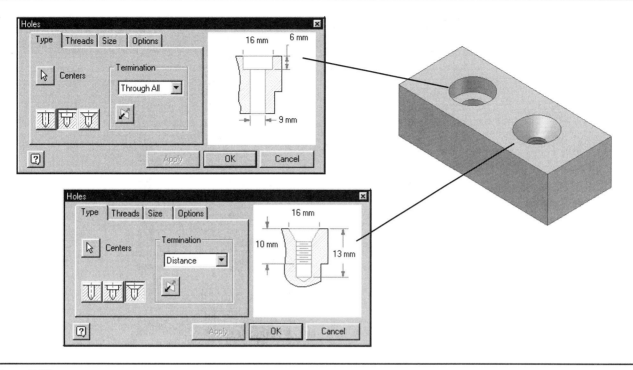

Figure 7-33

To dimension a countersunk hole

The countersunk hole has M8 threads and a thread depth of 10 mm.

1. Access the Drawing Annotation Panel bar and click the Hole/Thread Notes tool.

2. Click the edge of the circular view of the hole.
3. Drag the cursor away from the hole's edge and locate the dimension.
4. Right-click the mouse and select the Done option.

Note how the dimensions match exactly the values used to create the hole. See Figures 7-33 and 7-34.

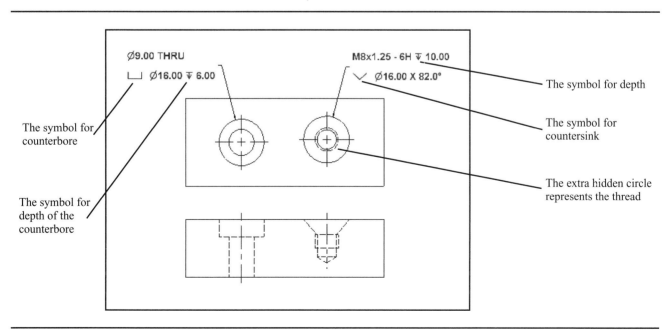

Figure 7-34

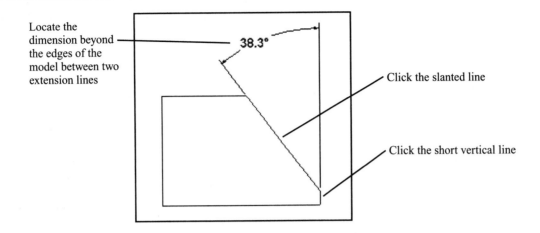

Figure 7-35

7-10 ANGULAR DIMENSIONS

Figure 7-35 shows a model that includes a slanted surface. The dimension value is located beyond the model between two extension lines. Locating dimensions between extension lines is preferred to locating the value between an extension line and the edge of the model.

To create an angular dimension

1. Access the Drawing Annotation Panel bar and click the General Dimension tool.
2. Click the slanted line, then click the short vertical line on the right side of the model.

3. Drag the cursor away from the hole's edge and locate the dimension.
4. Right-click the mouse and select the Done option.

Avoid overdimensioning

Figure 7-36 shows a shape dimensioned using an angular dimension. The shape is completely defined. Any additional dimension would be an error. It is tempting, in an effort to make sure a shape is completely defined, to add more dimensions, such as a horizontal dimension for the short horizontal edge at the top of the shape. This dimension is not needed and is considered *double dimensioning*.

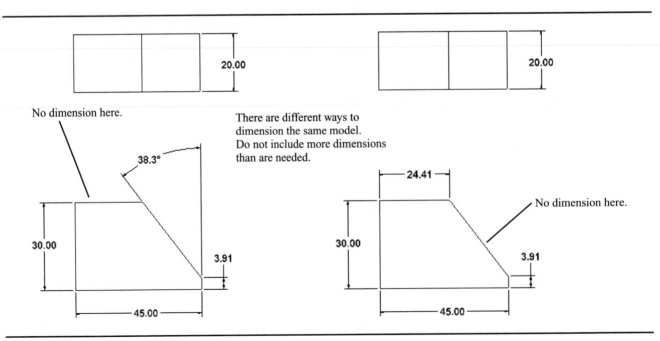

Figure 7-36

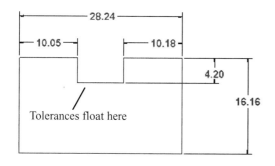

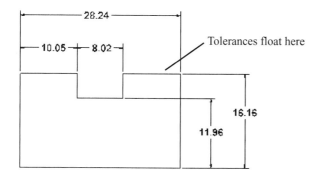

Figure 7-37

Figure 7-36 also shows the same front view dimensioned using only linear dimensions. The choice of whether to use angular or linear dimensions depends on the function of the model and which distances are more critical.

Figure 7-37 shows an object dimensioned two different ways. The dimensions used in the top example do not include a dimension for the width of the slot. This dimension is allowed to *float,* that is, allowed to accept any tolerance buildup. The dimensions used in the bottom example dimension the width of the slot but not the upper right edge. In this example the upper right edge is allowed to float or accept

any tolerance buildup. The choice of which edge to float depends on the function of the part. If the slot was to interface with a tab on another part, then it would be imperative that it be dimensioned and toleranced to match the interfacing part.

7-11 ORDINATE DIMENSIONS

Ordinate dimensions are dimensions based on an X,Y coordinate system. Ordinate dimensions do not include extension lines, dimension lines, or arrowheads, but simply horizontal and vertical leader lines drawn directly from the features of the object. Ordinate dimensions are particularly useful when dimensioning an object that includes many small holes.

Figure 7-38 shows a model that is to be dimensioned using ordinate dimensions. Ordinate dimensions values are calculated from the X,Y origin, which, in this example, is the lower left corner of the front view of the model.

To create ordinate dimensions

1. Access the Drawing Annotation Panel bar, then click on the Ordinate Dimension Set tool.

   See Figure 7-39.

2. Move the cursor into the drawing area and click the lower left corner of the model.

3. Drag the cursor away from the edge of the model and position the first ordinate dimension.

   See Figure 7-40.

4. Click the lower end of the vertical centerline of the first hole and position the dimension so that it is in line with the first dimension.

   See Figure 7-41.

   Note how the extension line from the first hole's centerline curves so that the dimension value may be located in

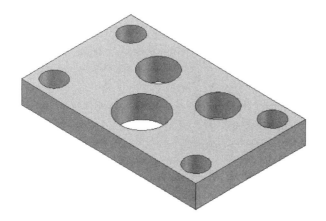

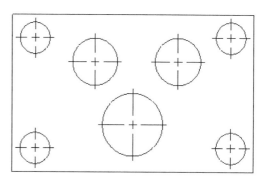

Figure 7-38

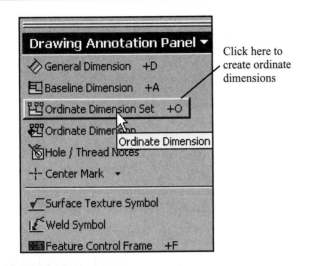

Figure 7-39

line with the first dimension. Inventor will automatically align ordinate dimensions.

5. Add the vertical dimensions.

Again start with the lower left corner of the model, then click the appropriate horizontal centerlines. See Figure 7-42.

To add hole dimensions

1. Click the Hole/Thread Notes on the Drawing Annotation Panel bar.
2. Add the appropriate hole dimensions.

Figure 7-40

See Figure 7-43.

7-12 BASELINE DIMENSIONS

Baseline dimensions are a series of dimensions that originate from a common baseline or datum line. Baseline dimensions are very useful because they help eliminate the tolerance buildup that is associated with chain-type dimensions. The Baseline Dimension tool is a flyout of the General Dimension tool located on the Drawing Annotation Panel bar. See Figure 7-44.

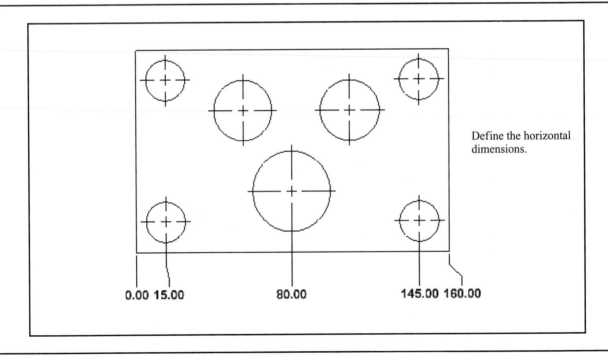

Figure 7-41

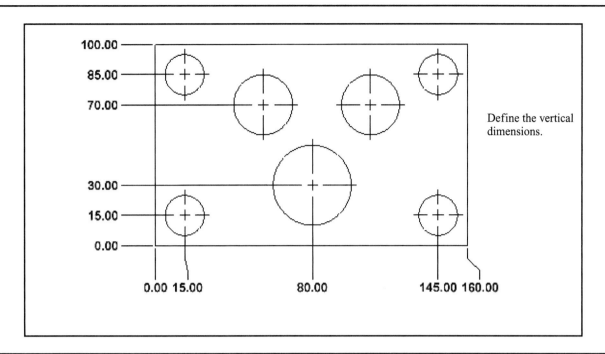

Figure 7-42

To use the Baseline Dimension command

See Figure 7-45.

1. Click the Baseline Dimension tool on the Drawing Annotation Panel bar.
2. Click the left vertical edge of the model.

3. Click the lower end on each vertical centerline and the right edge vertical line.
4. Right-click the mouse and select the Continue option.
5. Locate the baseline dimensions.

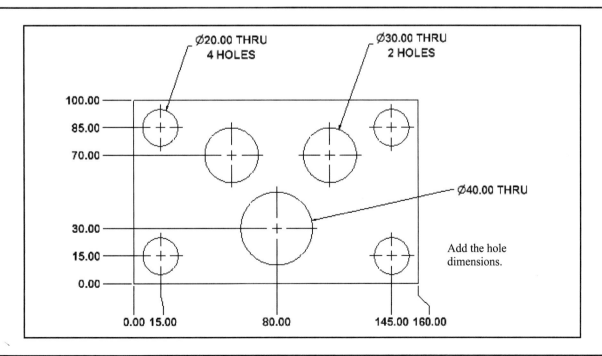

Figure 7-43

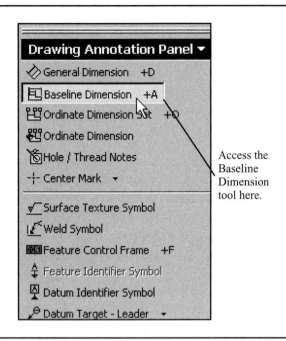

Access the Baseline Dimension tool here.

Figure 7-44

6. Press the right mouse button and select the Continue option, then create the vertical dimensions in a similar maner.
7. Add the hole dimensions using the Hole/Thread Notes tool.

7-13 HOLE TABLES

Inventor will generate hole tables that list holes' diameters and locations. There are two options: list all the holes, or list selected holes.

To list all holes in a table

1. Click the Hole Table - Selection tool on the Drawing Annotation Panel bar.
2. Select the View option.

See Figure 7-46.

3. Move the mouse into the drawing area and left-click the mouse.

A +-shaped cursor will appear.

4. Click on the lower left corner of the model.

This will define the origin for the table's X and Y values.

5. Right-click the mouse and select the Create option.

A rectangle representing the table will appear on the drawing screen. See Figure 7-47.

6. Locate the table on the drawing, then right-click the mouse and select the Done option.

7-14 LOCATING DIMENSIONS

There are eight general rules concerning the location of dimensions. See Figure 7-48.

1. Locate dimensions near the features they are defining.
2. Do not locate dimensions on the surface of the object.
3. Align and group dimensions so that they are neat and easy to understand.
4. Avoid crossing extension lines.
5. Do not cross dimension lines.

Sometimes it is impossible not to cross extension lines because of the complex shape of the object, but whenever possible, avoid crossing extension lines.

6. Locate shorter dimensions closer to the object than longer ones.
7. Always locate overall dimensions the farthest away from the object.
8. Do not dimension the same distance twice. This is called double dimensioning and will be discussed in Chapter 8.

7-15 FILLETS AND ROUNDS

Fillets and rounds may be dimensioned individually or by a note. In many design situations all the fillets and rounds are the same size, so a note as shown in Figure 7-49 is used. Any fillets or rounds that have a different radius from that specified by the note are dimensioned individually.

7-16 ROUNDED SHAPES – INTERNAL

Internal rounded shapes are called *slots*. Figure 7-50 shows three different methods for dimensioning slots. The end radii are indicated by the note R - 2 PLACES, but no numerical value is given. The width of the slot is dimensioned, and it is assumed that the radius of the rounded ends is exactly half of the stated width.

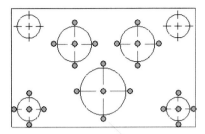

Click the left vertical edge, the lower end of each vertical centerline, and the right vertical edge.

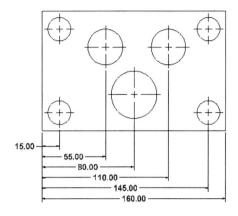

Press the right mouse button, select the Continue option and locate the dimensions.

Add the vertical dimensions

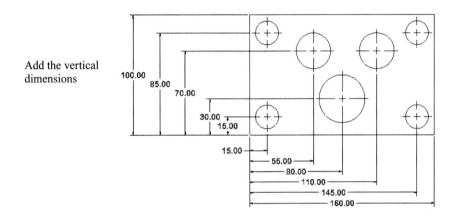

Add the hole dimensions

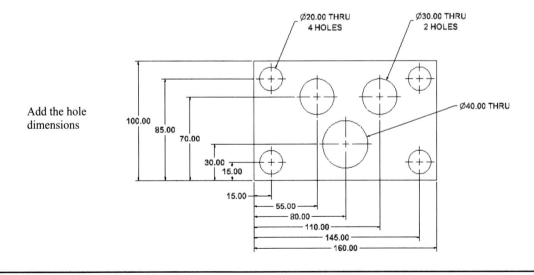

Figure 7-45

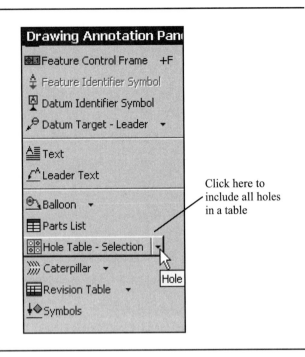

Figure 7-46

7-17 ROUNDED SHAPES – EXTERNAL

Figure 7-51 shows two shapes with external rounded ends. As with internal rounded shapes, the end radii are indicated but no value is given. The width of the object is given, and the radius of the rounded end is assumed to be exactly half of the stated width.

The second example shown in Figure 7-51 shows an object dimensioned using the object's centerline. This type of dimensioning is done when the distance between the hole is more important than the overall length of the object; that is, the tolerance for the distance between the holes is more exact than the tolerance for the overall length of the object.

The overall length of the object is given as a reference dimension (100). This means the object will be manufactured based on the other dimensions, and the 100 value will be used only for reference.

Objects with partially rounded edges should be dimensioned as shown in Figure 7-51. The radii of the end features are dimensioned. The centerpoint of the radii is implied to be on the object centerline. The overall dimension is given; it is not referenced unless specific radii values are included.

7-18 IRREGULAR SURFACES

There are three different methods for dimensioning irregular surfaces: tabular, baseline, and baseline with oblique extension lines. Figure 7-52 shows an irregular surface dimensioned using the tabular method. An XY axis is defined using the edges of the object. Points are then defined relative to the XY axis. The points are assigned reference numbers, and the reference numbers and XY coordinate values are listed in chart form as shown.

Figure 7-53 shows an irregular curve dimensioned using baseline dimensions. The baseline method references all dimensions to specified baselines. Usually there are two baselines, one horizontal and one vertical.

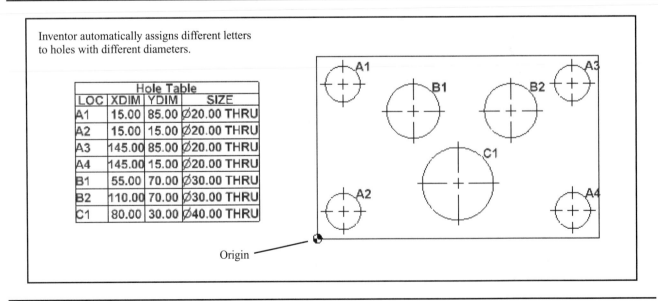

Figure 7-47

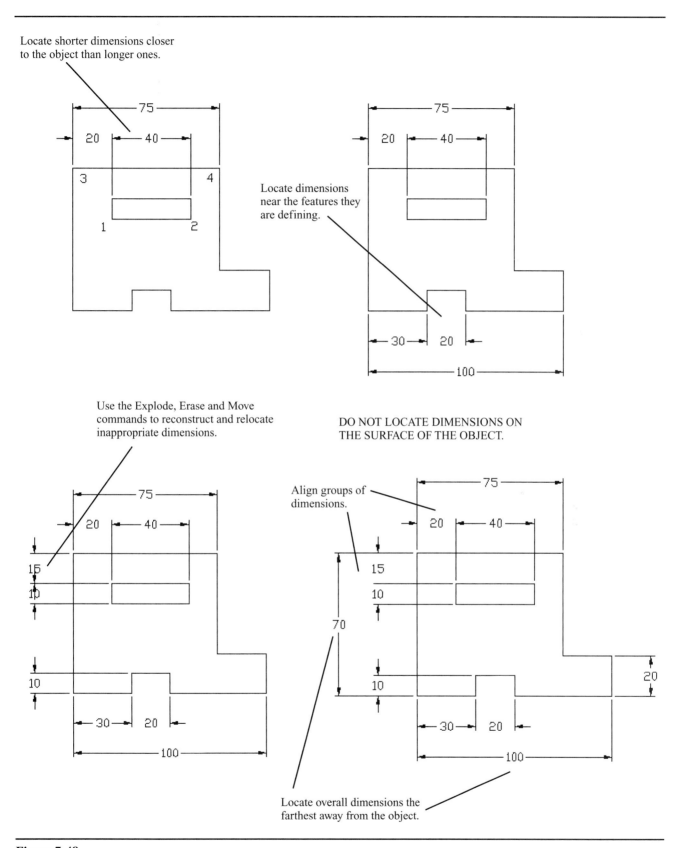

Locate shorter dimensions closer
to the object than longer ones.

Locate dimensions
near the features they
are defining.

Use the Explode, Erase and Move
commands to reconstruct and relocate
inappropriate dimensions.

DO NOT LOCATE DIMENSIONS ON
THE SURFACE OF THE OBJECT.

Align groups of
dimensions.

Locate overall dimensions the
farthest away from the object.

Figure 7-48

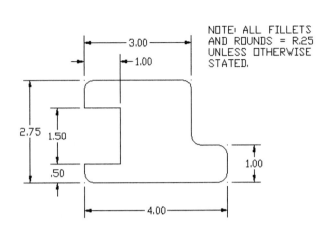

Figure 7-49

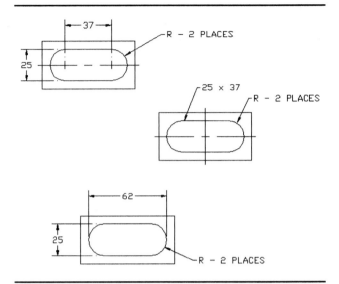

Figure 7-50

It is considered poor practice to use a centerline as a baseline. Centerlines are imaginary lines that do not exist on the object and would make it more difficult to manufacture and inspect the finished objects.

Baseline dimensioning is very common because it helps eliminate tolerance buildup (see Section 7-9) and is easily adaptable to many manufacturing processes. Inventor has a special Baseline Dimension command for use in creating baseline dimensions.

7-19 POLAR DIMENSIONS

Polar dimensions are similar to polar coordinates. A location is defined by a radius (distance) and an angle. Figure 7-54 shows an object that includes polar dimensions. The holes are located on a circular centerline, and their positions from the vertical centerline are specified using angles.

Figure 7-55 shows an example of a hole pattern dimensioned using polar dimensions.

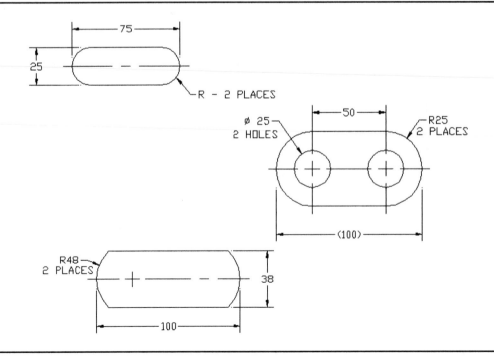

Figure 7-51

Station	1	2	3	4	5	6
X	0	20	40	55	62	70
Y	40	38	30	16	10	0

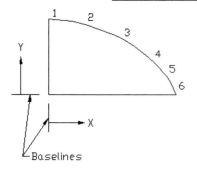

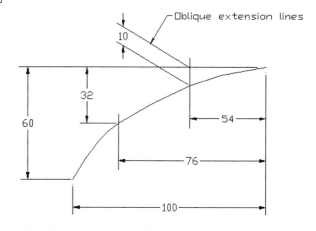

Figure 7-52

7-20 CHAMFERS

Chamfers are angular cuts made on the edges of objects. They are usually used to make it easier to fit two parts together. They are most often made at 45° angles but may be made at any angle. Figure 7-56 shows two objects with chamfers between surfaces 90° apart and two examples between surfaces that are not 90° apart. Either of the two types of dimensions shown for the 45° dimension may be used. If an angle other than 45° is used, the angle and setback distance must be specified.

Figure 7-57 shows two examples of internal chamfers. Both define the knurl using an angle and diameter. Internal chamfers are very similar to countersunk holes.

7-21 KNURLING

There are two types of knurls: diamond and straight. Knurls are used to make it easier to grip a shaft, or to rough a surface before it is used in a press fit.

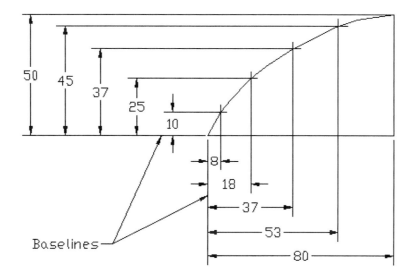

Figure 7-53

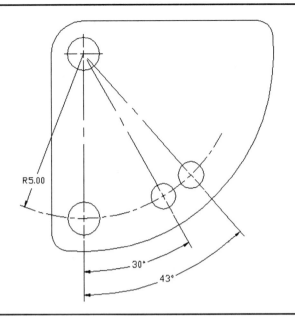

Figure 7-54

Knurls are defined by their pitch and diameter. See Figure 7-58. The pitch of a knurl is the ratio of the number of grooves on the circumference to the diameter. Standard knurling tools sized to a variety of pitch sizes are used to manufacture knurls for both English and metric units.

Diamond knurls may be represented by a double hatched pattern or by an open area with notes. The Hatch command is used to draw the double hatched lines. Straight knurls may be represented by straight lines in the pattern shown or by an open area with notes. The straight-line pattern is created by projecting lines from a construction circle. The construction points are evenly spaced on the circle.

7-22 KEYS AND KEYSEATS

Keys are small pieces of material used to transmit power. For example, Figure 7-59 shows how a key can be fitted between a shaft and a gear so that the rotary motion of the shaft can be transmitted to the gear.

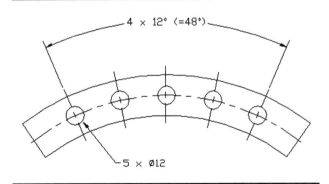

Figure 7-55

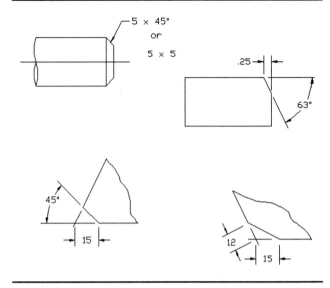

Figure 7-56

There are many different styles of keys. The key shown in Figure 7-59 has a rectangular cross section and is called a *square key*. Keys fit into grooves called *keyseats* or *keyways*.

Keyways are dimensioned from the bottom of the shaft or hole as shown.

7-23 SYMBOLS AND ABBREVIATIONS

Symbols are used in dimensioning to help accurately display the meaning of the dimension. Symbols also help eliminate language barriers when reading drawings. Figure 7-60 shows a list of dimensioning symbols available on the Dimension Styles dialog box under the Prefix/Suffix tab. See Section 7-8 for an explaination of how to apply symbols to dimensions.

Abbreviations should be used very carefully on drawings. Whenever possible, write out the full word including

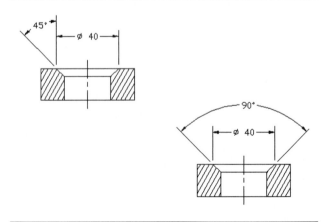

Figure 7-57

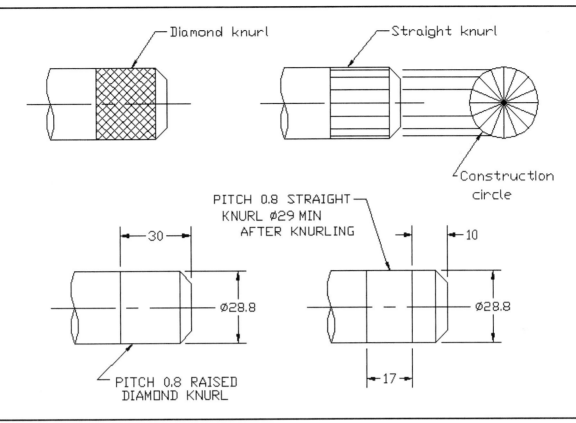

Figure 7-58

correct punctuation. Figure 7-61 shows several standard abbreviations used on technical drawings.

7-24 SYMMETRY AND CENTERLINE

An object is symmetrical about an axis when one side is an exact mirror image of the other. Figure 7-62 shows a symmetrical object. The two short parallel lines symbol or the note OBJECT IS SYMMETRICAL ABOUT THIS AXIS (centerline) may be used to designate symmetry.

If an object is symmetrical, only half the object need be dimensioned. The other dimensions are implied by the symmetry note or symbol.

Centerlines are slightly different from the axis of symmetry. An object may or may not be symmetrical about its centerline. See Figure 7-62. Centerlines are used to define the center of both individual features and entire objects. Use the centerline symbol when a line is a centerline, but do not use it in place of the symmetry symbol.

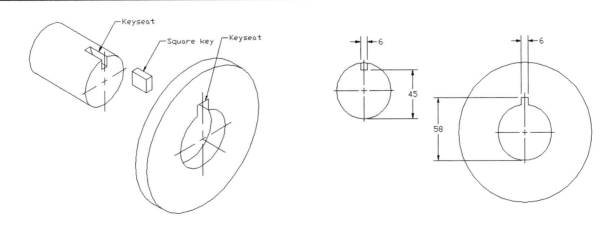

Figure 7-59

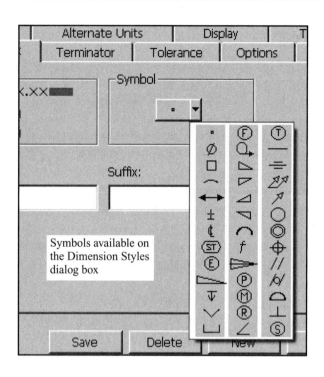

Figure 7-60

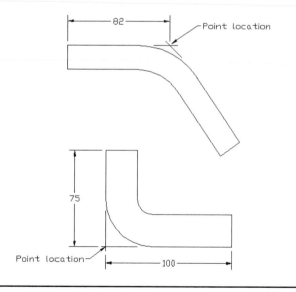

AL	=	Aluminum
C'BORE	=	Countebore
CRS	=	Cold Rolled Steel
CSK	=	Countersink
DIA	=	Diameter
EQ	=	Equal
HEX	=	Hexagon
MAT'L	=	Material
R	=	Radius
SAE	=	Society of Automotive Engineers
SFACE	=	Spotface
ST	=	Steel
SQ	=	Square
REQD	=	Required

Figure 7-61

7-25 DIMENSIONING TO A POINT

Curved surfaces can be dimensioned using theoretical points. See Figure 7-63. There should be a small gap between the surface of the object and the lines used to define the theoretical point. The point should be defined by the intersection of at least two lines.

There should also be a small gap between the extension lines and the theoretical point used to locate the point.

7-26 SECTIONAL VIEWS

Sectional views are dimensioned, as are orthographic views. See Figure 7-64. The sectional lines should be drawn at an angle that allows the viewer to clearly distinguish between the sectional lines and the extension lines.

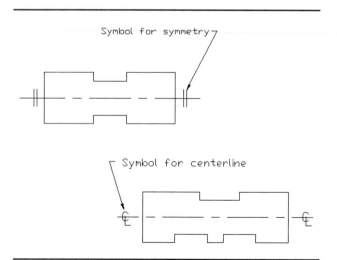

Figure 7-62

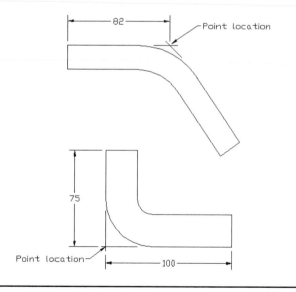

Figure 7-63

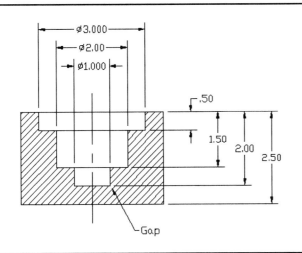

Figure 7-64

7-27 ORTHOGRAPHIC VIEWS

Dimensions should be added to orthographic views where the features appear in contour. Holes should be dimensioned in their circular views. Figure 7-65 shows three views of an object that has been dimensioned.

The hole dimensions are added to the top view, where the hole appears circular. The slot is also dimensioned in the top view because it appears in contour. The slanted surface is dimensioned in the front view.

The height of surface A is given in the side view rather than run along extension lines across the front view. The length of surface A is given in the front view. This is a contour view of the surface.

It is considered good practice to keep dimensions in groups. This makes it easier for the viewer to find dimensions.

Be careful not to double-dimension a distance. A distance should be dimensioned only once. If a 30 dimension were added above the 25 dimension on the right side view, it would be an error. The distance would be double-dimensioned: once with the 25 + 30 dimension and again with the 55 overall dimension. The 25 + 30 dimensions are mathematically equal to the 55 overall dimension, but there is a distinct difference in how they affect the manufacturing tolerances. Double dimensions are explained more fully in Chapter 9.

Dimensions using centerlines

Figure 7-66 shows an object dimensioned from its centerline. This type of dimensioning is used when the distance between the holes relative to each other is critical.

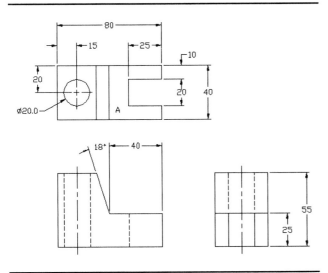

Figure 7-65

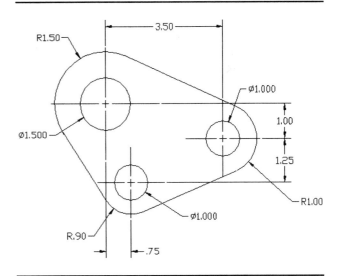

Figure 7-66

7-28 EXERCISE PROBLEMS

Measure and redraw the shapes in EX7-1 through EX7-18. The dotted grid background has either .50-in. or 10-mm spacing. All holes are through holes. Specify the units and scale of the drawing. Create a model by using the Extrude tool. Create a set of multiviews (front, top, side, and isometric views) using the .idw format and add the appropriate dimensions.

A. Measure using millimeters.
B. Measure using inches.

All dimensions are within either 0.25 in. or 5 mm. All fillets and rounds are R.50 in., R.25 in. or R10 mm, R5 mm.

EX7-1

THICKNESS:
40 mm
1.50 in.

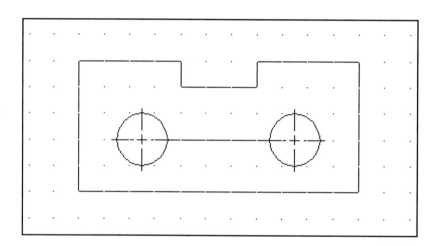

EX7-2

THICKNESS:
20 mm
0.75 in.

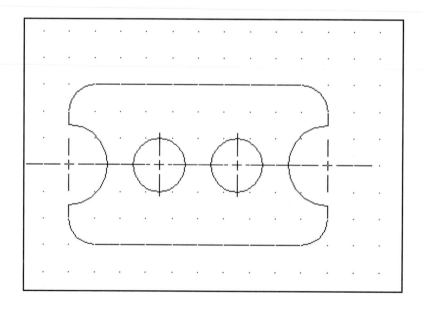

EX7-3

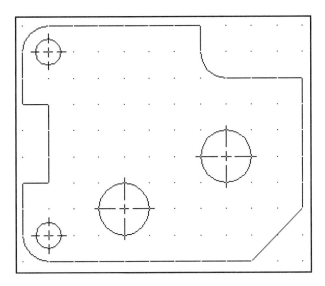

THICKNESS:
35 mm
1.25 in.

EX7-5

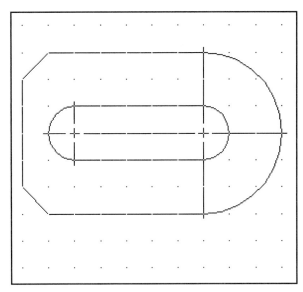

THICKNESS:
10 mm
.50 in.

EX7-4

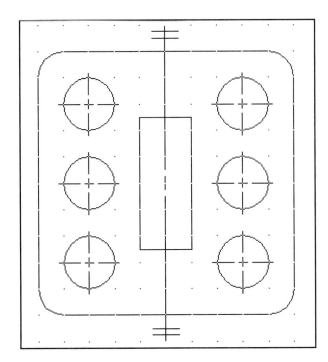

THICKNESS:
15 mm
0.50 in.

EX7-6

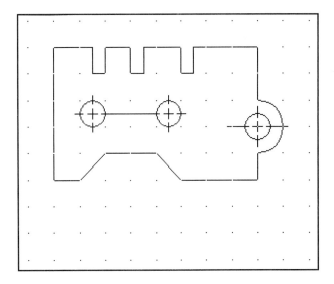

THICKNESS:
5 mm
.25 in.

EX7-7

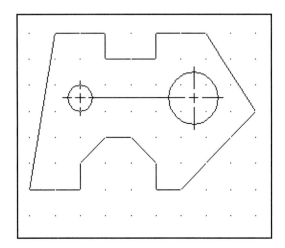

THICKNESS:
10 mm
0.25 in.

EX7-9

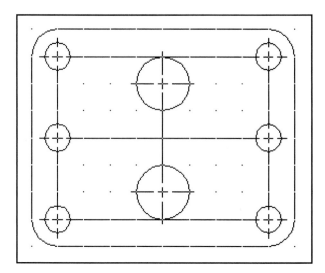

THICKNESS:
20 mm
.75 in.

EX7-8

THICKNESS:
8 mm
0.25 in.

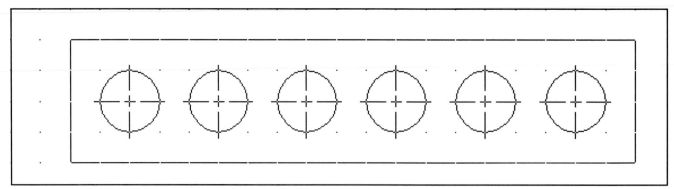

EX7-10

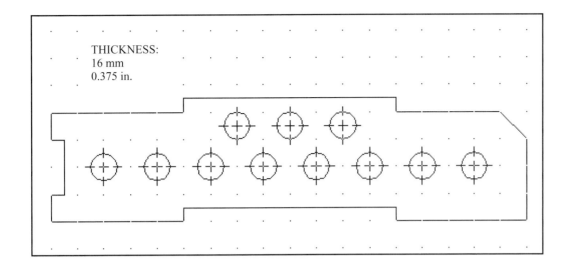

THICKNESS:
16 mm
0.375 in.

EX7-11

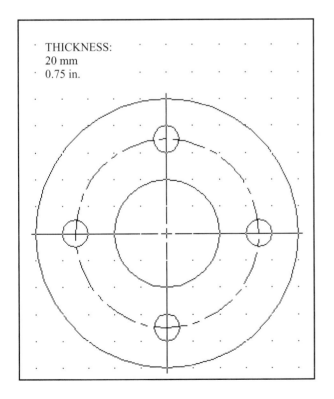

THICKNESS:
20 mm
0.75 in.

EX7-12

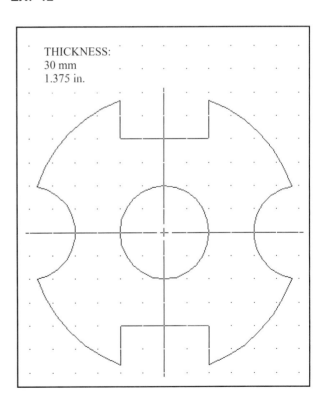

THICKNESS:
30 mm
1.375 in.

EX7-13

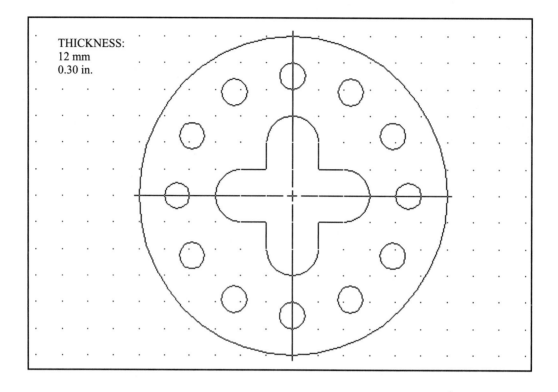

THICKNESS:
12 mm
0.30 in.

EX7-14

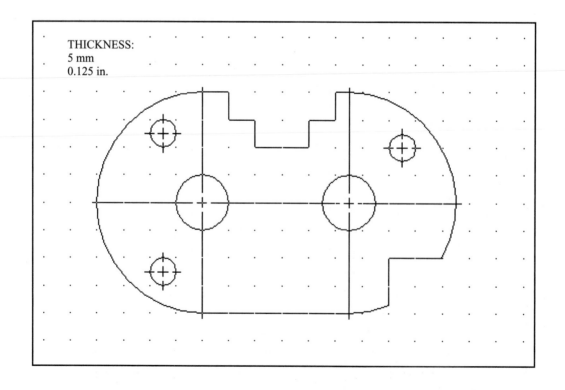

THICKNESS:
5 mm
0.125 in.

EX7-15

THICKNESS:
10 mm
0.25 in.

Dimension using
baseline dimensions

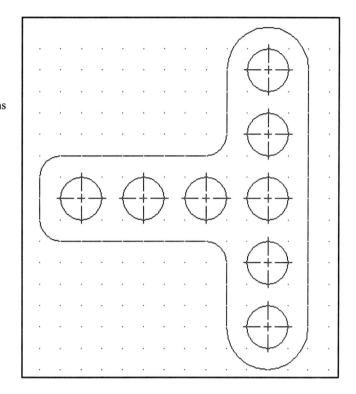

EX7-16

THICKNESS:
15 mm
0.50 in.

Dimension using
A. Baseline dimensions.
B. Ordinate dimensions.
C. Chain dimensions.
D. Hole table.

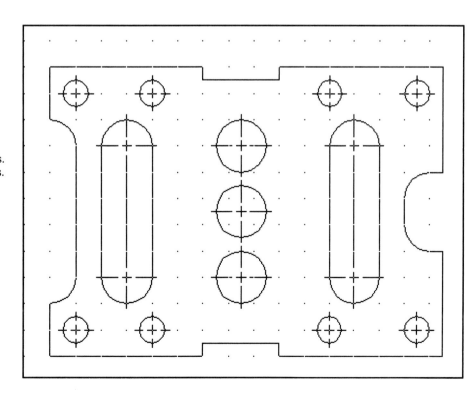

EX7-17

THICKNESS:
5 mm
.19 in.

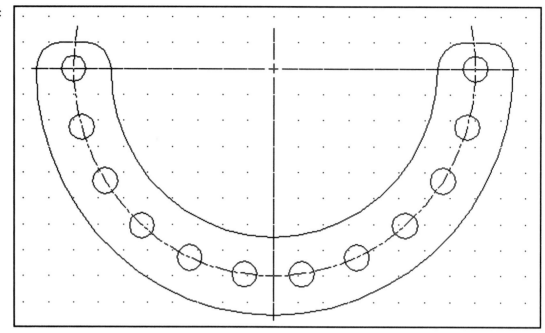

EX7-18

THICKNESS:
15 mm
.625 in.

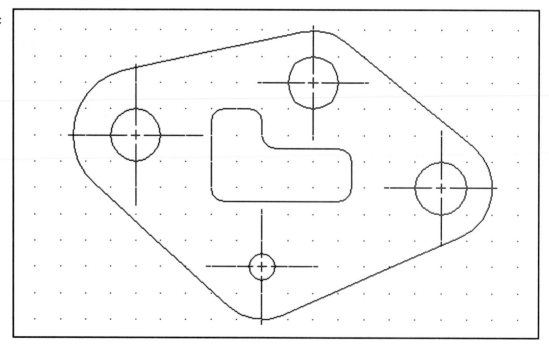

Draw models of the objects shown in exercise problems EX7-19 through EX7-36. Create and dimension multiview drawings of each.

EX7-19 MILLIMETERS

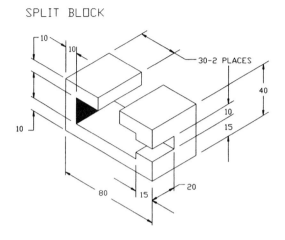

EX7-21 INCHES

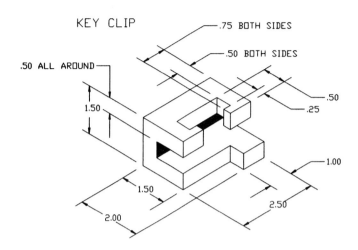

EX7-20 MILLIMETERS

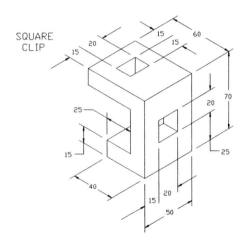

EX7-22 MILLIMETERS

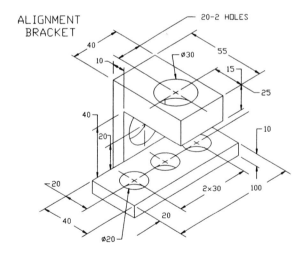

EX7-23 MILLIMETERS

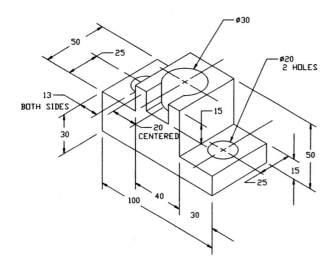

EX7-25 MILLIMETERS

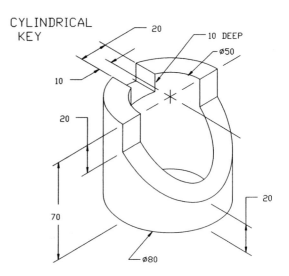

EX7-24 INCHES

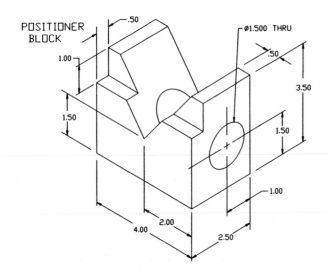

EX7-26 INCHES

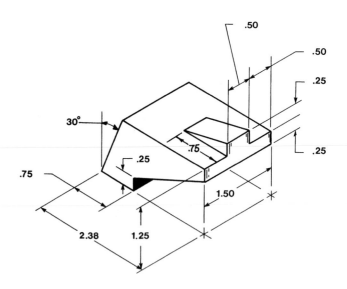

EX7-27 MILLIMETERS

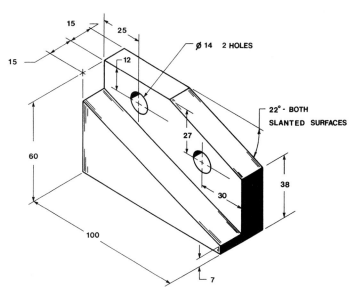

EX7-29 MILLIMETERS

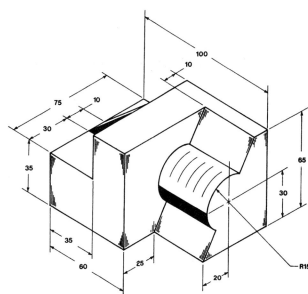

EX7-28 INCHES

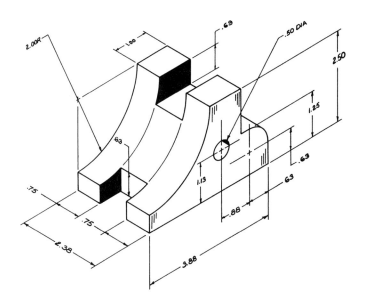

EX7-30 MILLIMETERS

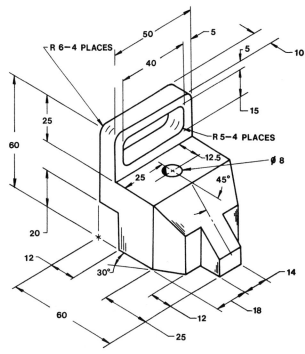

EX7-31 INCHES

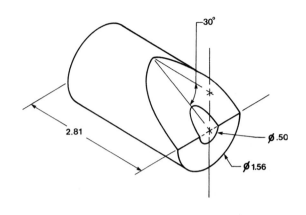

EX7-33 MILLIMETERS

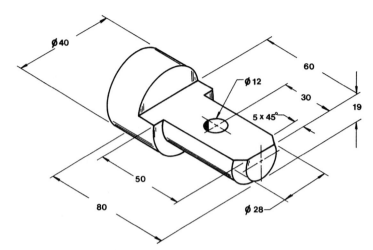

EX7-32 MILLIMETERS

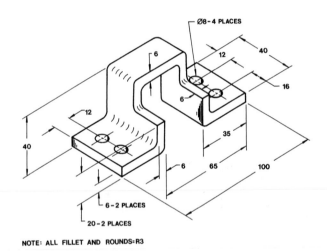

NOTE: ALL FILLET AND ROUNDS=R3

EX7-34 MILLIMETERS

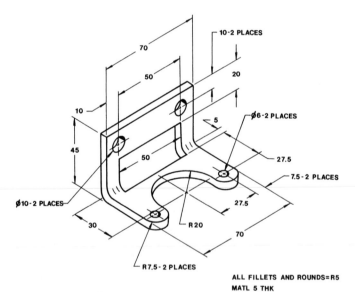

ALL FILLETS AND ROUNDS=R5
MATL 5 THK

EX7-35 MILLIMETERS

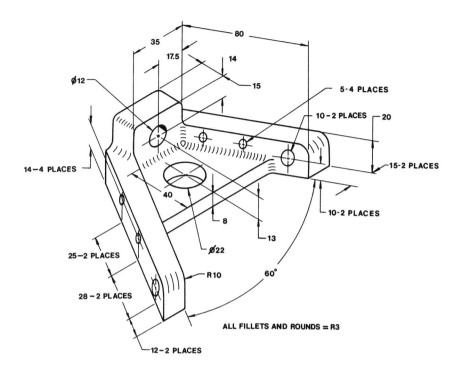

EX7-36 MILLIMETERS

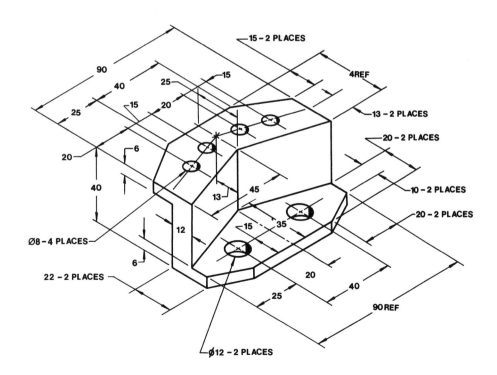

Tolerancing

8-1 INTRODUCTION

Tolerances define the manufacturing limits for dimensions. All dimensions have tolerances either written directly on the drawing as part of the dimension or implied by a predefined set of standard tolerances that apply to any dimension that does not have a stated tolerance.

This chapter explains general tolerance conventions and how they are applied using Inventor. It includes a sample tolerance study and an explanation of standard fits and surface finishes.

8-2 DIRECT TOLERANCE METHODS

There are two methods used to include tolerances as part of a dimension: *plus and minus,* and *limits.* Plus and minus tolerances can be expressed in either bilateral (deviation) or unilateral (symmetric) forms.

A *bilateral tolerance* has both a plus and a minus value, whereas a *unilateral tolerance* has either the plus or the minus value equal to 0. Figure 8-1 shows a horizontal dimension of 60 mm that includes a bilateral tolerance of plus or minus 1 and another dimension of 60.00 mm that includes a bilateral tolerance of plus 0.20 or minus 0.10. Figure 8-1

also shows a dimension of 65 mm that includes a unilateral tolerance of plus 1 or minus 0.

Plus or minus tolerances define a range for manufacturing. If inspection shows that all dimensioned distances on an object fall within their specified tolerance range, the object is considered acceptable; that is, it has been manufactured correctly.

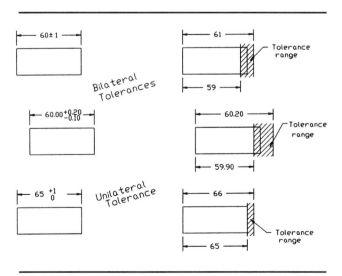

Figure 8-1

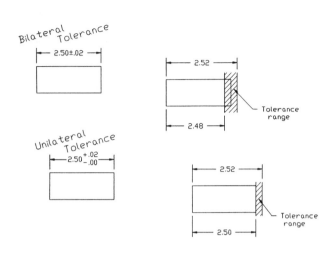

Figure 8-2

Figure 8-3

The dimension and tolerance of 60 ±0.1 means that the part must be manufactured within a range no greater than 60.1 or less than 59.9. The dimension and tolerance 65 +1, −0 defines the tolerance range as 65.0 to 66.0.

Figure 8-2 shows some bilateral and unilateral tolerances applied using decimal inch values. Inch dimensions and tolerances are written using a slightly different format than millimeter dimensions and tolerances, but they also define manufacturing ranges for dimension values. The horizontal bilateral dimension and tolerance 2.50 ±.02 defines the longest acceptable distance as 2.52 in. and the shortest as 2.48. The unilateral dimension 2.50 +.02, −.00 defines the longest acceptable distance as 2.52 and the shortest as 2.50.

8-3 TOLERANCE EXPRESSIONS

Dimension and tolerance values are written differently for inch and millimeter values. See Figure 8-3. Unilateral dimensions for millimeter values specify a zero limit with a single 0. A zero limit for inch values must include the same number of decimal places given for the dimension value. In the example shown in Figure 8-3, the dimension value .500 has a unilateral tolerance with minus zero tolerance. The zero limit is written as .000, three decimal places for both the dimension and the tolerance.

Both values in a bilateral tolerance for inch values must contain the same number of decimal places; for millimeter values the tolerance values need not include the

same number of decimal places as the dimension value. In Figure 8-3 the dimension value 32 is accompanied by tolerances of +0.25 and −0.10. This form is not acceptable for inch dimensions and tolerances. An equivalent inch dimension and tolerance would be written 32.00 +.25/−.10.

Degree values must include the same number of decimal places in both the dimension value and the tolerance values for bilateral tolerances. A single 0 may be used for unilateral tolerances.

8-4 UNDERSTANDING PLUS AND MINUS TOLERANCES

A millimeter dimension and tolerance of 12.0 +0.2/− 0.1 means the longest acceptable distance is 12.2000...0, and the shortest is 11.9000...0. The total range is 0.3000...0.

After an object is manufactured, it is inspected to ensure that the object has been manufactured correctly. Each dimensioned distance is measured and, if it is within the specified tolerance, is accepted. If the measured distance is not within the specified tolerance, the part is rejected. Some rejected objects may be reworked to bring them into the specified tolerance range, whereas others are simply scrapped.

Figure 8-4 shows a dimension with a tolerance. Assume that five objects were manufactured using the same 12.0 +0.2/−0.1 dimension and tolerance. The objects were then inspected and the results were as listed. Inspected

GIVEN (mm)

12 $^{+0.2}_{-0.1}$

MEANS

TOL MAX = 12.2
TOL MIN = 11.9
TOTAL TOL = 0.3

OBJECT	AS MEASURED	ACCEPTABLE?
1	12.160	OK
2	12.020	OK
3	12.203	TOO LONG
4	11.920	OK
5	11.895	TOO SHORT

Figure 8-4

GIVEN (inches)

3.50±.02

MEANS

TOL MAX = 3.52
TOL MIN = 3.48
TOTAL TOL = .04

OBJECT	AS MEASURED	ACCEPTABLE?
1	3.520	OK
2	3.486	OK
3	3.470	TOO SHORT
4	3.521	TOO LONG
5	3.515	OK

Figure 8-5

measurements are usually expressed to at least one more decimal place than that specified in the tolerance. Which objects are acceptable and which are not? Object 3 is too long and object 5 is too short because their measured distances are not within the specified tolerances.

Figure 8-5 shows a dimension and tolerance of 3.50 ±.02 in. Object 3 is not acceptable because it is too short, and object 4 is too long.

8-5 CREATING PLUS AND MINUS TOLERANCES

Plus and minus tolerances may be created using Inventor using the Tolerance option associated with existing dimensions, or by setting the plus and minus values using the Dimension Styles tool.

Set the drawing up for ANSI standards and metric units. Both model and working dimensions are toleranced in the same manner.

To create plus and minus tolerances

The example given is for a horizontal dimension, but the procedure is the same for any linear or radial dimension.

1. Click on the dimension that is to be toleranced.

Circles will appear on the dimension, indicating that it has been selected. A dialog box will appear. See Figure 8-6.

2. Select the Tolerance option.

The Dimension Tolerance dialog box will appear. See Figure 8-7.

3. Click the Symmetric option and enter an Upper value of 0.20, then click OK.

One value is needed for a symmetric tolerance. The value will automatically be made the ± value.

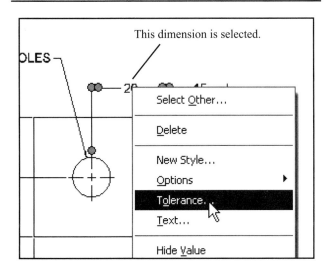

Figure 8-6

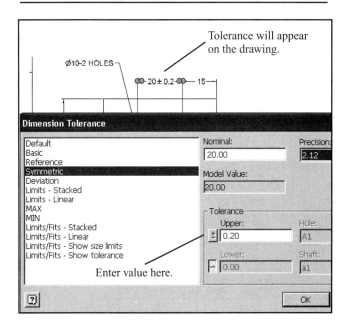

Figure 8-7

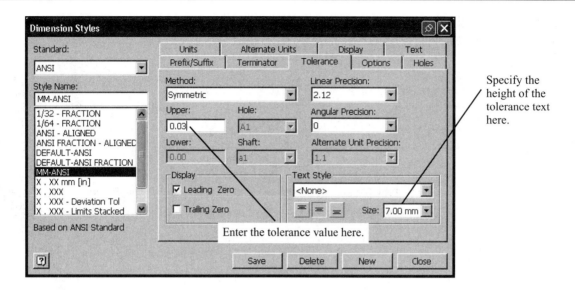

Figure 8-8

To create plus and minus tolerances using Dimension Styles

1. Click the Format heading at the top of the screen and select the Dimension Styles option.

The Dimension Styles dialog box will appear. See Figure 8-8.

2. Click the Tolerance tab, then select the Symmetric option in the Method box.

3. Set the Upper tolerance value for 0.03.

Inventor will automatically make this a ± tolerance.

4. Click Save, then Close.

Figure 8-9 shows the resulting tolerances. Note that all dimensions, except the hole value, have been changed to include a ±0.03 tolerance. If all dimensions require the same tolerance, it would be better to define a standard tolerance of that value and write the dimensions as whole numbers without a tolerance. Section 8-8 explains standard tolerances.

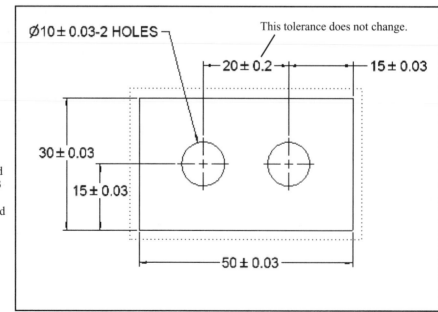

Figure 8-9

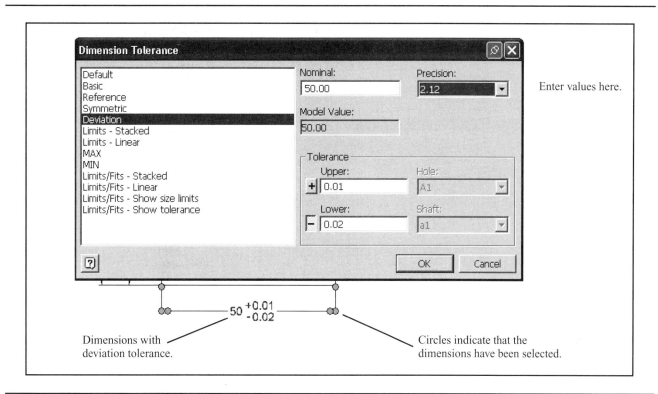

Figure 8-10

To create unequal plus and minus tolerances

1. Click on the dimension to be toleranced.

Circles will appear on the dimension, indicating that it has been selected. See Figure 8-6.

2. Right-click the mouse.

The Dimension Tolerance dialog box will appear. See Figure 8-10.

3. Select the Deviation option, then set the Upper value for 0.01 and the Lower value for 0.02.
4. Click OK.

Figure 8-10 shows the dimension with the unequal tolerances assigned.

8-6 LIMIT TOLERANCES

Figure 8-11 shows examples of limit tolerances. Limit tolerances replace dimension values. Two values are given: the upper and lower limits for the dimension value. The limit tolerance 62.1 and 61.9 is mathematically equal to 62 ± 0.1, but the stated limit tolerance is considered easier to read and understand.

Limit tolerances define a range for manufacture. Final distances on an object must fall within the specified range to be acceptable.

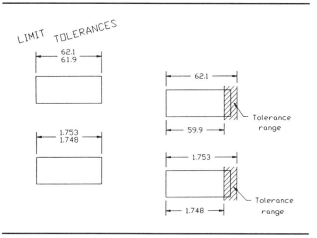

Figure 8-11

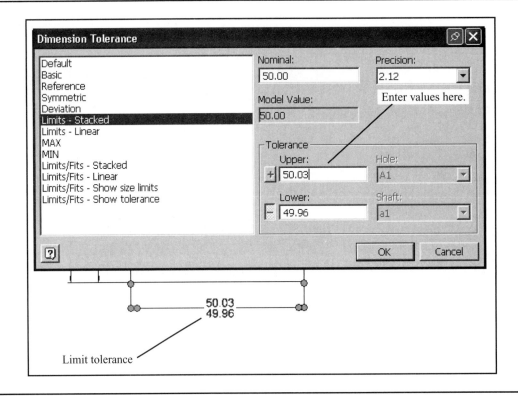

Figure 8-12

To create limit tolerances

1. Click on the dimension to be toleranced.

Circles will appear on the dimension, indicating that it has been selected. See Figure 8-6.

2. Right-click the mouse.

The Dimension Tolerance dialog box will appear. See Figure 8-12.

3. Select the Limits-Stacked option, then set the Upper value for 50.03 and the Lower value for 49.96.
4. Click OK.

Figure 8-12 shows the dimension with a limit tolerance assigned.

8-7 ANGULAR TOLERANCES

Figure 8-13 shows an example of an angular dimension with a symmetric tolerance. The procedures explained for applying different types of tolerances to linear dimensions also apply to angular dimensions.

To create angular tolerances

Figure 8-14 shows a model with a slanted surface. The model has been dimensioned, but no tolerances have been assigned. This example will assign a stacked limits tolerance to the angular dimension.

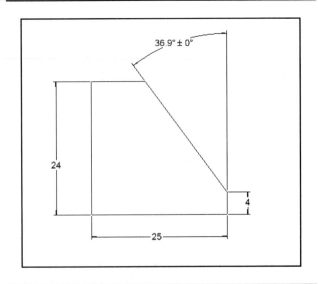

Figure 8-13

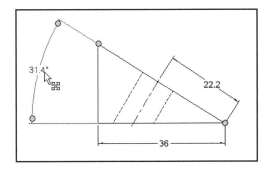
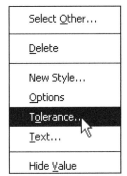
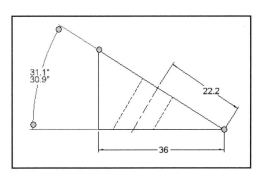

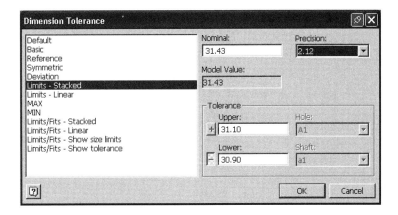

Figure 8-14

1. Right-click the angular dimension.

   Circles will appear, indicating that the dimension has been selected, and a dialog box will appear.

2. Select the Tolerance option.

   The Dimension Tolerance dialog box will appear.

3. Select the Limits-Stacked option and set the upper and lower values for the tolerance.
4. Click OK.

   Angular tolerances also can be assigned using the Dimension Styles tool. Figure 8-15 shows a Dimension Styles dialog box. Remember that if values are defined using Dimension Styles, all angular dimensions will have the assigned tolerance.

Enter values here.

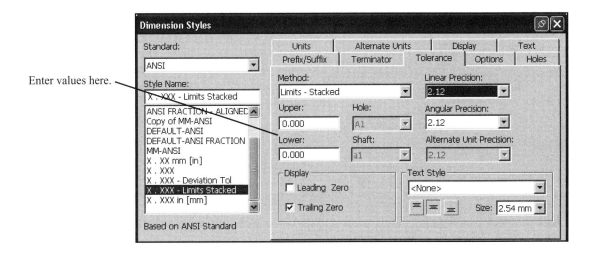

Figure 8-15

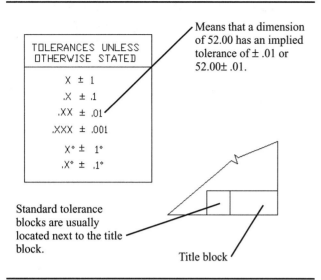

Means that a dimension of 52.00 has an implied tolerance of ± .01 or 52.00± .01.

Standard tolerance blocks are usually located next to the title block.

Title block

Figure 8-16

8-8 STANDARD TOLERANCES

Most manufacturers establish a set of standard tolerances that are applied to any dimension that does not include a specific tolerance. Figure 8-16 shows some possible standard tolerances. Standard tolerances vary from company to company. Standard tolerances are usually listed on the first page of a drawing to the left of the title block, but this location may vary.

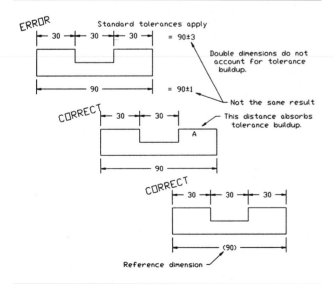

Figure 8-17

The X value used when specifying standard tolerances means any X stated in that format. A dimension value of 52.00 would have an implied tolerance of ±.01 because the stated standard tolerance is .XX ±.01, so any dimension value with two decimal places has a standard implied tolerance of ±.01. A dimension value of 52.000 would have an implied tolerance of ±.001.

8-9 DOUBLE DIMENSIONING

It is an error to dimension the same distance twice. This mistake is called *double dimensioning.* Double dimensioning is an error because it does not allow for tolerance buildup across a distance.

Figure 8-17 shows an object that has been dimensioned twice across its horizontal length, once using three 30-mm dimensions and a second time using the 90-mm overall dimension. The two dimensions are mathematically equal but are not equal when tolerances are considered. Assume that each dimension has a standard tolerance of ±1 mm. The three 30-mm dimensions could create an acceptable distance of 90 ±3 mm, or a maximum distance of 93 and a minimum distance of 87. The overall dimension of 90 mm allows a maximum distance of 91 and a minimum distance of 89. The two dimensions yield different results when tolerances are considered.

The size and location of a tolerance depends on the design objectives of the object, how it will be manufactured, and how it will be inspected. Even objects that have similar shapes may be dimensioned and toleranced very differently.

One possible solution to the double dimensioning shown in Figure 8-17 is to remove one of the 30-mm dimensions and allow that distance to "float", that is, absorb the cumulated tolerances. The choice of which 30 mm dimension to eliminate depends on the design objectives of the part. For this example the far-right dimension was eliminated to remove the double-dimensioning error.

Another possible solution to the double-dimensioning error is to retain the three 30-mm dimensions and to change the 90-mm overall dimension to a reference dimension. A reference dimension is used only for mathematical convenience. It is not used during the manufacturing or inspection process. A reference dimension is designated on a drawing using parentheses: (90).

If the 90-mm dimension was referenced, then only the three 30-mm dimensions would be used to manufacture and inspect the object. This would eliminate the double-dimensioning error.

8-10 CHAIN DIMENSIONS AND BASELINE DIMENSIONS

There are two systems for applying dimensions and tolerances to a drawing: chain and baseline. Figure 8-18 shows examples of both systems. *Chain dimensions* dimension each feature to the feature next to it. *Baseline dimensions* dimension all features from a single baseline or datum.

Chain and baseline dimensions may be used together. Figure 8-18 also shows two objects with repetitive features; one object includes two slots, and the other, three sets of three holes. In each example, the center of the repetitive feature is dimensioned to the left side of the object, which serves as a baseline. The sizes of the individual features are dimensioned using chain dimensions referenced to centerlines.

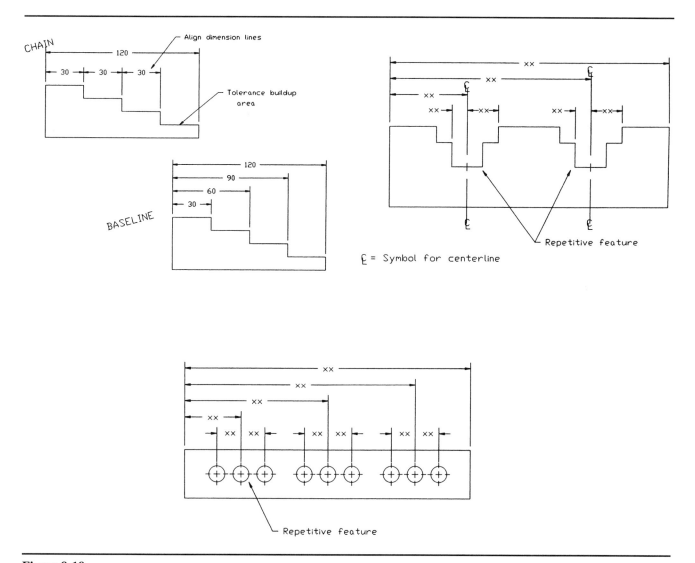

Figure 8-18

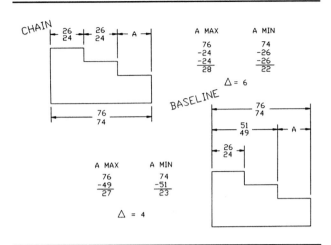

Figure 8-19

Figure 8-19 shows the same object dimensioned twice, once using chain dimensions and once using baseline dimensions. All distances are assigned a tolerance range of 2 mm, stated using limit tolerances. The maximum distance for surface A is 28 mm using the chain system and 27 mm using the baseline system. The 1-mm difference comes from the elimination of the first 26-24 limit dimension found on the chain example but not on the baseline.

The total tolerance difference is 6 mm for the chain and 4 mm for the baseline. The baseline reduces the tolerance variations for the object simply because it applies the tolerances and dimensions differently. So why not always use baseline dimensions? For most applications, the baseline system is probably better, but if the distance between the individual features is more critical than the distance from the feature to the baseline, use the chain system.

Baseline dimensions eliminate tolerance buildup and can be related directly to the reference axis of many machines. They tend to take up much more area on a drawing than do chain dimensions.

Chain dimensions are useful in relating one feature to another, such as the repetitive hole pattern shown in Figure 8-18. In this example the distance between the holes is more important than the individual hole's distance from the baseline.

Baseline dimensions created using Inventor

See Figure 8-20. See Section 7-14.

Baseline dimensions require a larger surface area on the drawing than do other styles of drawings. If necessary, use the Move View tool to make room for the baseline dimensions.

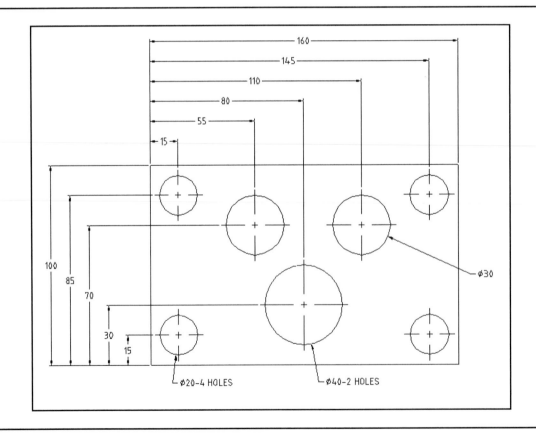

Figure 8-20

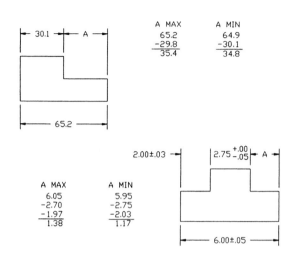

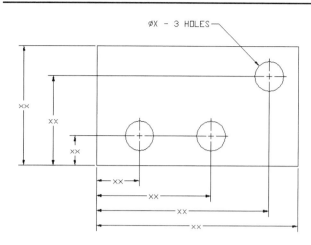

An example of rectangular coordinate dimensions.

Figure 8-21

Figure 8-22

8-11 TOLERANCE STUDIES

The term *tolerance study* is used when analyzing the effects of a group of tolerances on one another and on an object. Figure 8-21 shows an object with two horizontal dimensions. The horizontal distance A is not dimensioned. Its length depends on the tolerances of the two horizontal dimensions.

To calculate the maximum length of A

Distance A will be longest when the overall distance is at its longest and the other distance is at its shortest.

$$\begin{array}{r} 65.2 \\ -29.8 \\ \hline 35.4 \end{array}$$

To calculate the minimum length of A

Distance A will be shortest when the overall length is at its shortest and the other length is at its longest.

$$\begin{array}{r} 64.9 \\ -30.1 \\ \hline 34.8 \end{array}$$

8-12 RECTANGULAR DIMENSIONS

Figure 8-22 shows an example of rectangular dimensions referenced to baselines. Figure 8-23 shows a circular object on which dimensions are referenced to a circle's centerlines. Dimensioning to a circle's centerline is critical to accurate hole location.

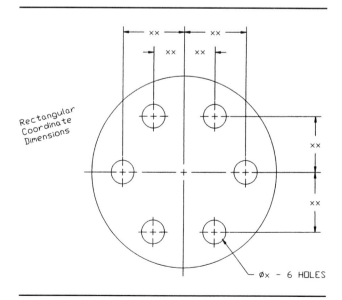

Figure 8-23

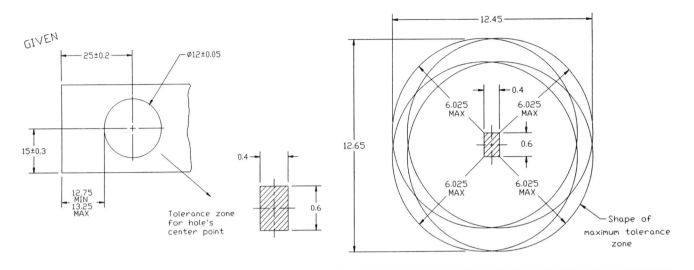

Figure 8-24

8-13 HOLE LOCATIONS

When rectangular dimensions are used, the location of a hole's center point is defined by two linear dimensions. The result is a rectangular tolerance zone whose size is based on the linear dimension's tolerances. The shape of the center point's tolerance zone may be changed to circular using positioning tolerancing as described in Section 8-42.

Figure 8-24 shows the location and size dimensions for a hole. Also shown are the resulting tolerance zone and the overall possible hole shape. The center point's tolerance is .2 by .3 based on the given linear locating tolerances.

The hole diameter has a tolerance of ±.05. This value must be added to the center point location tolerances to define the maximum overall possible shape of the hole. The maximum possible hole shape is determined by drawing the maximum radius from the four corner points of the tolerance zone.

This means that the left edge of the hole could be as close to the vertical baseline as 12.75 or as far as 13.25. The 12.75 value was derived by subtracting the maximum hole diameter value 12.05 from the minimum linear distance 24.80 (24.80 − 12.05 = 12.75). The 13.25 value was derived by subtracting the minimum hole diameter 11.95 from the maximum linear distance 25.20 (25.20 − 11.95 = 13.25).

Figure 8-25 shows a hole's tolerance zone based on polar dimensions. The zone has a sector shape, and the possible hole shape is determined by locating the maximum radius at the four corner points of the tolerance zone.

8-14 CHOOSING A SHAFT FOR A TOLERANCED HOLE

Given the hole location and size shown in Figure 8-24, what is the largest diameter shaft that will always fit into the hole?

Figure 8-26 shows the hole's center point tolerance zone based on the given linear locating tolerances. Four circles have been drawn centered at the four corners on the linear tolerance zone that represent the smallest possible hole diameter. The circles define an area that represents the maximum shaft size that will always fit into the hole, regardless of how the given dimensions are applied.

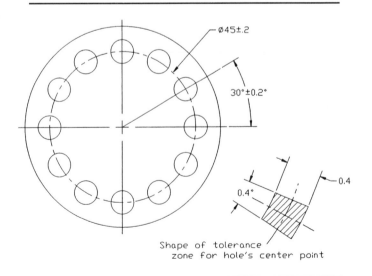

Figure 8-25

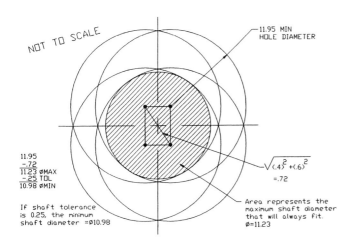

Figure 8-26

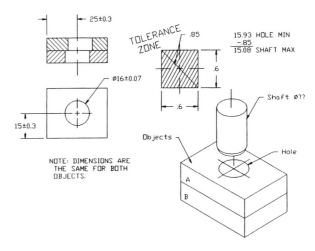

Figure 8-27

The diameter size of this circular area can be calculated by subtracting the maximum diagonal distance across the linear tolerance zone (corner to corner) from the minimum hole diameter.

The results can be expressed as a formula.

For linear dimensions and tolerances

$$S_{max} = H_{min} - DTZ$$

where

S_{max} = maximum shaft diameter

H_{min} = minimum hole diameter

DTZ = diagonal distance across the tolerance zone

In the example shown the diagonal distance is determined using the Pythagorean theorem:

$$DTZ = \sqrt{(.4)^2 + (.6)^2}$$
$$= \sqrt{.16 + .36}$$
$$DTZ = .72$$

This means that the maximum shaft diameter that will always fit into the given hole is 11.23.

$$S_{max} = H_{min} - DTZ$$
$$= 11.95 - .72$$
$$S_{max} = 11.23$$

This procedure represents a restricted application of the general formula presented later in the chapter for positioning tolerances. For a more complete discussion see Section 8-43.

Once the maximum shaft size has been established, a tolerance can be applied to the shaft. If the shaft had a total tolerance of .25, the minimum shaft diameter would be 11.23 − .25, or 10.98. Figure 8-26 shows a shaft dimensioned and toleranced using these values.

The formula presented is based on the assumption that the shaft is perfectly placed on the hole's center point. This assumption is reasonable if two objects are joined by a fastener and both objects are free to move. When both objects are free to move about a common fastener, they are called *floating objects*.

8-15 SAMPLE PROBLEM SP8-1

Parts A and B in Figure 8-27 are to be joined by a common shaft. The total tolerance for the shaft is to be .05. What are the maximum and minimum shaft diameters?

Both objects have the same dimensions and tolerances and are floating relative to each other.

$$S_{max} = H_{min} - DTZ$$
$$= 15.93 - .85$$
$$S_{max} = 15.08$$

The shaft's minimum diameter is found by subtracting the total tolerance requirement from the calculated maximum diameter:

$$15.08 - .05 = 15.03$$

Therefore,

Shaft max = 15.08

Shaft min = 15.03

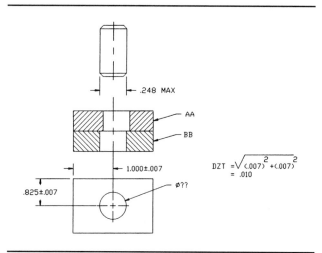

Figure 8-28

8-16 SAMPLE PROBLEM SP8-2

The procedure presented in Sample Problem SP8-1 can be worked in reverse to determine the maximum and minimum hole size based on a given shaft size.

Objects AA and BB as shown in Figure 8-28 are to be joined using a bolt whose maximum diameter is .248. What is the minimum hole size for objects that will always accept the bolt? What is the maximum hole size if the total hole tolerance is .005?

$$S_{max} = H_{min} - DZT$$

In this example H_{min} is the unknown factor, so the equation is rewritten as

$$H_{min} = S_{max} + DZT$$
$$= .248 + .010$$
$$H_{min} = .258$$

This is the minimum hole diameter, so the total tolerance requirement is added to this value:

$$.258 + .005 = .263$$

Therefore,

Hole max = .263

Hole min = .258

8-17 STANDARD FITS (METRIC VALUES)

Calculating tolerances between holes and shafts that fit together is so common in engineering design that a group of standard values and notations has been established. These values are listed in tables in the appendix.

There are three possible type of fits between a shaft and a hole: clearance, transitional, and interference. See Figure 8-29. There are several subclassifications within each of these categories.

A *clearance fit* always defines the maximum shaft diameter as smaller than the minimum hole diameter. The difference between the two diameters is the amount of clearance. It is possible for a clearance fit to be defined with zero clearance; that is, the maximum shaft diameter is equal to the minimum hole diameter.

An *interference fit* always defines the minimum shaft diameter as larger than the maximum hole diameter; that is, the shaft is always bigger than the hole. This definition means that an interference fit is the converse of a clearance fit. The difference between the diameter of the shaft and the hole is the amount of interference.

An interference fit is primarily used to assemble objects together. Interference fits eliminate the need for threads, welds, or other joining methods. Using an interference for joining two objects is generally limited to light load applications.

It is sometimes difficult to visualize how a shaft can be assembled into a hole with a diameter smaller than that of the shaft. It is sometimes done using a hydraulic press that slowly forces the two parts together. The joining process can be augmented by the use of lubricants or heat. The hole is heated, causing it to expand, the shaft is inserted, and the hole is allowed to cool and shrink around the shaft.

A *transition fit* may be either a clearance or an interference fit. It may have a clearance between the shaft and the hole or an interference.

Figure 8-30 shows two graphic representations of 20 different standard hole/shaft tolerance ranges. The figure shows ranges for hole tolerances, shaft tolerances, and the amount of clearance or interference for each classification. The notations are based on Standard International Tolerance values. A specific description for each category of fit is as follows.

Clearance fits

H11/c11 or C11/h11 = loose running fit
H8/d8 or D8/h8 = free running fit
H8/f7 or F8/h7 = close running fit
H7/g6 or G7/h6 = sliding fit
H7/h6 = locational clearance fit

Transitional fits

H7/k6 or K7/h6 = locational transition fit
H7/n6 or N7/h6 = locational transition fit

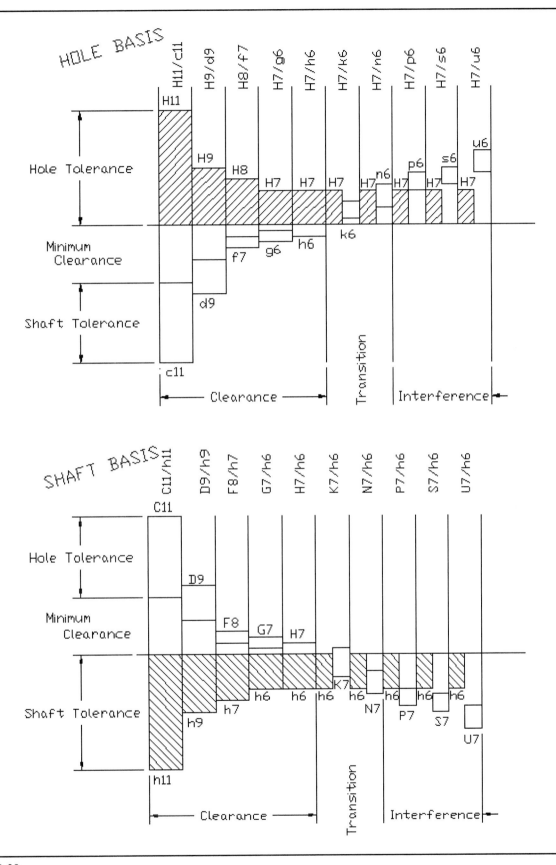

Figure 8-29

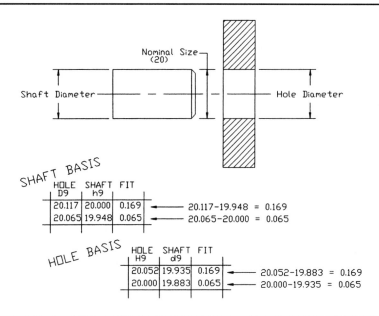

Figure 8-30

Interference fits

H7/p6 or P7/h6 = locational transition fit
H7/s6 or S7/h6 = medium drive fit
H7/u6 or U7/h6 = force fit

Not all possible sizes are listed in the tables in the appendix. Only preferred sizes are listed. Tolerances for sizes in between the stated sizes are derived by going to the next nearest given size; the values are not interpolated. A basic size of 27 would use the tolerance values listed for 25. Sizes that are exactly halfway between two stated sizes may use either set of values depending on the design requirements.

nominal sizes, called shaft basis tolerances. The choice of which set of values to use depends on the design application. In general, hole basis numbers are used more often because it is more difficult to vary hole diameters manufactured using specific drill sizes than shaft sizes manufactured using a lathe. Shaft sizes may be used when a specific fastener diameter is used to assemble several objects.

Figure 8-30 shows a hole, a shaft, and a set of sample values taken from the tables found in the appendix. One of the set of values is for hole basis tolerance and the other for shaft basis tolerance. The fit values are the same for both sets of values. The hole basis values were derived starting with a

8-18 NOMINAL SIZES

The term *nominal* refers to the approximate size of an object that matches a common fraction or whole number. A shaft with a dimension of 1.500±.003 is said to have a nominal size of "one and a half inches." A dimension of 1.500 +.000/−.005 is still said to have a nominal size of one and a half inches. In both examples 1.5 is the closest common fraction.

8-19 HOLE AND SHAFT BASIS

One of the charts shown in Figure 8-29 applies tolerances starting with the nominal hole sizes, called *hole basis tolerances;* the other applies tolerances starting with the *shaft*

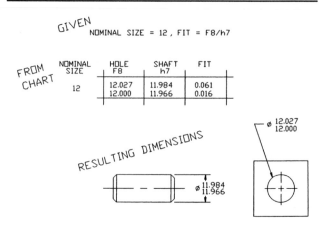

Figure 8-31

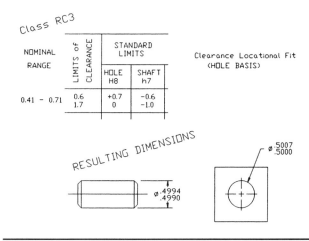

Figure 8-32

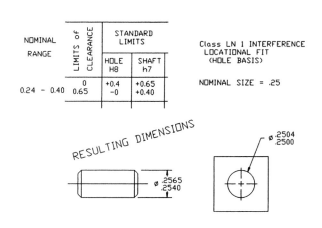

Figure 8-33

nominal hole size of 20.000, whereas the shaft basis values were derived starting with a shaft nominal size of 20.000. The letters used to identify holes are always written using capital letters, and the letters for shaft values use lowercase.

8-20 SAMPLE PROBLEM SP8-3

Dimension a hole and a shaft that are to fit together using a close running fit. Use hole basis values based on a nominal size of 12 mm.

Figure 8-31 shows values taken from the appropriate table in the appendix. The values may be applied directly to the shaft and hole as shown.

8-21 STANDARD FITS (INCH VALUES)

The appendix also includes tables of standard fit tolerances for inch values. The tables for inches are presented for a range of nominal values and are not for specific values, as are the metric value tables. The values may be hole or shaft basis.

Fits defined using inch values are classified as follows:

RC = running and sliding fits
LC = clearance locational fits
LT = transitional locational fits
LN = interference fits
FN = force fits

Each of these general categories has several subclassifications within it defined by a number, for example, Class RC1, Class RC2, through Class RC8. The letter designations are based on International Tolerance Standards, as are metric designations.

The values are listed in thousandths of an inch. A table value of 1.1 means .0011 in. A table value of .5 means .0005 in.

Figure 8-32 shows a set of values for a Class RC3 clearance fit hole basis taken from the appendix. If the values are applied to a nominal size of .5 in., the resulting hole and shaft sizes will be as shown. Plus table values are added to the nominal value, and minus values are subtracted from the nominal value.

Nominal values that are common to two nominal ranges (0.71) may use values from either range.

8-22 SAMPLE PROBLEM SP8-4

Dimension a hole and shaft for a Class LN1 interference fit based on a nominal diameter of .25 in. Use hole basis values.

Figure 8-33 shows the values for the .24–.40 nominal range as listed in the appendix. The values are in thousandths of an inch. Plus values are added to the nominal size. The resulting shaft and hole dimensions are as shown. The shaft's diameter is larger than the hole's diameter because this example calls for an interference fit.

PREFERRED SIZES			
First Choice	Second Choice	First Choice	Second Choice
1		12	
	1.1		14
1.2		16	
	1.4		18
1.6		20	
	1.8		22
2		25	
	2.2		28
2.5		30	
	2.8		35
3		40	
	3.5		45
4		50	
	4.5		55
5		60	
	5.5		70
6		80	
	7		90
8		100	
	9		110
10		120	
	11		140

Figure 8-34

Fraction	Decimal Equivalent	Fraction	Decimal Equivalent	Fraction	Decimal Equivalent
7/64	.1094	21/64	.3281	11/16	.6875
1/8	.1250	11/32	.3438	3/4	.7500
9/64	.1406	23/64	.3594	13/16	.8125
5/32	.1562	3/8	.3750	7/8	.8750
11/64	.1719	25/64	.3906	15/16	.9375
3/16	.1875	13/32	.4062	1	1.0000
13/64	.2031	27/64	.4219		
7/32	.2188	7/16	.4375	PARTIAL LIST	
1/4	.2500	29/64	.4531	of standard	
17/64	.2656	15/32	.4688	Twist Drill Sizes	
9/32	.2812	1/2	.5000	(Fractional	
19/64	.2969	9/16	.5625	Sizes)	
5/16	.3125	5/8	.6250		

Figure 8-35

8-23 PREFERRED AND STANDARD SIZES

It is important that designers always consider preferred and standard sizes when selecting sizes for designs. Most tooling is set up to match these sizes, so manufacturing is greatly simplified when preferred and standard sizes are specified. Figure 8-34 shows a listing of preferred sizes for metric values.

Consider the case of design calculations that call for a 42-mm-diameter hole. A 42-mm-diameter hole is not a preferred size. A diameter of 40 mm is the closest preferred size, and a 45-mm-diameter is a second choice. A 42-mm hole could be manufactured but would require an unusual drill size that may not be available. It would be wise to reconsider the design to see if a 40-mm-diameter hole could be used, and if not, possibly a 45-mm-diameter hole.

A very large quantity production run could possibly justify the cost of special tooling, but for smaller runs it is probably better to use preferred sizes. Machinists will have the required drills, and maintenance people will have the appropriate tools for these sizes.

Figure 8-35 shows a listing of standard fractional drill sizes. Most companies now specify metric units or decimal inches; however, many standard items are still available in fractional sizes, and many older objects may still require fractional-sized tools and replacement parts. A more complete listing is available in the appendix.

8-24 SURFACE FINISHES

The term *surface finish* refers to the accuracy (flatness) of a surface. Metric values are measured using micrometers (μm), and inch values are measured in microinches (μin.).

The accuracy of a surface depends on the manufacturing process used to produce the surface. Figure 8-36 shows a listing of manufacturing processes and the quality of the surface finish they can be expected to produce.

Surface finishes have several design applications. *Datum surfaces,* or surfaces used for baseline dimensioning, should have fairly accurate surface finishes to help assure accurate measurements. Bearing surfaces should have good-quality surface finishes for better load distribution, and parts that operate at high speeds should have smooth finishes to help reduce friction. Figure 8-37 shows a screw head sitting on a very wavy surface. Note that the head of the screw is actually in contact with only two wave peaks, meaning all the bearing load is concentrated on the two peaks. This situation could cause stress cracks and greatly weaken the surface. A better-quality surface finish would increase the bearing contact area.

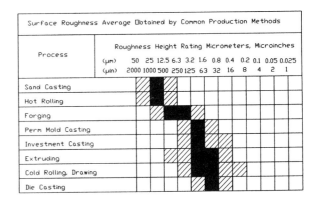

Figure 8-36

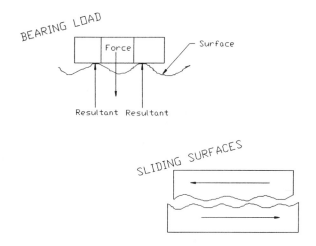

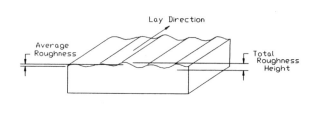

Figure 8-37

Figure 8-38

Figure 8-37 also shows two very rough surfaces moving in contact with each another. The result will be excess wear to both surfaces because the surfaces touch only on the peaks, and these peaks will tend to wear faster than flatter areas. Excess vibration can also result when interfacing surfaces are too rough.

Surface finishes are classified into three categories: surface texture, roughness, and lay. *Surface texture* is a general term that refers to the overall quality and accuracy of a surface.

Roughness is a measure of the average deviation of a surface's peaks and valleys. See Figure 8-38.

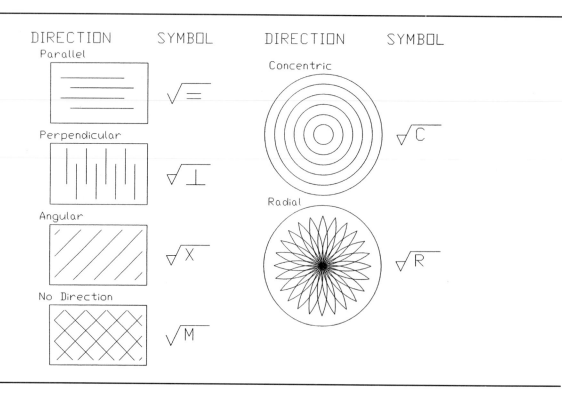

Figure 8-39a

Figure 8-39b

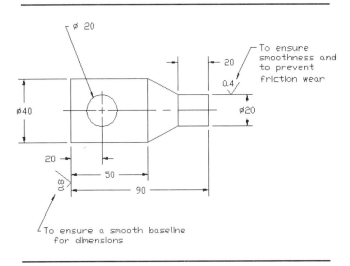

Figure 8-40

Lay refers to the direction of machine marks on a surface. See Figure 8-39. The lay of a surface is particularly important when two moving objects are in contact with each other, especially at high speeds.

8-25 SURFACE CONTROL SYMBOLS

Surface finishes are indicated on a drawing using surface control symbols. See Figure 8-40. The general surface control symbol looks like a check mark. Roughness values may be included with the symbol to specify the required accuracy. Surface control symbols can also be used to specify the manufacturing process that may or may not be used to produce a surface.

Figure 8-40 shows two applications of surface control symbols. In the first example, a 0.8-μm (32 μin.) surface finish is specified on the surface that serves as a datum for several horizontal dimensions. A 0.8-μm surface finish is generally considered the minimum acceptable finish for datums.

A second finish mark with a value of 0.4 μm is located on an extension line that refers to a surface that will be in contact with a moving object. The extra flatness will help prevent wear between the two surfaces.

It is suggested that a general finish mark be drawn and saved as a WBlock so it can be inserted as needed on future drawings. Add the machine mark WBlock to any prototype drawings created.

To apply surface control symbols using Inventor

Figure 8-42 shows a dimensioned view of a model. Surface symbols are to be added to the sides of the slot.

1. Access the Drawing Annotation Panel bar and click the Surface Texture Symbol tool.

   See Figure 8-41.

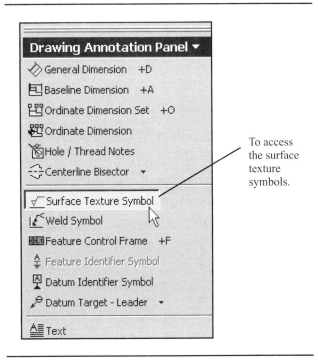

Figure 8-41

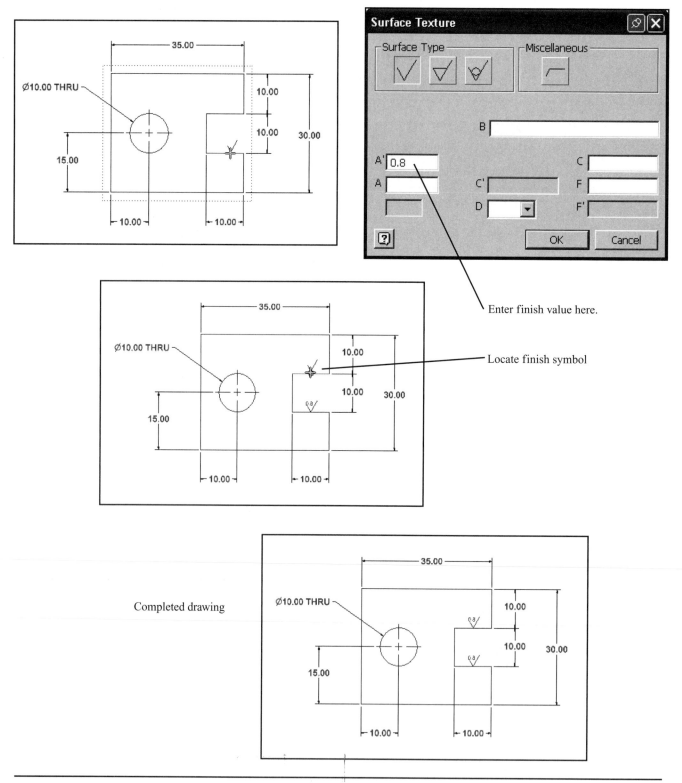

Enter finish value here.

Locate finish symbol

Completed drawing

Figure 8-42

2. Move the cursor to the drawing area and click the lower horizontal edge of the slot.

This step locates the surface texture symbol on the drawing. See Figure 8-42.

3. Right-click the mouse, and select the Continue option.

The Surface Texture diaolg box will appear.

4. Enter a surface texture value of 0.8.
5. Click OK.

The surface symbol will be added to the drawing.

The cursor will remain in Surface Texture mode so that other symbols may be applied.

6. Add a second symbol to the upper edge of the slot.
7. Right-click the mouse and select the Done option.

Available lay symbols

Inventor includes a group of lay symbols that can be added to the drawing using the Surface Texture Symbol tool. The definition of the symbols is located on the Drafting Standards dialog box.

1. Click the Format heading at the top of the screen, then the Standards option.

The Drafting Standards dialog box will appear. See Figure 8-43.

To add lay symbols

1. Click on the Surface Texture Symbol tool, locate a symbol, then right-click the mouse.

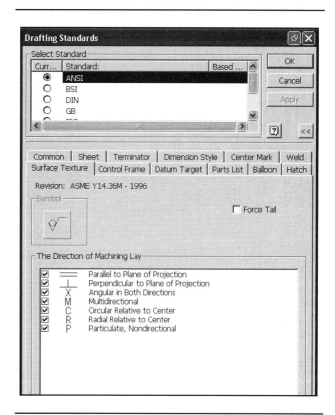

Figure 8-43

A dialog box will appear. See Figure 8-44.

2. Scroll down the available symbols in box D and select an appropriate symbol.

In this example the symbol M for Multidirectional was selected.

3. Click OK.

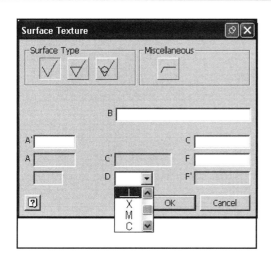

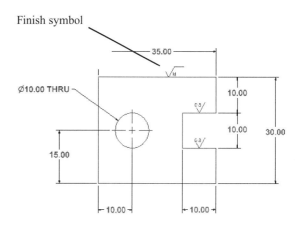

Figure 8-44

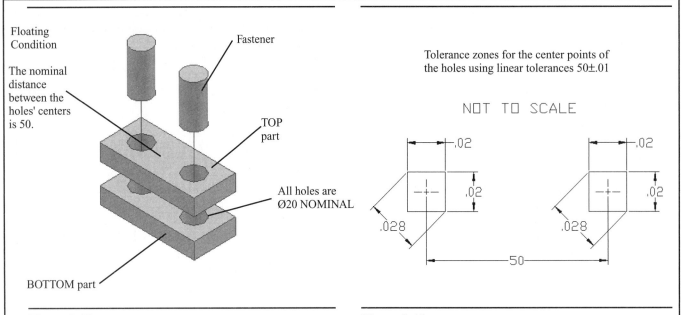

Floating Condition

The nominal distance between the holes' centers is 50.

Fastener

TOP part

All holes are Ø20 NOMINAL

BOTTOM part

Figure 8-45

Tolerance zones for the center points of the holes using linear tolerances 50±.01

NOT TO SCALE

.02 .02 .028 .02 .02 .028 50

Figure 8-46

8-26 DESIGN PROBLEMS

Figure 8-45 shows two objects that are to be fitted together using a fastener such as a screw-and-nut combination. For this example a cylinder will be used to represent a fastener. Only two nominal dimensions are given. The dimensions and tolerances were derived as follows.

The distance between the centers of the holes is given as 50 nominal. The term *nominal* means that the stated value is only a starting point. The final dimensions will be close to the given value but do not have to equal it.

Assigning tolerances is an iteration process; that is, a tolerance is selected and other tolerance values are calculated from the selected initial values. If the results are not satisfactory, go back and modify the initial value and calculate the other values again. As your experience grows you will become better at selecting realistic initial values.

In the example shown in Figure 8-45, start by assigning a tolerance of ±.01 to both the top and bottom parts for both the horizontal and vertical dimensions used to locate the holes. This means that there is a possible center point variation of .02 for both parts. The parts must always fit together, so tolerances must be assigned based on the worst-case condition, or when the parts are made at the extreme ends of the assigned tolerances.

Figure 8-46 shows a greatly enlarged picture of the worst-case condition created by a tolerance of ±.01. The center points of the holes could be as much as .028 apart if the two center points were located at opposite corners of the tolerance zones. This means that the minimum hole diameter must always be at least .028 larger than the maximum stud

diameter. In addition, there should be a clearance tolerance assigned so that the hole and stud are never exactly the same size. Figure 8-47 shows the resulting tolerances.

Floating condition

The top and bottom parts shown in Figure 8-45 are to be joined by two independent fasteners; that is, the location of one fastener does not depend on the location of the other. This situation is called a *floating condition*.

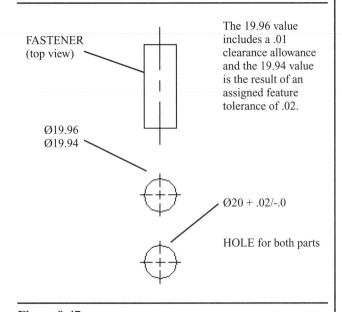

FASTENER (top view)

The 19.96 value includes a .01 clearance allowance and the 19.94 value is the result of an assigned feature tolerance of .02.

Ø19.96
Ø19.94

Ø20 + .02/-.0

HOLE for both parts

Figure 8-47

This means that the tolerance zones for both the top and bottom parts can be assigned the same values and that a fastener diameter selected to fit one part will also fit the other part.

The final tolerances were developed by first defining a minimum hole size of 20.00. An arbitrary tolerance of .02 was assigned to the hole and was expressed as $20.00 +.02/-0$ so that the hole can never be any smaller than 20.00.

The 20.00 minimum hole diameter dictates that the maximum fastener diameter can be no greater than 19.97, or .03 (the rounded-off diagonal distance across the tolerance zone—.028) less than the minimum hole diameter. A .01 clearance was assigned. The clearance ensures that the hole and fastener are never exactly the same diameter. The resulting maximum allowable diameter for the fastener is 19.96. Again an arbitrary tolerance of .02 was assigned to the fastener. The final fastener dimensions are therefore 19.96 to 19.94.

The assigned tolerances ensure that there will always be at least .01 clearance between the fastener and the hole. The other extreme condition occurs when the hole is at its largest possible size (20.02) and the fastener is at its smallest (19.94). This means that there could be as much as .08 clearance between the parts. If this much clearance is not acceptable, then the assigned tolerances will have to be reevaluated.

Figure 8-48 shows the top and bottom parts dimensioned and toleranced. Any dimensions that do not have assigned tolerances are assumed to have standard tolerances.

Note, in Figure 8-48, that the top edge of each part has been assigned a surface finish. This was done to help ensure the accuracy of the $20 \pm .01$ dimension. If this edge surface was rough, it could affect the tolerance measurements.

This example will be done again in Section 8-50 using geometric tolerances. Geometric tolerance zones are circular rather than rectangular.

Fixed condition

Figure 8-49 shows the same nominal conditions presented in Figure 8-45, but the fasteners are now fixed to the top part. This situation is called the *fixed condition*. In analyzing the tolerance zones for the fixed condition, two position tolerances must be considered; the positional tolerances for the holes in the bottom part, and the positional tolerances for the fixed fasteners in the top part. This relationship may be expressed in an equation as follows.

$$S_{max} + DTSZ = H_{min} - DTZ$$

where:

S_{max} = maximum shaft (fastener) diameter

H_{min} = minimum hole diameter

$DTSZ$ = diagonal distance across the shaft's center point tolerance zone

DTZ = diagonal distance across the hole's center point tolerance zone

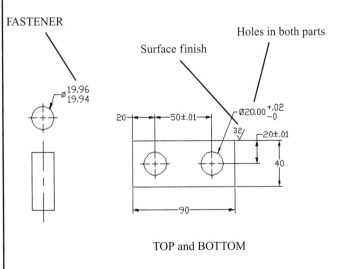

FASTENER

Surface finish

Holes in both parts

$\varnothing$ 19.96 / 19.94

20 — $50\pm.01$ — $\varnothing20.00 \, {}^{+.02}_{-0}$

32 — $20\pm.01$

40

90

TOP and BOTTOM

All dimensions not assigned a tolerance will be assumed to have a standard tolerance.

Figure 8-48

Fixed condition

Fasteners are fixed to TOP part

All holes are $\varnothing20$ nominal.

The nominal distance between the holes' centers is 50.

Figure 8-49

If a dimension and tolerance of 50 ±.01 is assigned to both the center distance between the holes and the center distance between the fixed fasteners, the values for DTSZ and DTZ will be equal. The formula can then be simplified as follows.

$$S_{max} = H_{min} - 2(DTZ)$$

where DTZ equals the diagonal distance across the tolerance zone. If a hole tolerance of 20.00 + .02/−0 is also defined, the resulting maximum shaft size can be determined, assuming that the calculated distance of .028 is rounded off to .03. See Figure 8-50.

$$S_{max} = 20.00 - 2(0.03)$$
$$= 19.94$$

This means that 19.94 is the largest possible shaft diameter that will just fit. If a clearance tolerance of .01 is assumed to ensure that the shaft and hole are never exactly the same size, the maximum shaft diameter becomes 19.93. See Figure 8-51.

A feature tolerance of .02 on the shaft will result in a minimum shaft diameter of 19.91. Note that the .01 clearance tolerance and the .02 feature tolerance were arbitrarily chosen. Other values could have been used.

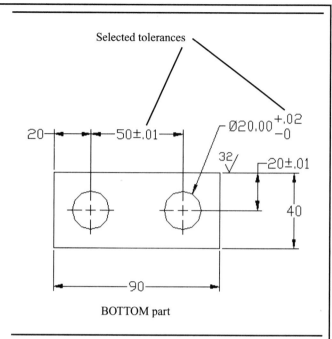

Figure 8-50

The shaft values were derived as follows:

20.00	The selected value for the minimum hole diameter.
−.03	The rounded-off value for the hole positional tolerance
−.03	The rounded-off value for the shaft positional tolerance
−.01	The selected clearance value
19.93	The maximum shaft value
−.02	The selected tolerance value
19.91	The minimum shaft diameter

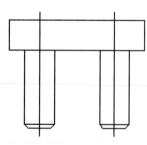

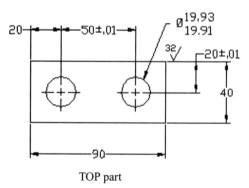

Figure 8-51

To design a hole given a fastener size

The previous two examples started by selecting a minimum hole diameter and then calculating the resulting fastener size. Figure 8-52 shows a situation in which the fastener size is defined, and the problem is to determine the appropriate hole sizes. Figure 8-53 shows the dimensions and tolerances for both top and bottom parts.

Requirements:

Clearance, minimum = .003
Hole tolerances = .005
Positional tolerance = .002

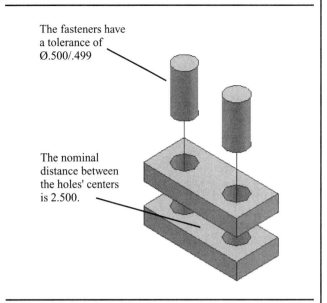

The fasteners have a tolerance of Ø.500/.499

The nominal distance between the holes' centers is 2.500.

Figure 8-52

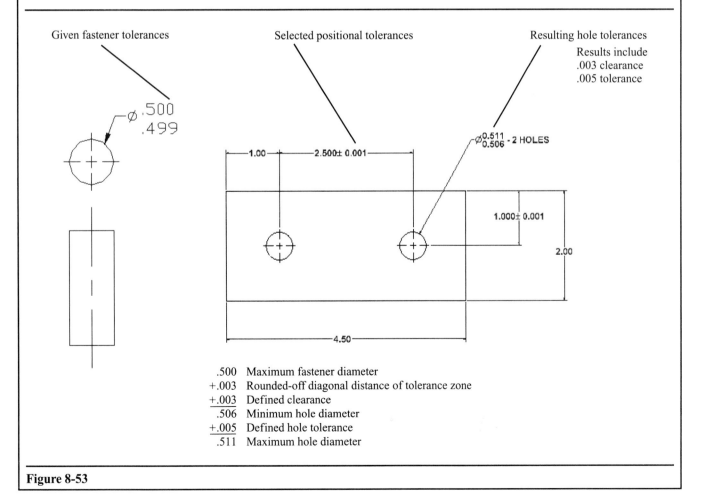

Given fastener tolerances

Selected positional tolerances

Resulting hole tolerances
Results include
.003 clearance
.005 tolerance

Ø.500
.499

1.00 — 2.500± 0.001

Ø0.511
0.506 - 2 HOLES

1.000± 0.001

2.00

4.50

.500	Maximum fastener diameter
+.003	Rounded-off diagonal distance of tolerance zone
+.003	Defined clearance
.506	Minimum hole diameter
+.005	Defined hole tolerance
.511	Maximum hole diameter

Figure 8-53

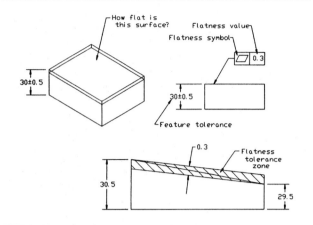

Figure 8-54

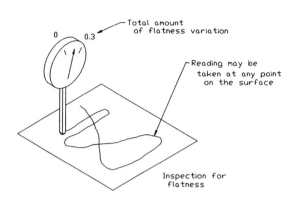

Figure 8-55

8-27 GEOMETRIC TOLERANCES

Geometric tolerancing is a dimensioning and tolerancing system based on the geometric shape of an object. Surfaces may be defined in terms of their flatness or roundness, or in terms of how perpendicular or parallel they are to other surfaces.

Geometric tolerances allow a more exact definition of the shape of an object than do conventional coordinate-type tolerances. Objects can be toleranced in a manner more closely related to their design function or so that their features and surfaces are more directly related to each other.

8-28 TOLERANCES OF FORM

Tolerances of form are used to define the shape of a surface relative to itself. There are four classifications: flatness, straightness, roundness, and cylindricity. Tolerances of form are not related to other surfaces but apply only to an individual surface.

8-29 FLATNESS

Flatness tolerances are used to define the amount of variation permitted in an individual surface. The surface is thought of as a plane not related to the rest of the object.

Figure 8-54 shows a rectangular object. How flat is the top surface? The given plus or minus tolerances allow a variation of (± 0.5) across the surface. Without additional tolerances the surface could look like a series of waves varying between 30.5 and 29.5.

If the example in Figure 8-54 was assigned a flatness tolerance of 0.3, the height of the object—the feature tolerance—could continue to vary based on the 30 ± 0.5 toler-

ance, but the surface itself could not vary by more than 0.3. In the most extreme condition, one end of the surface could be 30.5 above the bottom surface and the other end 29.5, but the surface would still be limited to within two parallel planes 0.3 apart as shown.

To better understand the meaning of flatness, consider how the surface would be inspected. The surface would be acceptable if a gauge could be moved all around the surface and never vary by more than 0.3. See Figure 8-55. Every point in the plane must be within the specified tolerance.

8-30 STRAIGHTNESS

Straightness tolerances are used to measure the variation of an individual feature along a straight line in a specified direction. Figure 8-56 shows an object with a straightness tolerance applied to its top surface. Straightness differs from flatness because straightness measurements are checked by moving a gauge directly across the surface in a single direction. The gauge is not moved randomly about the surface, as is required by flatness.

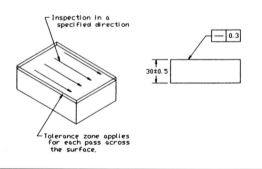

Figure 8-56

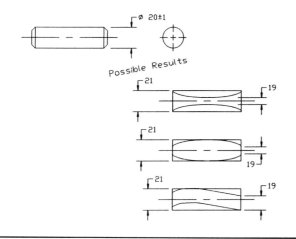

Figure 8-57

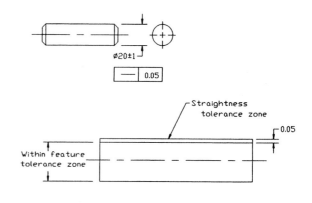

Figure 8-58

Straightness tolerances are most often applied to circular or matching objects to help ensure that the parts are not barreled or warped within the given feature tolerance range and, therefore, do not fit together well. Figure 8-57 shows a cylindrical object dimensioned and toleranced using a standard feature tolerance. The surface of the cylinder may vary within the specified tolerance range as shown.

Figure 8-58 shows the same object shown in Figure 8-57 dimensioned and toleranced using the same feature tolerance but also including a 0.05 straightness tolerance. The straightness tolerance limits the surface variation to 0.05 as shown.

8-31 STRAIGHTNESS (RFS AND MMC)

Figure 8-59 again shows the same cylinder shown in Figures 8-57 and 8-58. This time the straightness tolerance is applied about the cylinder's centerline. This type of tolerance permits the feature tolerance and geometric tolerance to be used together to define a *virtual condition*. A virtual condition is used to determine the maximum possible size variation of the cylinder or the smallest diameter hole that will always accept the cylinder. See Section 8-19.

The geometric tolerance specified in Figure 8-59 is applied to any circular segment along the cylinder, regardless of the cylinder's diameter. This means that the 0.05 tolerance is applied equally when the cylinder's diameter measures 19 or when it measures 21. This application is called RFS, *regardless of feature size*. RFS conditions are specified in a tolerance either by an S with a circle around it or implied tacitly when no other symbol is used. In Figure 8-59 no symbol is listed after the 0.05 value, so it is assumed to be applied RFS.

Figure 8-60 shows the cylinder dimensioned with an MMC condition applied to the straightness tolerance. MMC stands for *maximum material condition* and means that the specified straightness tolerance (0.05) is applied only at the MMC condition or when the cylinder is at its maximum diameter size (21).

A shaft is an external feature, so its largest possible size or MMC occurs when it is at its maximum diameter. A hole is an internal feature. A hole's MMC condition occurs when it is at its smallest diameter. The MMC condition for holes will be discussed later in the chapter along with positional tolerances.

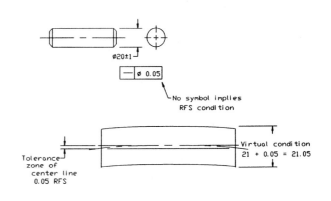

Figure 8-59

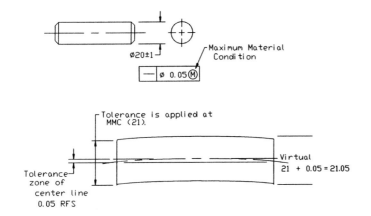

Measured Size	Allowable Tolerance Zone	Virtual Condition
21.0	0.05	21.05
20.9	0.15	21.15
20.8	0.25	21.25
.	.	.
.	.	.
20.0	1.05	22.05
.	.	.
.	.	.
19.0	2.05	23.05

Figure 8-60

Applying a straightness tolerance at MMC allows for a variation in the resulting tolerance zone. Because the 0.05 flatness tolerance is applied at MMC, the virtual condition is still 21.05, the same as with the RFS condition; however, the tolerance is applied only at MMC. As the cylinder's diameter varies within the specified feature tolerance range the acceptable tolerance zone may vary to maintain the same virtual condition.

The table in Figure 8-60 shows how the tolerance zone varies as the cylinder's diameter varies. When the cylinder is at its largest size or MMC, the tolerance zone equals 0.05, or the specified flatness variation. When the cylinder is at its smallest diameter, the tolerance zone equals 2.05, or the total feature size plus the total flatness size. In all variations the virtual size remains the same, so at any given cylinder diameter value, the size of the tolerance zone can be determined by subtracting the cylinder's diameter value from the virtual condition.

Figure 8-61 shows a comparison between different methods used to dimension and tolerance a .750 shaft. The first example uses only a feature tolerance. This tolerance sets an upper limit of .755 and a lower limit of .745. Any variations within that range are acceptable.

The second example in Figure 8-61 sets a straightness tolerance of .003 about the cylinder's centerline. No conditions are defined, so the tolerance is applied RFS. This limits the variations in straightness to .003 at all feature sizes. For example, when the shaft is at its smallest possible feature size of .745, the .003 still applies. This means that a shaft measuring .745 that had a straightness variation greater than .003 would be rejected. If the tolerance had been applied at MMC, the part would be accepted. This does not mean that straightness tolerances should always be applied at MMC. If straightness is critical to the design integrity or function of the part, then straightness should be applied in the RFS condition.

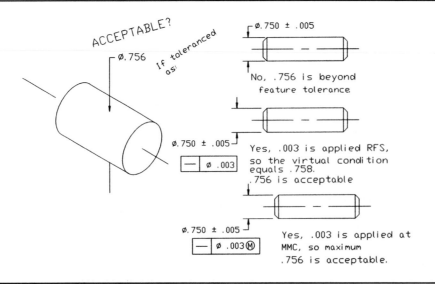

The RFS condition does not allow the tolerance zone to grow, as does the same tolerance applied at MMC.

Figure 8-61

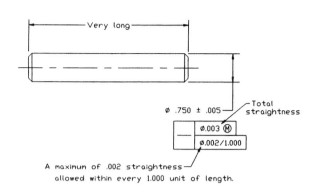

Figure 8-62

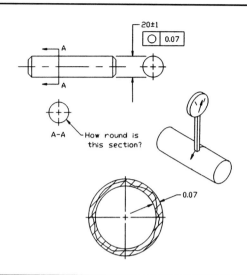

Figure 8-63

The third example in Figure 8-61 applies the straightness tolerance about the center line at MMC. This tolerance creates a virtual condition of .758. The MMC condition allows the tolerance to vary as the feature tolerance varies, so when the shaft is at its smallest feature size, .745, a straightness tolerance of .003 is acceptable (.005 feature tolerance + .003 straightness tolerance).

If the tolerance specification for the cylinder shown in Figure 8-61 was 0.000 applied at MMC, it would mean that the shaft would have to be perfectly straight at MMC or when the shaft was at its maximum value (.755); however, the straightness tolerance can vary as the feature size varies, as discussed for the other tolerance conditions. A 0.000 tolerance means that the MMC and the virtual conditions are equal.

Figure 8-62 shows a very long .750 diameter shaft. Its straightness tolerance includes a length qualifier that serves to limit the straightness variations over each inch of the shaft length and to prevent excess waviness over the full length. The tolerance Ø.002/1.000 means that the total straightness may vary over the entire length of the shaft by .003 but that the variation is limited to .002 per 1.000 of shaft length.

8-32 CIRCULARITY

A *circularity tolerance* is used to limit the amount of variation in the roundness of a surface of revolution. It is measured at individual cross sections along the length of the object. The measurements are limited to the individual cross sections and are not related to other cross sections. This means that in extreme conditions the shaft shown in Figure 8-63 could actually taper from a diameter of 21 to a diameter

of 19 and never violate the circularity requirement. It also means that qualifications such as MMC can- not be applied.

Figure 8-63 shows a shaft that includes a feature tolerance and a circularity tolerance of 0.07. To understand circularity tolerances, consider an individual cross section or slice of the cylinder. The actual shape of the outside edge of the slice varies around the slice. The difference between the maximum diameter and the minimum diameter of the slice can never exceed the stated circularity tolerance.

Circularity tolerances can be applied to tapered sections and spheres, as shown in Figure 8-64. In both applications, circularity is measured around individual cross sections, as it was for the shaft shown in Figure 8-63.

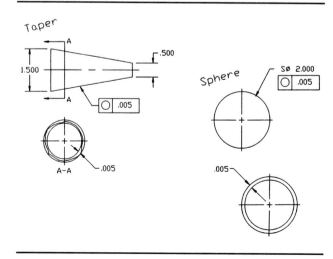

Figure 8-64

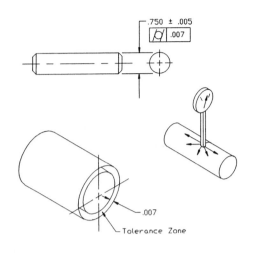

Figure 8-65

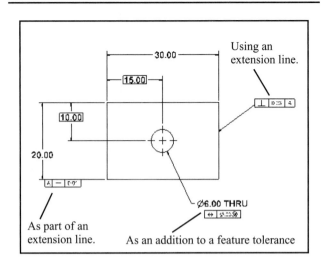

Figure 8-66

8-33 CYLINDRICITY

Cylindricity tolerances are used to define a tolerance zone both around individual circular cross sections of an object and also along its length. The resulting tolerance zone looks like two concentric cylinders.

Figure 8-65 shows a shaft that includes a cylindricity tolerance that establishes a tolerance zone of .007. This means that if the maximum measured diameter is determined to be .755, the minimum diameter cannot be less than

.748 anywhere on the cylindrical surface. Cylindricity and circularity are somewhat analogous to flatness and straightness. Flatness and cylindricity are concerned with variations across an entire surface or plane. In the case of cylindricity, the plane is shaped like a cylinder. Straightness and circularity are concerned with variations of a single element of a surface: a straight line across the plane in a specified direction for straightness, and a path around a single cross section for circularity.

	TYPE OF TOLERANCE	CHARACTERISTIC	SYMBOL
FOR INDIVIDUAL FEATURES	FORM	STRAIGHTNESS	—
		FLATNESS	▱
		CIRCULARITY	○
		CYLINDRICITY	⌭
INDIVIDUAL OR RELATED FEATURES	PROFILE	PROFILE OF A LINE	⌒
		PROFILE OF A SURFACE	⌓
RELATED FEATURES	ORIENTATION	ANGULARITY	∠
		PERPENDICULARITY	⊥
		PARALLELISM	//
	LOCATION	POSITION	⌖
		CONCENTRICITY	◎
	RUNOUT	CIRCULAR RUNOUT	↗
		TOTAL RUNOUT	↗↗

TERM	SYMBOL
AT MAXIMUM MATERIAL CONDITION	Ⓜ
REGARDLESS OF FEATURE SIZE	Ⓢ
AT LEAST MATERIAL CONDITION	Ⓛ
PROJECTED TOLERANCE ZONE	Ⓟ
DIAMETER	∅
SPHERICAL DIAMETER	S∅
RADIUS	R
SPHERICAL RADIUS	SR
REFERENCE	( )
ARC LENGTH	⌒

Figure 8-67

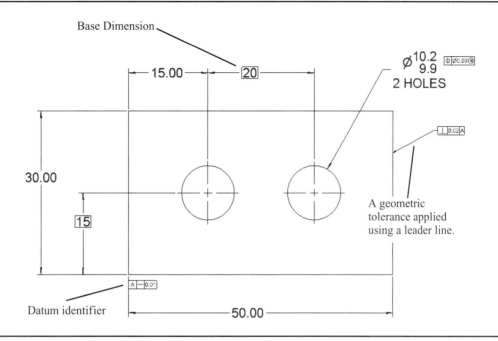

Figure 8-68

8-34 GEOMETRIC TOLERANCES USING INVENTOR

Geometric tolerances are tolerances that limit dimensional variations based on the geometric properties. Figure 8-66 shows three different ways geometric tolerance boxes can be added to a drawing.

Figure 8-67 shows lists geometric tolerance symbols. Figure 8-68 shows an object dimensioned using geometric tolerances. The geometric tolerances were created as follows.

To define a datum— Tolerance tool

1. Access the Drawing Annotation Panel bar.

   See Figure 8-69.

2. Click the Datum Identifier Symbol tool and move the cursor into the drawing area.

   A datum box will appear on the cursor. See Figure 8-70.

3. Position the datum identifier and press the left mouse button, then the right button.

   A dialog box will appear.

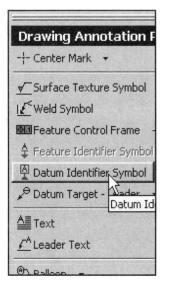

The Drawing annotation panel bar.

Figure 8-69

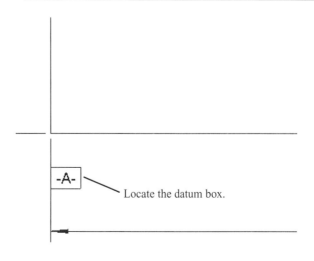

Locate the datum box.

Figure 8-70

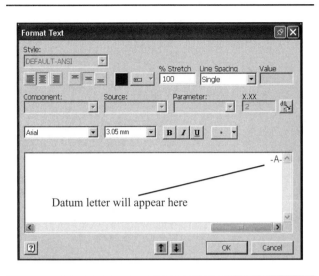

Figure 8-71

4. Click the Continue option.

The Format Text dialog box will appear. See Figure 8-71. The letter A will automatically be selected. If another letter is required, backspace out the existing letter and type in a new one. Flank the letter with dashes.

5. Click OK, then right-click the mouse and select the Done option.

The symbol may be moved after it is created if needed.

To define a perpendicular tolerance

1. Access the Drawing Annotation panel bar.
2. Click the Feature Control Frame tool, move the cursor into the drawing area, and select a location for the control frame.
3. Left-click the mouse, then right-click it. Select the Continue option.

The Feature Control Frame dialog box will appear. See Figure 8-72.

4. Click the Sym box and select the perpendicular symbol.
5. Set the Tolerance 1 value for .02 and the Datum 1 value for A.

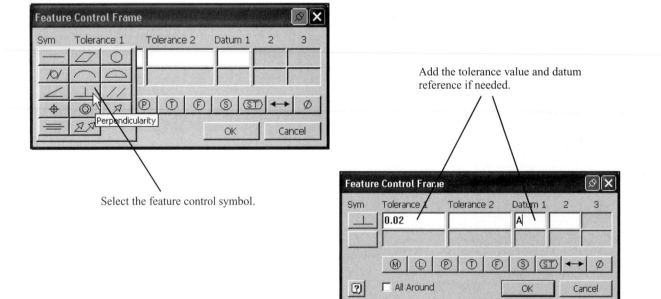

Figure 8-72

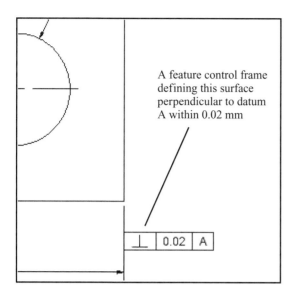

A feature control frame defining this surface perpendicular to datum A within 0.02 mm

Figure 8-73

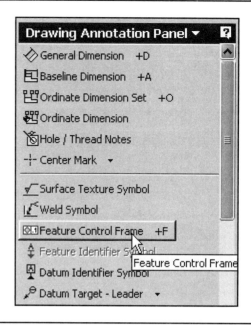

Figure 8-74

Figure 8-73 shows the resulting feature control frame.

To define a straightness tolerance with a leader line

1. Access the Drawing Annotation Panel bar and select the Feature Control Frame tool.

   See Figure 8-74.

2. Move the cursor into the drawing area.

   A feature control box will appear.

3. Left-click the right vertical edge line of the part, then move the frame away from the edge.
4. Select a location for the frame, left-click, then right-click the mouse and select the Continue option.
5. Edit the Feature Control Frame dialog box as shown in Figure 8-75.

   Figure 8-75 shows the resulting feature control.

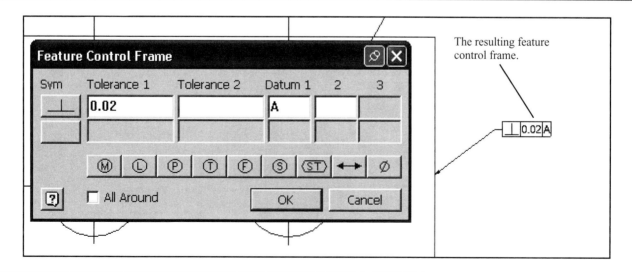

The resulting feature control frame.

Figure 8-75

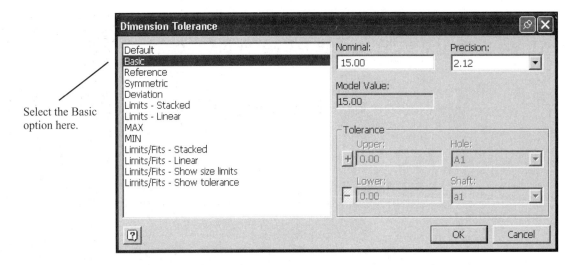

Figure 8-76

Positional tolerance

A *positional tolerance* is used to locate and tolerance a hole in an object. Positional tolerances require basic locating dimensions for the hole's center point. Positional tolerances also require a feature tolerance to define the diameter tolerances of the hole, and a geometric tolerance to define the position tolerance for the hole's center point.

To create a basic dimension

The 15 and 20 dimensions in Figure 8-68 used to locate the center position of the hole are basic dimensions. *Basic dimensions* are dimensions enclosed in rectangles.

1. Create dimensions using the General Dimension tool.
2. Right-click the existing dimension and select the Tolerance option.

The Dimension Tolerance dialog box will appear. See Figure 8-76.

3. Select the Basic option, then click OK.

The selected dimension will be enclosed in a rectangle. This is a basic dimension. See Figure 8-77.

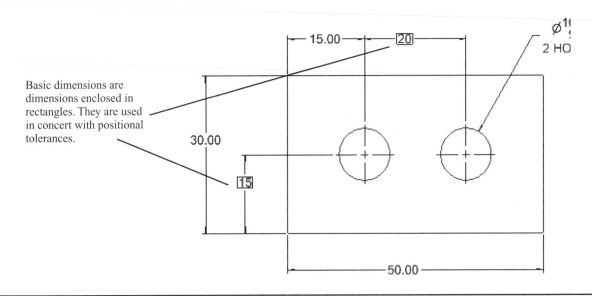

Basic dimensions are dimensions enclosed in rectangles. They are used in concert with positional tolerances.

Figure 8-77

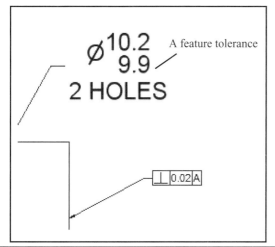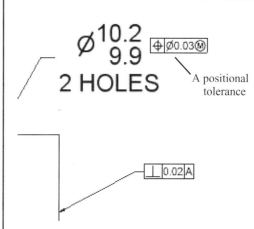

Figure 8-78

To add a positional tolerance to a hole's feature tolerance

Figure 8-78 shows a feature tolerance for a hole. A feature tolerance defines the hole's size limits. In the example shown in Figure 8-78 the hole is defined as 10.2 to 9.9. These values define the tolerance of the hole. In addition, the location of the hole's center point must be defined and toleranced.

1. Click the Feature Control Frame tool.
2. Move the cursor and locate the feature control frame next to the hole's feature control dimensions.
3. Click the left mouse button, then the right mouse button. Select the Continue option.

The Feature Control Frame dialog box will appear. See Figure 8-79.

4. Select the positional symbol, then move to the Tolerance 1 box and locate the cursor to the left of the 0.000 default tolerance value.
5. Select the centerline symbol, Ø.
6. Create a tolerance value of 0.03, then click the maximum material condition symbol, an M with a circle around it.
7. Click OK.
8. Right-click the mouse and select the Done option.

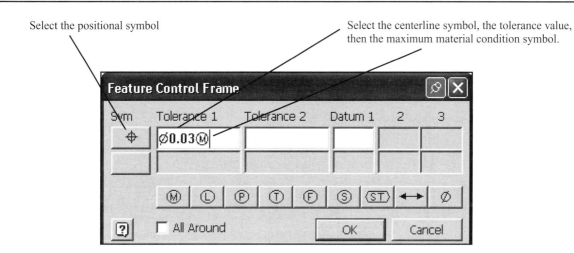

Figure 8-79

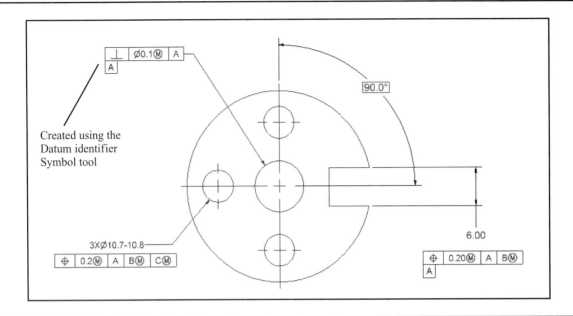

Figure 8-80

To create more complex geometric tolerance drawing callouts

Figure 8-80 shows a model that has several, more complex, geometric tolerance drawing callouts. These callouts are created using the procedure and the same Feature Control Frame dialog box as shown in Figure 8-79, but data must be entered. The individual datum callouts are created using the Datum Identifier Symbol tool.

Figure 8-81 shows the Feature Control Frame used to create the slot dimension in Figure 8-80. Figure 8-81 also shows the Feature Control Fame dialog box for a multilined tolerance callout.

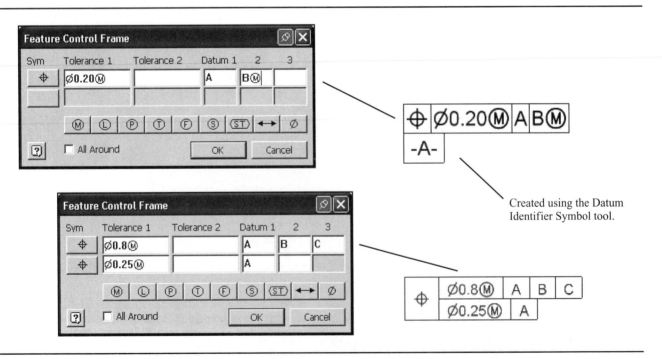

Figure 8-81

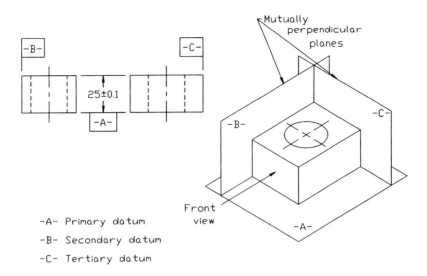

Figure 8-82

8-35 TOLERANCES OF ORIENTATION

Tolerances of orientation are used to relate a feature or surface to another feature or surface. Tolerances of orientation include perpendicularity, parallelism, and angularity. They may be applied using RFS or MMC conditions, but they cannot be applied to individual features by themselves. To define a surface as parallel to another surface is very much like assigning a flatness value to the surface. The difference is that flatness applies only within the surface; every point on the surface is related to a defined set of limiting parallel planes. Parallelism defines every point in the surface relative to another surface. The two surfaces are therefore directly related to each other, and the condition of one affects the other.

Orientation tolerances are used with locational tolerances. A feature is first located, then it is oriented within the locational tolerances. This means that the orientation tolerance must always be less than the locational tolerances. The next four sections will further explain this requirement.

8-36 DATUMS

A *datum* is a point, axis, or surface used as a starting reference point for dimensions and tolerances. Figure 8-82 and Figure 8-83 show a rectangular object with three datum planes labeled –A–, –B–, and –C–. The three datum planes are called the primary, secondary, and tertiary datums, respectively. The three datum planes are, by definition, exactly 90° to one another.

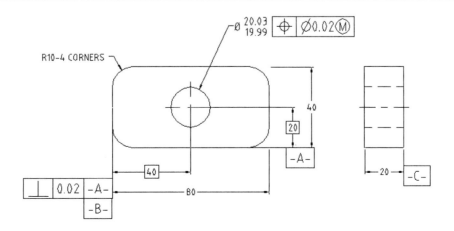

Figure 8-83

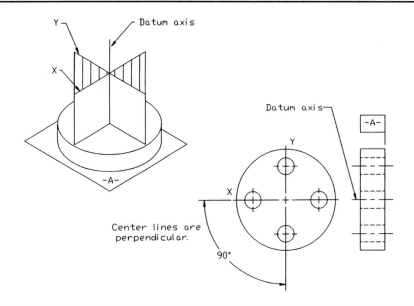

Figure 8-84

Figure 8-84 shows a cylindrical datum frame that includes three datum planes. The X and Y planes are perpendicular to each other, and the base A plane is perpendicular to the datum axis between the X and Y planes.

Datums are defined on a drawing by letters enclosed in rectangular boxes, as shown. The defining letters are written flanked by dashes: –A–, –B–, and –C–.

Datum planes are assumed to be perfectly flat. When assigning a datum status to a surface, be sure that the surface is reasonably flat. This means that datum surfaces should be toleranced using surface finishes, or created using machine techniques that produce flat surfaces.

8-37 PERPENDICULARITY

Perpendicularity tolerances are used to limit the amount of variation for a surface or feature within two planes perpendicular to a specified datum. Figure 8-85 shows a rectangular object. The bottom surface is assigned as datum –A–, and the right vertical edge is toleranced so that it must be perpendicular within a limit of 0.05 to datum –A–. The perpendicularity tolerance defines a tolerance zone 0.05 wide between two parallel planes that are perpendicular to datum –A–.

The object also includes a horizontal dimension and tolerance of 40±1. This tolerance is called a *locational tolerance* because it serves to locate the right edge of the object. As with rectangular coordinate tolerances, discussed

earlier in the chapter, the 40±1 controls the location of the edge—how far away or how close it can be to the left edge—but does not directly control the shape of the edge. Any shape that falls within the specified tolerance range is acceptable. This may in fact be sufficient for a given design, but if a more controlled shape is required, a perpendicularity tolerance must be added. The perpendicularity tolerance works within the locational tolerance to ensure that the edge is not only within the locational tolerance but is also perpendicular to datum –A–.

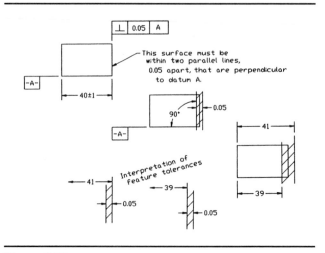

Figure 8-85

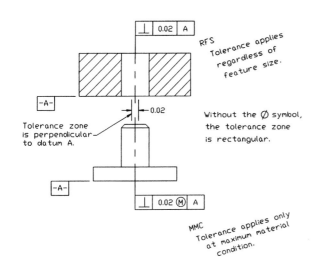

Figure 8-86

If tolerance is

Ø20±0.03

| ⊥ | 0.02 | A |

Feature Tolerance	Allowable Tolerance
20.03	.02
20.02	.02
20.01	.02
20.00	.02
19.99	.02
19.98	.02
19.97	.02

Tolerance zone shape

| 0.02 SQUARE |

If tolerance is

Ø20±0.03

| ⊥ | Ø0.02 Ⓜ | A |

Feature Tolerance	Allowable Tolerance
20.03	.02
20.02	.03
20.01	.04
20.00	.05
19.99	.06
19.98	.07
19.97	.08

Tolerance zone shape

Ø0.02

Figure 8-87

Figure 8-85 shows the two extreme conditions for the 40±1 locational tolerance. The perpendicularity tolerance is applied by first measuring the surface and determining its maximum and minimum lengths. The difference between these two measurements must be less than 0.05. So if the measured maximum distance is 41, then no other part of the surface may be less then 41 − 0.05 = 40.95.

Tolerances of perpendicularity serve to complement locational tolerances, to make the shape more exact, so tolerances of perpendicularity must always be smaller than tolerances of location. It would be of little use, for example, to assign a perpendicularity tolerance of 1.5 for the object shown in Figure 8-86. The locational tolerance would prevent the variation from ever reaching the limits specified by such a large perpendicularity tolerance.

Figure 8-87 shows a perpendicularity tolerance applied to cylindrical features: a shaft and a hole. The figure includes examples of both RFS and MMC applications. As with straightness tolerances applied at MMC, perpendicularity tolerances applied about a hole or shaft's centerline allow the tolerance zone to vary as the feature size varies.

The inclusion of the Ø symbol in a geometric tolerance is critical to its interpretation. See Figure 8-88. If the Ø symbol is not included, the tolerance applies only to the view in which it is written. This means that the tolerance zone is shaped like a rectangular slice, not a cylinder, as would be the case if the Ø symbol were included. In general it is better to always include the Ø symbol for cylindrical features because it generates a tolerance zone more like that used in positional tolerancing.

Figure 8-88 shows a perpendicularity tolerance applied to a slot, a noncylindrical feature. Again, the MMC specification is always for variations in the tolerance zone.

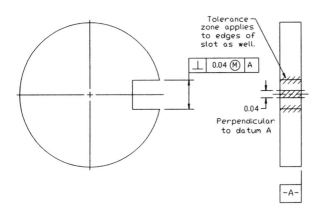

Figure 8-88

8-38 PARALLELISM

Parallelism is used to ensure that all points within a plane are within two parallel planes that are parallel to a referenced datum plane. Figure 8-89 shows a rectangular object that is toleranced so that its top surface is parallel to the bottom surface within 0.02. This means that every point on the top surface must be within a set of parallel planes 0.02 apart. These parallel tolerancing planes are located by determining the maximum and minimum distances from the datum surface. The difference between the maximum and minimum values may not exceed the stated 0.02 tolerance.

In the extreme condition of maximum feature size, the top surface is located 40.5 above the datum plane. The parallelism tolerance is then applied, meaning that no point on the surface may be closer than 40.3 to the datum. This is an RFS condition. The MMC condition may also be applied, thereby allowing the tolerance zone to vary as the feature size varies.

8-39 ANGULARITY

Angularity tolerances are used to limit the variance of surfaces and axes that are at an angle relative to a datum. Angularity tolerances are applied like perpendicularity and parallelism tolerances as a way to better control the shape of locational tolerances.

Figure 8-90 shows an angularity tolerance and several ways it is interpreted at extreme conditions.

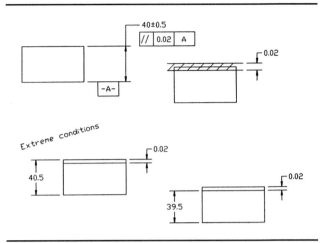

Figure 8-89

8-40 PROFILES

Profile tolerances are used to limit the variations of irregular surfaces. They may be assigned as either bilateral or unilateral tolerances. There are two types of profile tolerances: surface and line. Surface profile tolerances limit the variation of an entire surface, whereas a line profile tolerance limits the variations along a single line across a surface.

Figure 8-91 shows an object that includes a surface profile tolerance referenced to an irregular surface. The tolerance is considered a bilateral tolerance because no other specification is given. This means that all points on the surface must be located between two parallel planes 0.08 apart that are centered about the irregular surface. The measurements are taken perpendicular to the surface.

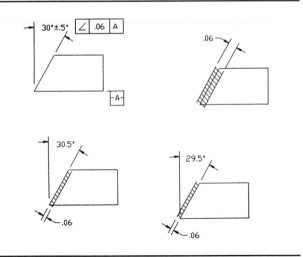

Figure 8-90

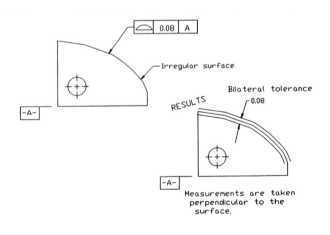

Figure 8-91

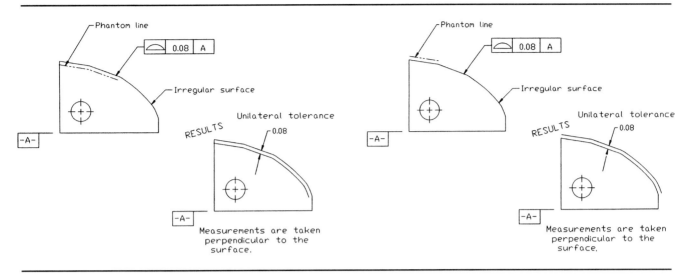

Figure 8-92

Unilateral applications of surface profile tolerances must be indicated on the drawing using phantom lines. The phantom line indicates the side of the true profile line of the irregular surface on which the tolerance is to be applied. A phantom line above the irregular surface indicates that the tolerance is to be applied using the true profile line as 0 and then the specified tolerance range is to be added above that line. See Figures 8-92 and 8-93.

Profiles of line tolerances are applied to irregular surfaces, as shown in Figure 8-93. Profiles of line tolerances are particularly helpful when tolerancing an irregular surface that is constantly changing, such as the surface of an airplane wing.

Surface and line profile tolerances are somewhat analogous to flatness and straightness tolerances. Flatness and surface profile tolerances are applied across an entire surface, whereas straightness and line profile tolerances are applied only along a single line across the surface.

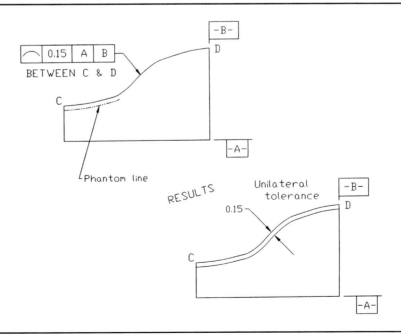

Figure 8-93

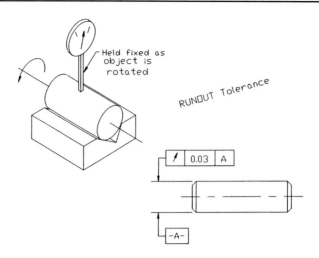

Figure 8-94

8-41 RUNOUTS

A *runout tolerance* is used to limit the variations between features of an object and a datum. More specifically they are applied to surfaces around a datum axis such as a cylinder or to a surface constructed perpendicular to a datum axis. There are two types of runout tolerances: circular and total.

Figure 8-94 shows a cylinder that includes a circular runout tolerance. The runout requirements are checked by rotating the object about its longitudinal axis or datum axis while holding an indicator gauge in a fixed position on the object's surface.

Runout tolerances may be either bilateral or unilateral. A runout tolerance is assumed to be bilateral unless otherwise indicated. If a runout tolerance is to be unilateral, a phantom line is used to indicate the side of the object's true surface to which the tolerance is to be applied. See Figure 8-95.

Runout tolerances may be applied to tapered areas of cylindrical objects, as shown in Figure 8-96. The tolerance is checked by rotating the object about a datum axis while holding an indicator gauge in place.

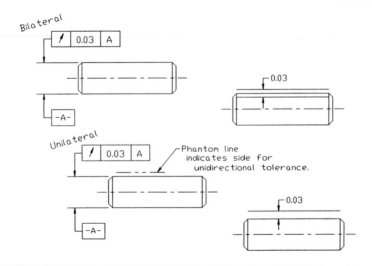

Figure 8-95

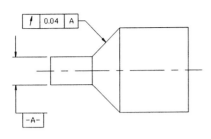

Figure 8-96

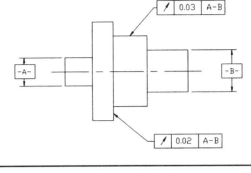

Figure 8-98

A total runout tolerance limits the variation across an entire surface. See Figure 8-97. An indicator gauge is not held in place while the object is rotated, as it is for circular runout tolerances, but is moved about the rotating surface.

Figure 8-98 shows a circular runout tolerance that references two datums. The two datums serve as one datum. The object can then be rotated about both datums simultaneously as the runout tolerances are checked.

8-42 POSITIONAL TOLERANCES

Positional tolerances are used to locate and tolerance holes. Positional tolerances create a circular tolerance zone for hole center point locations, in contrast with the rectangular-shaped tolerance zone created by linear coordinate dimensions. See Figure 8-99. The circular tolerance zone allows for an increase in acceptable tolerance variation without compromising the design integrity of the object. Note how some of the possible hole center points fall in an area outside the rectangular tolerance zone but are still within the circular tolerance zone. If the hole had been located using linear coordinate dimensions, center points located beyond the rectangular tolerance zone would have been rejected as beyond tolerance, and yet holes produced using these locations would function correctly from a design standpoint. The center point locations would be acceptable if positional tolerances had been specified. The finished hole is round, so a round tolerance zone is appropriate. The rectangular tolerance zone rejects some holes unnecessarily.

Holes are dimensioned and toleranced using geometric tolerances by a combination of locating dimensions, feature dimensions and tolerances, and positional tolerances.

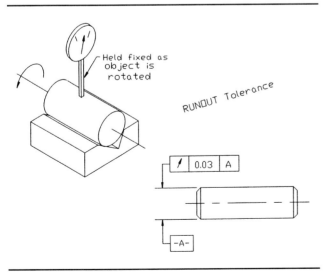

Figure 8-97

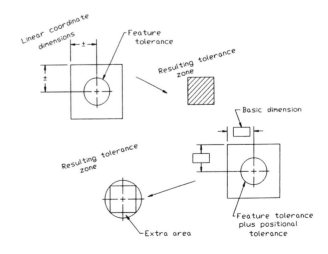

Figure 8-99

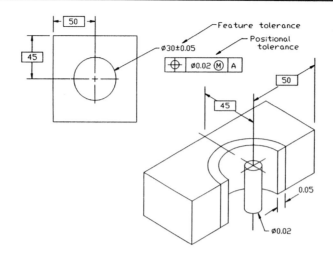

Figure 8-100

See Figure 8-100. The locating dimensions are enclosed in rectangular boxes and are called *basic dimensions*. Basic dimensions are assumed to be exact.

The feature tolerances for the hole are as presented earlier in the chapter. They can be presented using plus or minus or limit-type tolerances. In the example shown in Figure 8-100 the diameter of the hole is toleranced using a plus and minus 0.05 tolerance.

The basic locating dimensions of 45 and 50 are assumed to be exact. The tolerances that would normally accompany linear locational dimensions are replaced by the positional tolerance. The positional tolerance also specifies that the tolerance be applied at the centerline at maximum material condition. The resulting tolerance zones are as shown in Figure 8-100.

Figure 8-101 shows an object containing two holes that are dimensioned and toleranced using positional tolerances. There are two consecutive horizontal basic dimensions. Because basic dimensions are exact, they do not have tolerances that accumulate; that is, there is no tolerance buildup.

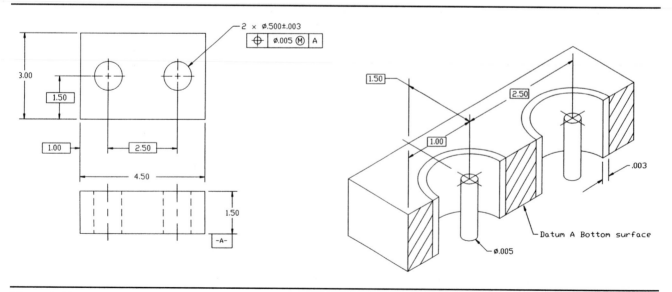

Figure 8-101

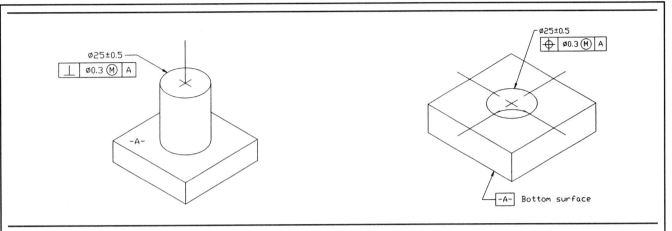

Figure 8-102

8-43 VIRTUAL CONDITION

Virtual condition is a combination of a feature's MMC and its geometric tolerance. For external features (shafts) it is the MMC plus the geometric tolerance; for internal features (holes) it is the MMC minus the geometric tolerance.

The following calculations are based on the dimensions shown in Figure 8-102.

To calculate the virtual condition for a shaft

25.5	MMC for shaft—maximum diameter
+0.3	Geometric tolerance
25.8	Virtual condition

To calculate the virtual condition for a hole

24.5	MMC for hole—minimum diameter
−0.3	Geometric tolerance
24.2	Virtual condition

8-44 FLOATING FASTENERS

Positional tolerances are particularly helpful when dimensioning matching parts. Because basic locating dimensions are considered exact, the sizing of mating parts is dependent only on the hole and shaft's MMC and the geometric tolerance between them.

The relationship for floating fasteners and holes in objects may be expressed as a formula:

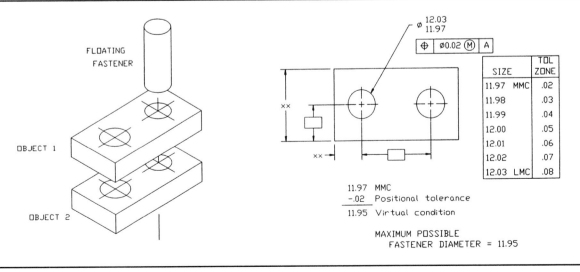

Figure 8-103

$$H - T = F$$

where

H = hole at MMC

T = geometric tolerance

F = shaft at MMC

A *floating fastener* is one that is free to move in either object. It is not attached to either object and it does not screw into either object. Figure 8-103 shows two objects that are to be joined by a common floating shaft, such as a bolt or screw. The feature size and tolerance and the positional geometric tolerance are both given. The minimum size hole that will always just fit is determined using the preceding formula.

$$H - T = F$$
$$11.97 - .02 = 11.95$$

Therefore, the shaft's diameter at MMC, the shaft's maximum diameter, equals 11.95. Any required tolerance would have to be subtracted from this shaft size.

The 0.02 geometric tolerance is applied at the hole's MMC, so as the hole's size expands within its feature tolerance, the tolerance zone for the acceptable matching parts also expands.

8-45 SAMPLE PROBLEM SP8-5

The situation presented in Figure 8-103 can be worked in reverse; that is, hole sizes can be derived from given shaft sizes.

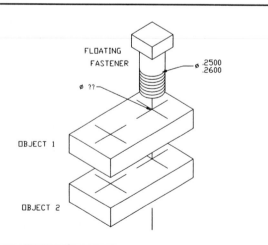

Figure 8-104

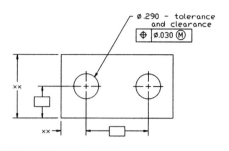

Figure 8-105

The two objects shown in Figure 8-104 are to be joined by a .250-in. bolt. The parts are floating; that is, they are both free to move, and the fastener is not joined to either object. What is the MMC of the holes if the positional tolerance is to be .030?

A manufacturer's catalog specifies that the tolerance for .250 bolts is .2500 to .2600.

Rewriting the formula

$$H - T = F$$

to isolate the H yields

$$H = F + T$$
$$= .260 + .030$$
$$= .290$$

The .290 value represents the minimum hole diameter, MMC, for all four holes that will always accept the .250 bolt. Figure 8-105 shows the resulting drawing callout.

Any clearance requirements or tolerances for the hole would have to be added to the .290 value.

8-46 SAMPLE PROBLEM SP8-6

Repeat the problem presented in SP8-5 but be sure that there is always a minimum clearance of .002 between the hole and the shaft, and assign a hole tolerance of .008.

Sample problem SP8-5 determined that the maximum hole diameter that will always accept the .250 bolt was .290 based on the .030 positioning tolerance. If the minimum clearance is to be .002, the maximum hole diameter is found as follows:

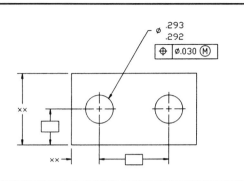

Figure 8-106

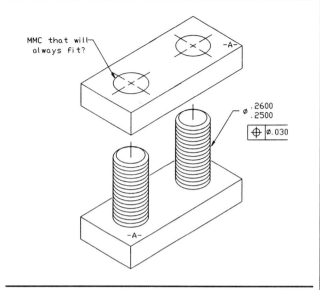

Figure 8-108

.290 Minimum hole diameter that will always accept the bolt (0 clearance at MMC)
+.002 Minimum clearance
.292 Minimum hole diameter including clearance

Now, assign the tolerance to the hole:

.292 Minimum hole diameter
+.001 Tolerance
.293 Maximum hole diameter

See Figure 8-106 for the appropriate drawing callout. The choice of clearance size and hole tolerance varies with the design requirements for the objects.

8-47 FIXED FASTENERS

A *fixed fastener* is one that is attached to one of the mating objects. See Figure 8-107. Because the fastener is fixed to one of the objects, the geometric tolerance zone must be smaller than that used for floating fasteners. The fixed fastener cannot move without moving the object it is attached to. The relationship between fixed fasteners and holes in mating objects is defined by the formula

$$H - 2T = F$$

The tolerance zone is cut in half. This can be demonstrated by the objects shown in Figure 8-108. The same feature sizes that were used in Figure 8-103 are assigned, but in this example the fasteners are fixed. Solving for the geometric tolerance yields a value as follows:

$$H - F = 2T$$
$$11.97 - 11.95 = 2T$$
$$.02 = 2T$$
$$.01 = T$$

The resulting positional tolerance is half that obtained for floating fasteners.

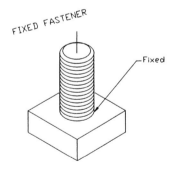

Figure 8-107

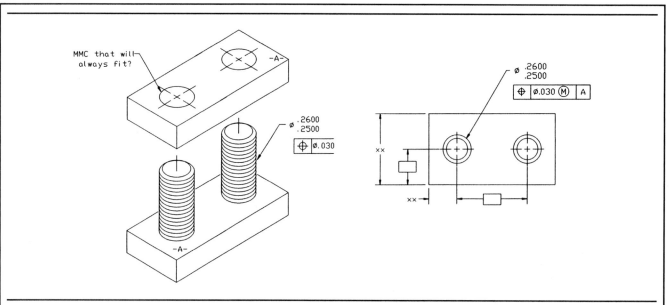

Figure 8-109

8-48 SAMPLE PROBLEM SP8-7

This problem is similar to sample problem SP8-5, but the given conditions are applied to fixed fasteners rather than floating fasteners. Compare the resulting shaft diameters for the two problems. See Figure 8-109.

A. What is the minimum diameter hole that will always accept the fixed fasteners?

B. If the minimum clearance is .005 and the hole is to have a tolerance of .002, what are the maximum and minimum diameters of the hole?

$$H - 2T = F$$

$$H = F + 2T$$
$$= .260 + 2(.030)$$
$$= .260 + .060$$
$$= .320 \text{ Minimum diameter that will always accept the fixed fastener}$$

If the minimum clearance is .005 and the hole tolerance is .002,

```
 .320  Virtual condition
+.005  Clearance
 .325  Minimum hole diameter
```

```
 .325  Minimum hole diameter
+.002  Tolerance
 .327  Maximum hole diameter
```

The maximum and minimum values for the hole's diameter can then be added to the drawing of the object that fits over the fixed fasteners. See Figure 8-110.

8-49 DESIGN PROBLEMS

This problem was originally done in Section 8-27 using rectangular tolerances. It is done in this section using positional geometric tolerances so that the two systems can be compared. It is suggested that Section 8-27 be reviewed before reading this section.

Figure 8-111 shows top and bottom parts that are to be joined in the floating condition. A nominal distance of 50 between hole centers and 20 for the holes has been assigned.

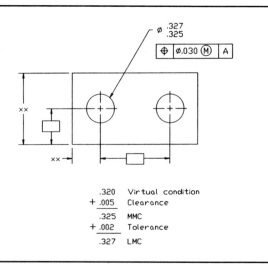

```
 .320  Virtual condition
+.005  Clearance
 .325  MMC
+.002  Tolerance
 .327  LMC
```

Figure 8-110

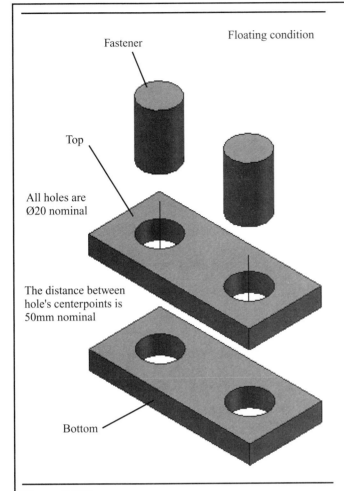

Figure 8-111

Fastener

Floating condition

Top

All holes are
Ø20 nominal

The distance between
hole's centerpoints is
50mm nominal

Bottom

In Section 8-27 a rectangular tolerance of ±.01 was selected, and there was a minimum hole diameter of 20.00. Figure 8-112 shows the resulting tolerance zones.

The diagonal distance across the rectangular tolerance zone is .028 and was rounded off to .03 to yield a maximum possible fastener diameter of 19.97. If the same .03 value is used to calculate the fastener diameter using positional tolerance, the results are as follows:

$$H - T = F$$
$$20.00 - .03 = 19.97$$

The results seem to be the same, but because of the circular shape of the positional tolerance zone, the manufactured results are not the same. The minimum distance between the inside edges of the rectangular zones is 49.98, or .01 from the center point of each hole. The minimum distance from the innermost points of the circular tolerance zones is 49.97, or .015 (half of the rounded-off .03 value) from the center point of each hole. The same value difference also occurs for the maximum distance between center points, where 50.02 is the maximum distance for the rectangular tolerances, and 50.03 is the maximum distance for the circular tolerances. The size of the circular tolerance zone is larger because the hole tolerances are assigned at MMC. Figure 8-112 shows a comparison between the tolerance zones, and Figure 8-113 shows how the positional tolerances would be presented on a drawing of either the top or bottom part.

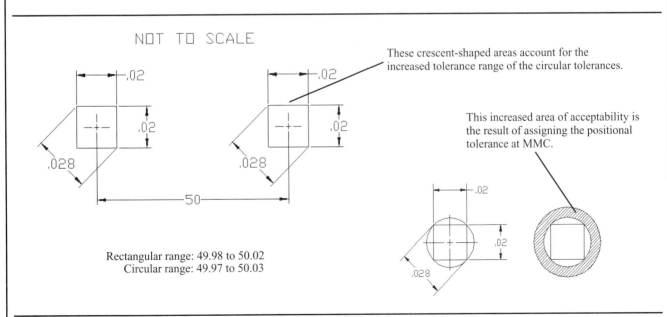

NOT TO SCALE

These crescent-shaped areas account for the
increased tolerance range of the circular tolerances.

This increased area of acceptability is
the result of assigning the positional
tolerance at MMC.

Rectangular range: 49.98 to 50.02
Circular range: 49.97 to 50.03

Figure 8-112

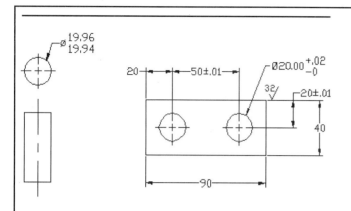

Figure 8-113

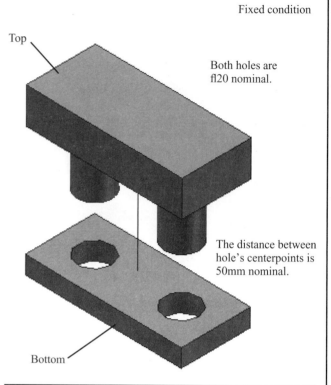

Fixed condition

Both holes are
fl20 nominal.

The distance between
hole's centerpoints is
50mm nominal.

Top

Bottom

Figure 8-114

Figure 8-114 shows the same top and bottom parts joined together in the fixed condition. The initial nominal values are the same. If the same .03 diagonal value is assigned as a positional tolerance, the results are as follows:

$$H - 2T = F$$
$$20.00 - .06 = 19.94$$

These results appear to be the same as those generated by the rectangular tolerance zone, but the circular tolerance zone allows a greater variance in acceptable manufactured parts. Figure 8-115 shows how the positional tolerance would be presented on a drawing.

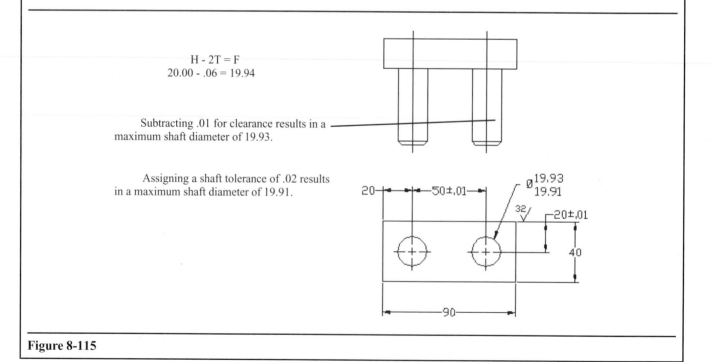

$$H - 2T = F$$
$$20.00 - .06 = 19.94$$

Subtracting .01 for clearance results in a maximum shaft diameter of 19.93.

Assigning a shaft tolerance of .02 results in a maximum shaft diameter of 19.91.

Figure 8-115

8-50 EXERCISE PROBLEMS

Draw a model of the objects shown in EX8-1 to EX8-4 using the given dimensions and tolerances. Create a drawing layout with a view of the model as shown. Add the specified dimensions and tolerances.

EX8-1 MILLIMETERS

1. 38±0.05
2. 10±0.1
3. 5±0.05
4. $45.50°$ / $44.50°$
5. 40±0.1
6. 22±0.1
7. $12 \; {}^{+0}_{-.1}$
8. $25 \; {}^{+.05}_{-0}$
9. 51.50 / 50.75
10. 76±0.1

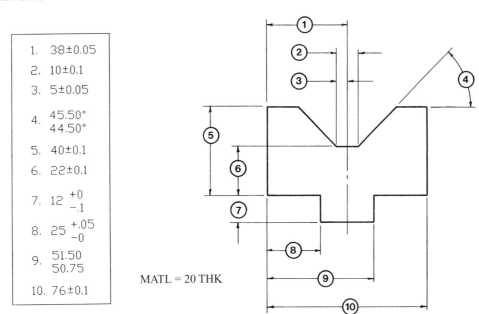

MATL = 20 THK

EX8-2 MILLIMETERS

1. 34±0.25
2. 17±0.25
3. 25±0.05
4. 15.00 / 14.80
5. 50±0.05
6. 80±0.1
7. R5±0.1-8 PLACES
8. 45±0.25
9. 60±0.1
10. Ø14 - 3 HOLES
11. 15.00 / 14.80
12. 30.00 / 29.80

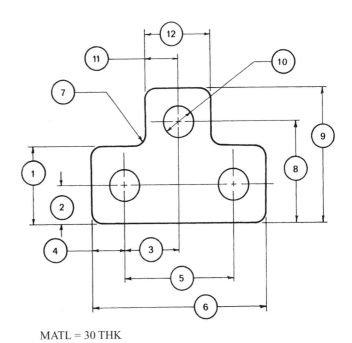

MATL = 30 THK

EX8-3 INCHES

1. 3.00±.01
2. 1.56±.01
3. 46.50°
   45.50°
4. .750±.005
5. 2.75
   2.70
6. 3.625±.010
7. 45°±.5°
8. 2.250±.005

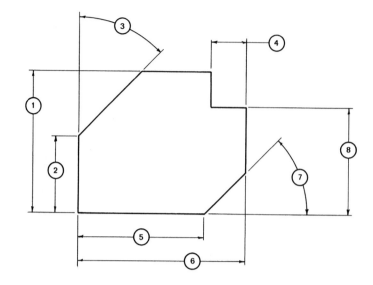

MATL = .75 THK

EX8-4 MILLIMETERS

1. 50 $^{+.2}_{\ 0}$
2. R45±.1 – 2 PLACES
3. 63.5 $^{\ 0}_{-.2}$
4. 76±.1
5. 38±.1
6. Ø12.00 $^{+.05}_{\ 0}$ – 3 HOLES
7. 30±.03
8. 30±.03
9. 100 $^{+.4}_{\ 0}$

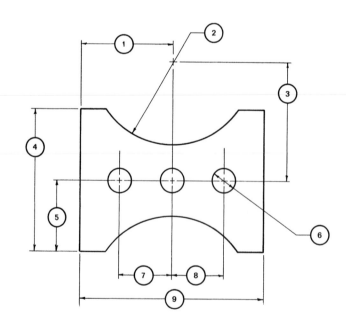

MATL = 10 THK

EX8-5 MILLIMETERS

Redraw the following object, including the given dimensions and tolerances. Calculate and list the maximum and minimum distances for surface A.

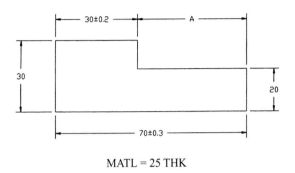

MATL = 25 THK

EX8-7 MILLIMETERS

Redraw the following object, including the dimensions and tolerances. Calculate and list the maximum and minimum distances for surfaces D and E.

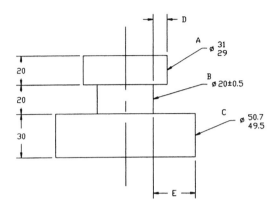

EX8-6 INCHES

A. Redraw the following object, including the dimensions and tolerances. Calculate and list the maximum and minimum distances for surface A.
B. Redraw the given object and dimension it using baseline dimensions. Calculate and list the maximum and minimum distances for surface A.

EX8-8 MILLIMETERS

Dimension the following object twice, once using chain dimensions and once using baseline dimensions. Calculate and list the maximum and minimum distances for surface D for both chain and baseline dimensions. Compare the results.

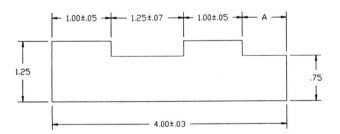

MATL = 1.25 THK

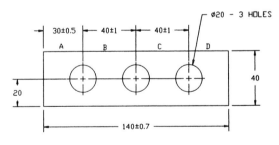

MATL = 20 THK

Redraw the following shapes, including the dimensions and tolerances. Also list the required minimum and maximum values for the specified distances.

EX8-9 INCHES

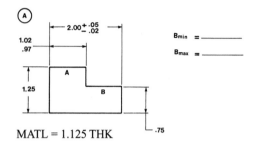

MATL = 1.125 THK

B_{min} = _____
B_{max} = _____

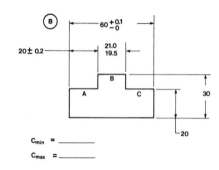

C_{min} = _____
C_{max} = _____

MATL = 45 THK

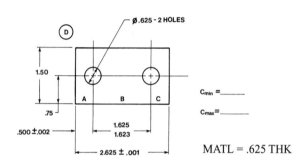

C_{min} = _____
C_{max} = _____

MATL = .625 THK

EX8-10 MILLIMETERS

Redraw and complete the following inspection report. Under the Results column classify each "AS MEASURED" value as OK if the value is within the stated tolerances, REWORK if the value indicates that the measured value is beyond the stated tolerance but can be reworked to bring it into the acceptable range, or SCRAP if the value is not within the tolerance range and cannot be reworked to make it acceptable.

INSPECTION REPORT

PART NAME AND NO: 1075-5002

1.00 3 PLACES

INSPECTOR:

DATE:

BASE DIMENSION	TOLERANCES		AS MEASURED	RESULTS
	MAX	MIN		
① 100 ± 0.5			99.8	
② $\phi\,^{57}_{56}$			57.01	
③ 22 ± 0.3			21.72	
④ $^{40.05}_{39.95}$			39.98	
⑤ 22 ± 0.3			21.68	
⑥ $R52 ^{+\ 0}_{-0.2}$			51.99	
⑦ $35 ^{+0.2}_{-0.3}$			35.20	
⑧ $30 ^{+0.4}_{0}$			30.27	
⑨ $6.0 ^{+.1}_{-.2}$			5.85	
⑩ 12.0 ± 0.2			11.90	

.50 —10 PLACES

EX8-11 MILLIMETERS

Redraw the following charts and complete them based on the following information. All values are in millimeters.

A. Nominal = 16, Fit = H8/d8
B. Nominal = 30, Fit = H11/c11
C. Nominal = 22, Fit = H7/g6
D. Nominal = 10, Fit = C11/h11
E. Nominal = 25, Fit = F8/h7
F. Nominal = 12, Fit = H7/k6
G. Nominal = 3, Fit = H7/p6
H. Nominal = 18, Fit = H7/s6
I. Nominal = 27, Fit = H7/u6
J. Nominal = 30, Fit = N7/h6

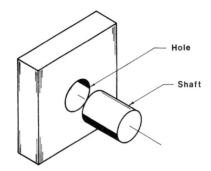

half space

3.75
6 equal
spaces

NOMINAL	HOLE		SHAFT		CLEARANCE	
	MAX	MIN	MAX	MIN	MAX	MIN
A						
B						
C						
D						
E						

|← 1.5 →|← 6.0 – 6 equal spaces →|

NOMINAL	HOLE		SHAFT		INTERFERENCE	
	MAX	MIN	MAX	MIN	MAX	MIN
F						
G						
H						
I						
J						

Use the same dimensions given above

EX8-12 INCHES

Redraw the following charts and complete them based on the following information. All values are in inches.

A. Nominal = 0.25, Fit = Class LC5
B. Nominal = 1.00, Fit = Class LC7
C. Nominal = 1.50, Fit = Class LC10
D. Nominal = 0.75, Fit = Class RC3
E. Nominal = 1.75, Fit = Class RC6
F. Nominal = .500, Fit = Class LT2
G. Nominal = 1.25, Fit = Class LT5
H. Nominal = 1.38, Fit = Class LN3
I. Nominal = 1.625, Fit = Class FN1
J. Nominal = 2.00, Fit = Class FN4

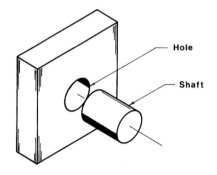

half space

NOMINAL	HOLE		SHAFT		CLEARANCE	
	MAX	MIN	MAX	MIN	MAX	MIN
A						
B						
C						
D						
E						

3.75
6 equal
spaces

1.5   6.0 – 6 equal spaces

NOMINAL	HOLE		SHAFT		INTERFERENCE	
	MAX	MIN	MAX	MIN	MAX	MIN
F						
G						
H						
I						
J						

Use the same dimensions given above

Draw the chart shown and add the appropriate values based on the dimensions and tolerances given in exercise problems EX8-13 to EX8-16.

EX8-13 MILLIMETERS

PART NO: 9–M53A

A. 20±0.1

B. 30±0.2

C. Ø20±0.05

D. 40

E. 60

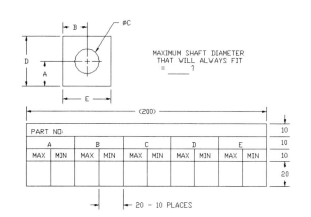

MAXIMUM SHAFT DIAMETER
THAT WILL ALWAYS FIT
= _____ ?

EX8-14 MILLIMETERS

PART NO: 9–M53B

A. 32.02
    31.97

B. 47.52
    47.50

C. Ø18 $^{+0.05}_{0}$

D. 64±0.05

E. 100±0.05

EX8-15 MILLIMETERS

PART NO: 9–E47A

A. 2.00±.02

B. 1.75±.03

C. Ø.750±.005

D. 4.00±.05

E. 3.50±.05

EX8-16 MILLIMETERS

PART NO: 9–E47B

A. 18 $^{+0}_{-0.02}$

B. 26 $^{+0}_{-0.04}$

C. Ø 24.03
      23.99

D. 52±0.04

E. 36±0.02

EX8-17 MILLIMETERS

Prepare front and top views of Parts 4A and 4B based on the given dimensions. Add tolerances to produce the stated clearances.

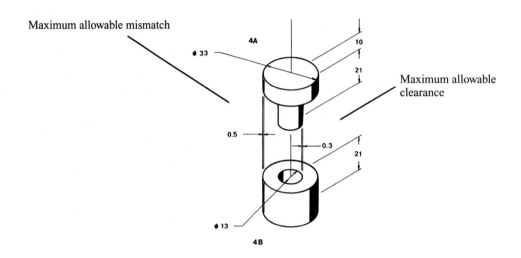

EX8-18 INCHES

Redraw parts A and B and dimensions and tolerances to meet the "UPON ASSEMBLY" requirements.

BOX, TOP

MAT'L=1.00 THK

Nominal dimensions

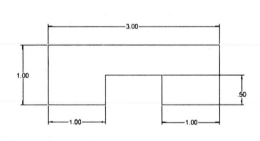

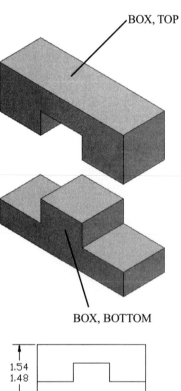

BOX, BOTTOM

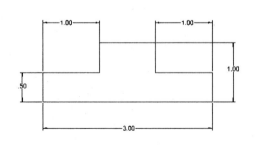

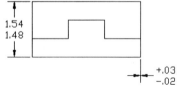

EX8-19 MILLIMETERS

Draw a front and top view of both given objects. Add dimensions and tolerances to meet the "FINAL CONDITION" requirements.

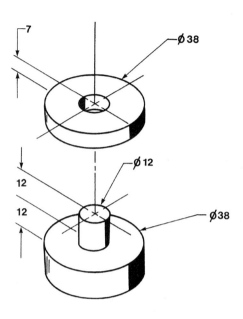

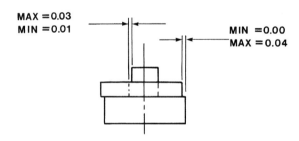

FINAL CONDITIONAL

MAX =0.03
MIN =0.01

MIN =0.00
MAX =0.04

EX8-20 MILLIMETERS

Given the following nominal sizes, dimension and tolerance parts AM-311 and AM-312 so that they always fit together regardless of orientation. Further, dimension the overall lengths of each part so that in the assembled condition they will always pass through a clearance gauge with an opening of 80.00±.02.

In the assembled condition, both parts must always pass through the clearance gauge.

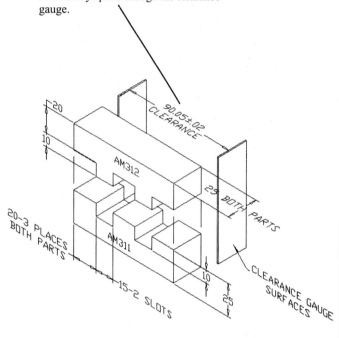

All given dimensions, except for the the clearance gauge, are nominal.

EX8-21 MILLIMETERS

Design a bracket that will support the three Ø100 wheels shown. The wheels will utilize three Ø5.00±.01 shafts attached to the bracket. The bottom of the bracket must have a minimum of 10 mm from the ground. The wall thickness of the bracket must always be at least 5 mm, and the minimum bracket opening must be at least 15 mm.

1. Prepare a front and a side view of the bracket.
2. Draw the wheels in their relative positions using phantom lines.
3. Add all appropriate dimensions and tolerances.

ALL SIZES ARE NOMINAL, UNLESS OTHERWISE STATED.

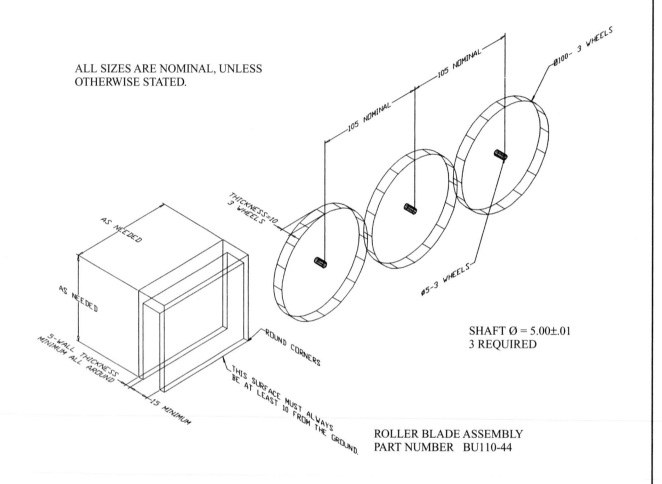

SHAFT Ø = 5.00±.01
3 REQUIRED

ROLLER BLADE ASSEMBLY
PART NUMBER BU110-44

Given a TOP and a BOTTOM part in the floating condition as shown in Figure 8-116, satisfy the requirements given in exercise problems EX8-22 to EX8-25 so that the parts always fit together regardless of orientation. Prepare drawings of each part including dimensions and tolerances.

A. Use linear tolerances
B. Use positional tolerances.

EX8-22

A. The distance between the holes' center points is 2.00 nominal.
B. The holes are Ø.375 nominal.
C. The fasteners have a tolerance of .001.
D. The holes have a tolerance of .002.
E. The minimum allowable clearance between the fasteners and the holes is .003.

EX8-23

A. The distance between the holes' center points is 80 nominal.
B. The holes are Ø12 nominal.
C. The fasteners have a tolerance of 0.05.
D. The holes have a tolerance of 0.03.
E. The minimum allowable clearance between the fasteners and the holes is 0.02.

EX8-24

A. The distance between the holes' center points is 3.50 nominal.
B. The holes are Ø.625 nominal.
C. The fasteners have a tolerance of .005.
D. The holes have a tolerance of .003.
E. The minimum allowable clearance between the fasteners and the holes is .002.

EX8-25

A. The distance between the holes' center points is 65 nominal.
B. The holes are Ø16 nominal.
C. The fasteners have a tolerance of 0.03.
D. The holes have a tolerance of 0.04.
E. The minimum allowable clearance between the fasteners and the holes is 0.03.

Given a top and a bottom part in the fixed condition as shown in Figure 8-117, satisfy the requirements given in exercise problems EX8-26 through EX8-29 so that the parts fit

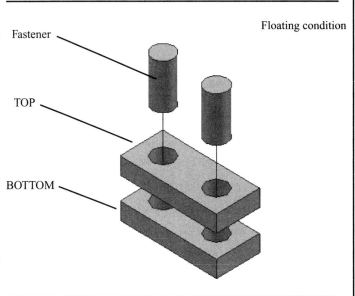

Floating condition

Figure 8-116

together regardless of orientation. Prepare drawings of each part including dimensions and tolerances.

A. Use linear tolerances
B. Use positional tolerances.

EX8-26

A. The distance between the holes' center points is 60 nominal.
B. The holes are Ø10 nominal.
C. The fasteners have a tolerance of 0.04.
D. The holes have a tolerance of .0.02.
E. The minimum allowable clearance between the fasteners and the holes is 0.02.

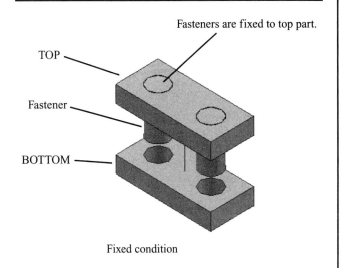

Fixed condition

Figure 8-117

EX8-27

A. The distance between the holes' center points is 3.50 nominal.
B. The holes are Ø.563 nominal.
C. The fasteners have a tolerance of .005.
D. The holes have a tolerance of .003.
E. The minimum allowable clearance between the fasteners and the holes is .002.

EX8-28

A. The distance between the holes' center points is 100 nominal.
B. The holes are Ø18 nominal.
C. The fasteners have a tolerance of 0.02.
D. The holes have a tolerance of 0.01.
E. The minimum allowable clearance between the fasteners and the holes is 0.03

EX8-29

A. The distance between the holes' center points is 1.75 nominal.
B. The holes are Ø.250 nominal.
C. The fasteners have a tolerance of .002.
D. The holes have a tolerance of .003.
E. The minimum allowable clearance between the fasteners and the holes is .001.

EX8-30

Dimension and tolerance the rotator assembly shown in Figure 8-118. Use the given dimensions as nominal and add sleeve bearings between the LINKs and both the CROSS-LINK and the PLATE. Create drawings of each part. Modify the dimensions as needed and add the appropriate tolerances. Specify the selected sleeve bearing.

CROSS-LINK P/N AM311-2, SAE 1020

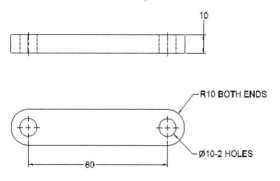

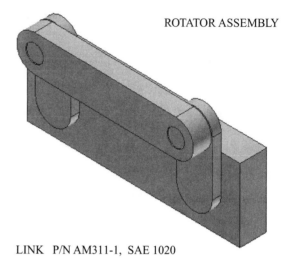

ROTATOR ASSEMBLY

LINK P/N AM311-1, SAE 1020

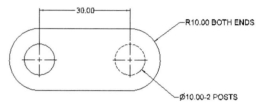

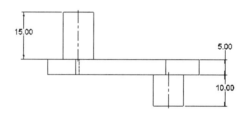

PLATE P/N AM311-3, SAE 1020

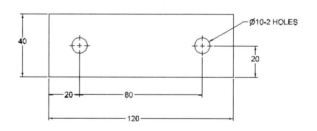

Figure 8-118

EX8-31

Dimension and tolerance the rocker assembly shown in Figure 8-119. Use the given dimensions as nominal, and add sleeve bearings between all moving parts. Create drawings of each part. Modify the dimensions as needed and add the appropriate tolerances. Specify the selected sleeve bearing.

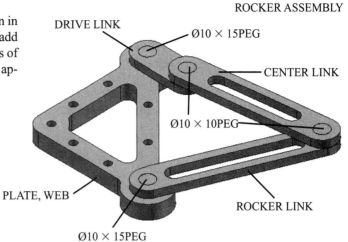

ROCKER ASSEMBLY

DRIVE LINK
Ø10 × 15PEG
CENTER LINK
Ø10 × 10PEG
PLATE, WEB
Ø10 × 15PEG
ROCKER LINK

DRIVE LINK
AM312-2
SAE 1040
5 mm THK

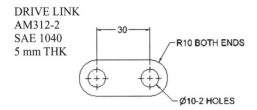

30
R10 BOTH ENDS
Ø10-2 HOLES

PLATE, WEB   AM312-1, SAE 1040, 10 mm THK

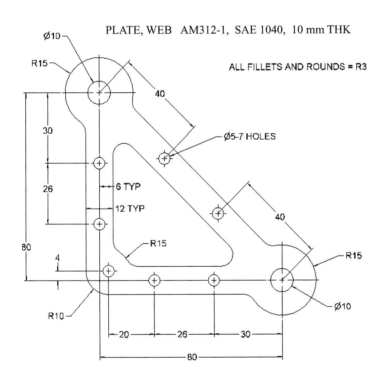

ALL FILLETS AND ROUNDS = R3

Ø10
R15
40
30
26
6 TYP
12 TYP
80
4
R10
R15
Ø5-7 HOLES
40
R15
Ø10
20
26
30
80

ROCKER LINK
AM312-4
SAE 1040
5 mm THK

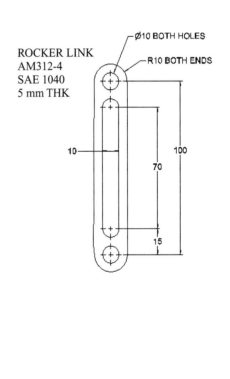

Ø10 BOTH HOLES
R10 BOTH ENDS
10
100
70
15

Ø10 × 10PEG AM312-5
SAE 1020

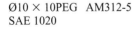
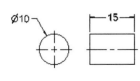

Ø10
15

CENTER LINK   AM312-3, SAE 1040, 5 mm THK

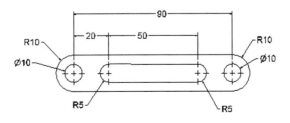

90
20
50
R10
Ø10
R10
Ø10
R5
R5

Ø10 × 15PEG AM312-6
SAE 1020

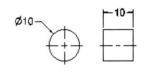

Ø10
10

Figure 8-119

EX8-32

Draw the following model, create a drawing layout with the appropriate views, and add the specified dimensions and tolerances.

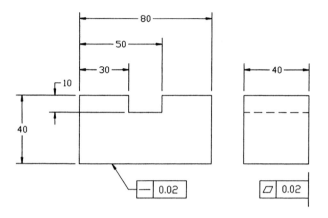

EX8-33

Redraw the following shaft, create a drawing layout with the appropriate views, and add a feature dimension and tolerance of 36±0.1 and a straightness tolerance of 0.07 about the centerline at MMC.

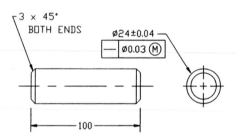

EX8-34

A. Given the shaft shown, what is the minimum hole diameter that will always accept the shaft?

B. If the minimum clearance between the shaft and a hole is equal to 0.02, and the tolerance on the hole is to be 0.6, what are the maximum and minimum diameters for the hole?

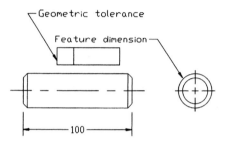

EX8-35

A. Given the shaft shown, what is the minimum hole diameter that will always accept the shaft?

B. If the minimum clearance between the shaft and a hole is equal to .005, and the tolerance on the hole is to be .007, what are the maximum and minimum diameters for the hole?

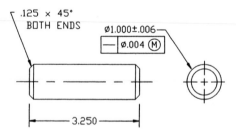

EX8-36

Draw a front and a right-side view of the object shown in Figure 8-120 and add the appropriate dimensions and tolerances based on the following information. Numbers located next to an edge line indicate the length of the edge.

A. Define surfaces A, B, and C as primary, secondary, and tertiary datums, respectively.
B. Assign a tolerance of ±0.5 to all linear dimensions.
C. Assign a feature tolerance of 12.07 − 12.00 to the protruding shaft.
D. Assign a flatness tolerance of 0.01 to surface –A–.
E. Assign a straightness tolerance of 0.03 to the protruding shaft.
F. Assign a perpendicularity tolerance to the centerline of the protruding shaft of 0.02 at MMC relative to datum –A–.

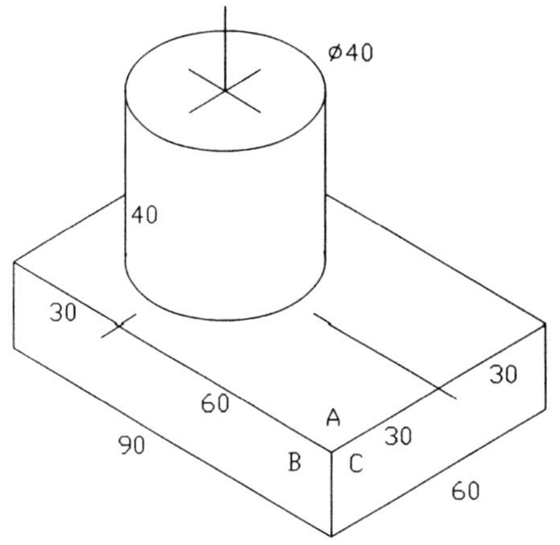

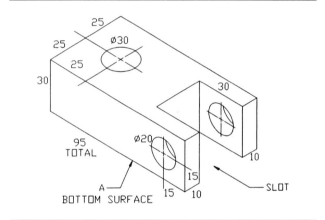

Figure 8-120

EX8-37

Draw a front and a right-side view of the object shown in Figure 8-121 and add the following dimensions and tolerances.

A. Define the bottom surface as datum –A–.
B. Assign a perpendicularity tolerance of 0.4 to both sides of the slot relative to datum –A–.
C. Assign a perpendicularity tolerance of 0.2 to the centerline of the 30 diameter hole at MMC relative to datum –A–.
D. Assign a feature tolerance of ±0.8 to all three holes.
E. Assign a parallelism tolerance of 0.2 to the common centerline between the two 20 diameter holes relative to datum –A–.
F. Assign a tolerance of ±0.5 to all linear dimensions.

Figure 8-121

EX8-38

Draw a circular front and the appropriate right-side view of the object shown in Figure 8-122 and add the following dimensions and tolerances.

A. Assign datum –A– as indicated.
B. Assign the object's longitudinal axis as datum –B–.
C. Assign the object's centerline through the slot as datum –C–.
D. Assign a tolerance of ±0.5 to all linear tolerances.
E. Assign a tolerance of ±0.5 to all circular shaped features.
F. Assign a parallelism tolerance of 0.01 to both edges of the slot.
G. Assign a perpendicularity tolerance of 0.01 to the outside edge of the protruding shaft.

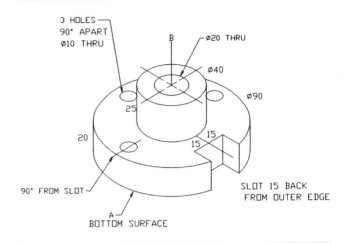

Figure 8-122

EX8-39

Given the two objects shown in Figure 8-123, draw a front and a side view of each. Assign a tolerance of ±0.5 to all linear dimensions. Assign a feature tolerance of ±0.4 to the shaft, and also assign a straightness tolerance of 0.2 to the shaft's centerline at MMC.

Tolerance the hole so that it will always accept the shaft with a minimum clearance of 0.1 and a feature tolerance of 0.2. Assign a perpendicularity tolerance of 0.05 to the centerline of the hole at MMC.

EX8-40

Given the two objects shown in Figure 8-124, draw a front and a side view of each. Assign a tolerance of ±0.005 to all linear dimensions. Assign a feature tolerance of ±0.004 to the shaft, and also assign a straightness tolerance of 0.002 to the shaft's centerline at MMC.

Tolerance the hole so that it will always accept the shaft with a minimum clearance of 0.001 and a feature tolerance of 0.002.

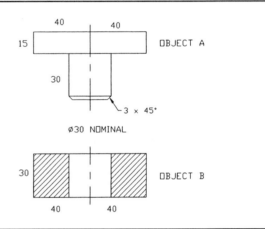

Figure 8-123

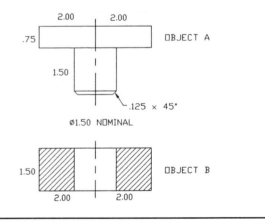

Figure 8-124

EX8-41

Draw a model of the object shown in Figure 8-125, then create a drawing layout including the specified dimensions. Add the following tolerances and specifications to the drawing.

A. Surface 1 is datum –A–.

B. Surface 2 is datum –B– and is perpendicular to datum –A– within 0.1 mm.

C. Surface 3 is datum –C– and is parallel to datum A within 0.3 mm.

D. Locate a 16-mm diameter hole in the center of the front surface that goes completely through the object. Use positional tolerances to locate the hole. Assign a positional tolerance of 0.02 at MMC perpendicular to datum –A–.

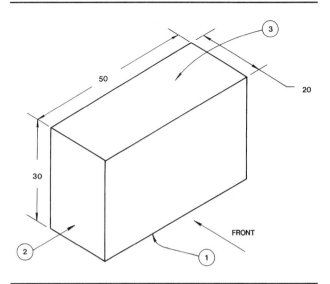

Figure 8-125

EX8-42

Draw a model of the object shown in Figure 8-126, then create a drawing layout including the specified dimensions. Add the following tolerances and specifications to the drawing.

A. Surface 1 is datum –A–.

B. Surface 2 is datum –B– and is perpendicular to datum –A– within .003 in.

C. Surface 3 is parallel to datum –A– within .005 in.

D. The cylinder's longitudinal centerline is to be straight within .001 in. at MMC.

E. Surface 2 is to have circular accuracy within .002 in.

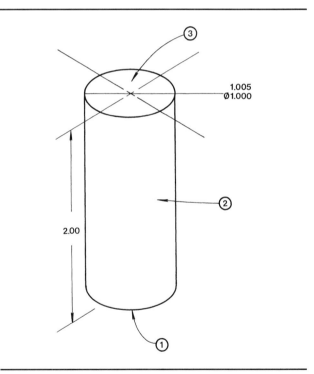

Figure 8-126

EX8-43

Draw a model of the object shown in Figure 8-127, then create a drawing layout including the specified dimensions. Add the following tolerances and specifications to the drawing.

A. Surface 1 is datum –A–.
B. Surface 4 is datum –B– and is perpendicular to datum A within 0.08 mm.
C. Surface 3 is flat within 0.03 mm.
D. Surface 5 is parallel to datum A within 0.01 mm.
E. Surface 2 has a runout tolerance of 0.2 mm relative to surface 4.
F. Surface 1 is flat within 0.02 mm.
G. The longitudinal centerline is to be straight within 0.02 at MMC and perpendicular to datum –A–.

EX8-44

Draw a model of the object shown in Figure 8-128, then create a drawing layout including the specified dimensions. Add the following tolerances and specifications to the drawing.

A. Surface 2 is datum –A–.
B. Surface 6 is perpendicular to datum –A– with .000 allowable variance at MMC but with a .002 in. MAX variance limit beyond MMC.
C. Surface 1 is parallel to datum –A– within .005.
D. Surface 4 is perpendicular to datum –A– within .004 in.

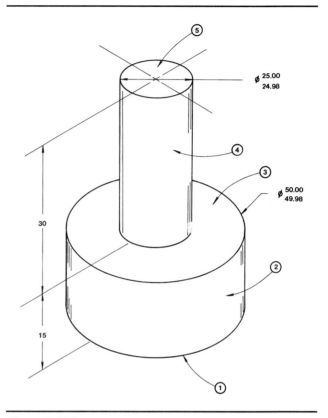

Figure 8-127

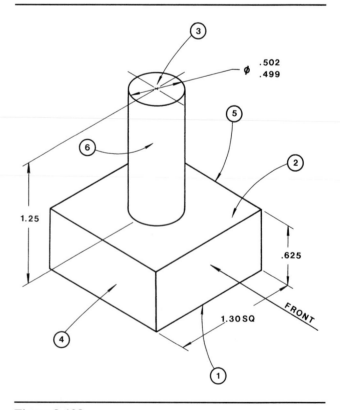

Figure 8-128

EX8-45

Draw a model of the object shown in Figure 8-129, then create a drawing layout including the specified dimensions. Add the following tolerances and specifications to the drawing.

A. Surface 1 is datum –A–.
B. Surface 2 is datum –B–.
C. The hole is located using a true position tolerance value of 0.13 mm at MMC. The true position tolerance is referenced to datums –A– and –B–.
D. Surface 1 is to be straight within 0.02 mm.
E. The bottom surface is to be parallel to datum –A– within 0.03 mm.

EX8-46

Draw a model of the object shown in Figure 8-130, then create a drawing layout including the specified dimensions. Add the following tolerances and specifications to the drawing.

A. Surface 1 is datum –A–.
B. Surface 2 is datum –B–.
C. Surface 3 is perpendicular to surface 2 within 0.02 mm.
D. The four holes are to be located using a positional tolerance of 0.07 mm at MMC referenced to datums –A– and –B–.
E. The centerlines of the holes are to be straight within 0.01 mm at MMC.

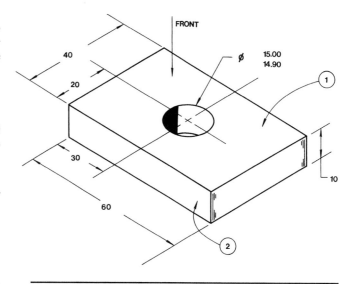

Figure 8-129

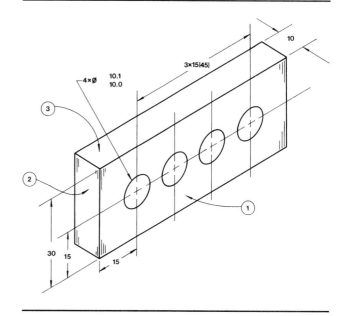

Figure 8-130

EX8-47

Draw a model of the object shown in Figure 8-131 then create a drawing layout including the specified dimensions. Add the following tolerances and specifications to the drawing.

A. Surface 1 has a dimension of .378–.375 in. and is datum –A–. The surface has a dual primary runout with datum –B– to within .005 in. The runout is total.

B. Surface 2 has a dimension of 1.505–1.495 in. Its runout relative to the dual primary datums –A– and –B– is .008 in. The runout is total.

C. Surface 3 has a dimension of 1.000 ±.005 and has no geometric tolerance.

D. Surface 4 has no circular dimension but has a total runout tolerance of .006 in. relative to the dual datums –A– and –B–.

E. Surface 5 has a dimension of .500–.495 in. and is datum –B–. It has a dual primary runout with datum –A– within .005 in. The runout is total.

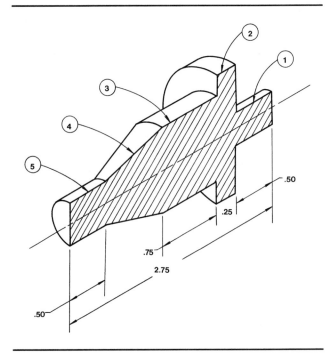

Figure 8-131

EX8-48

Draw a model of the object shown in Figure 8-132, then create a drawing layout including the specified dimensions. Add the following tolerances and specifications to the drawing.

A. Hole 1 is datum –A–.

B. Hole 2 is to have its circular centerline parallel to datum –A– within 0.2 mm at MMC when datum –A– is at MMC.

C. Assign a positional tolerance of 0.01 to each hole's centerline at MMC.

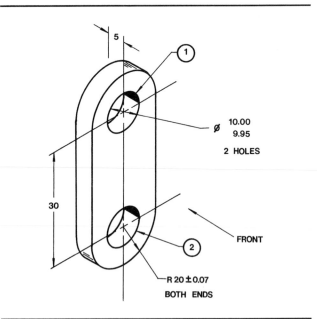

Figure 8-132

EX8-49

Draw a model of the object shown in Figure 8-133, then create a drawing layout including the specified dimensions. Add the following tolerances and specifications to the drawing.

A. Surface 1 is datum –A–.
B. Surface 2 is datum –B–.
C. The six holes have a diameter range of .502–.499 in. and are to be located using positional tolerances so that their centerlines are within .005 in. at MMC relative to datums –A– and –B–.
D. The back surface is to be parallel to datum –A– within .002 in.

EX8-50

Draw a model of the object shown in Figure 8-134, then create a drawing layout including the specified dimensions. Add the following tolerances and specifications to the drawing.

A. Surface 1 is datum –A–.
B. Hole 2 is datum –B–.
C. The eight holes labeled 3 have diameters of 8.4–8.3 mm with a positional tolerance of 0.15 mm at MMC relative to datums –A– and –B–. Also, the eight holes are to be counterbored to a diameter of 14.6–14.4 mm and to a depth of 5.0 mm.
D. The large center hole is to have a straightness tolerance of 0.2 at MMC about its centerline.

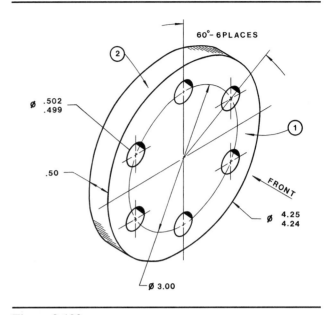

Figure 8-133

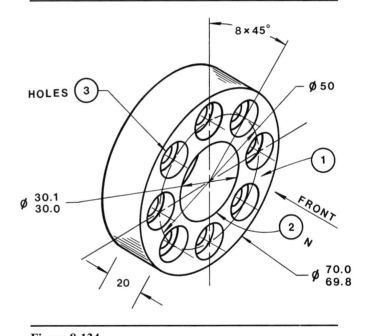

Figure 8-134

EX8-51

Draw a model of the object shown in Figure 8-135, then create a drawing layout including the specified dimensions. Add the following tolerances and specifications to the drawing.

A. Surface 1 is datum –A–.
B. Surface 2 is datum –B–.
C. Surface 3 is datum –C–.
D. The four holes labeled 4 have a dimension and tolerance of 8 +0.3, −0 mm. The holes are to be located using a positional tolerance of 0.05 mm at MMC relative to datums –A–, –B–, and –C–.
E. The six holes labeled 5 have a dimension and tolerance of 6 +0.2, −0 mm. The holes are to be located using a positional tolerance of 0.01 mm at MMC relative to datums –A–, –B–, and –C–.

EX8-52

The objects on page 349, labeled A and B, are to be toleranced using four different tolerances as shown. Redraw the charts shown in Figure 8-136 and list the appropriate allowable tolerance for "as measured" increments of 0.1 mm or .001 in. Also include the appropriate geometric tolerance drawing called out above each chart.

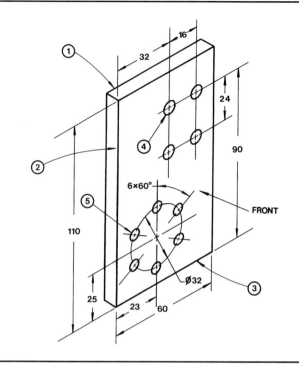

Figure 8-135

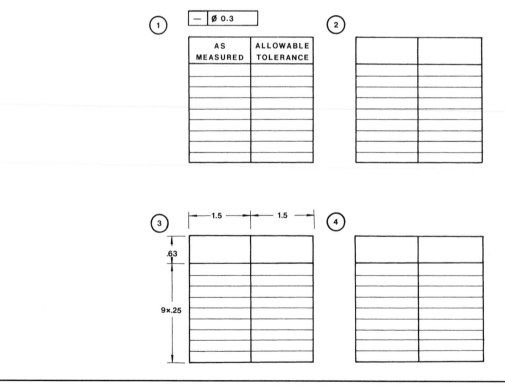

Figure 8-136

A. MILLIMETERS

1. $\boxed{-\ \boxed{\varnothing 0.3}}$

2. $\boxed{-\ \boxed{\varnothing 0.3 \textcircled{M}}}$

3. $\boxed{-\ \boxed{\varnothing 0.0 \textcircled{M}}}$

4. $\boxed{-\ \boxed{\varnothing 0.0 \textcircled{M}}\ \boxed{\varnothing 0.04\ \text{MAX}}}$

20.5
⌀19.7

B. INCHES

1. $\boxed{\bigcirc\ \boxed{\varnothing .003}}$

2. $\boxed{\bigcirc\ \boxed{\varnothing .003 \textcircled{M}}}$

3. $\boxed{\bigcirc\ \boxed{\varnothing .000 \textcircled{M}}}$

4. $\boxed{\bigcirc\ \boxed{\varnothing .000 \textcircled{M}}\ \boxed{\varnothing .000\ \text{MAX}}}$

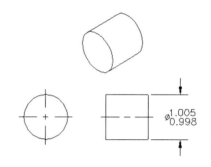

$\varnothing {}^{1.005}_{0.998}$

EX8-53 MILLIMETERS

Assume that there are two copies of the part in Figure 8-137 and that these parts are to be joined together using four fasteners in the floating condition. Draw front and top views of the object, including dimensions and tolerances. Add the following tolerances and specifications to the drawing, then draw front and top views of a shaft that can be used to join the two objects. The shaft should be able to fit into any of the four holes.

A. Surface 1 is datum –A–.
B. Surface 2 is datum –B–.
C. Surface 3 is perpendicular to surface 2 within 0.02 mm.
D. Specify the positional tolerance for the four holes applied at MMC.
E. The centerlines of the holes are to be straight within 0.01 mm at MMC.
F. The clearance between the shafts and the holes is to be 0.05 minimum and 0.10 maximum.

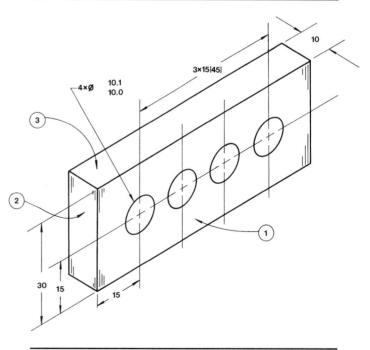

Figure 8-137

EX8-54

Dimension and tolerance parts 1 and 2 of Figure 8-138 so that part 1 always fits into part 2 with a minimum clearance of .005 in. The tolerance for part 1's outer matching surface is .006 in.

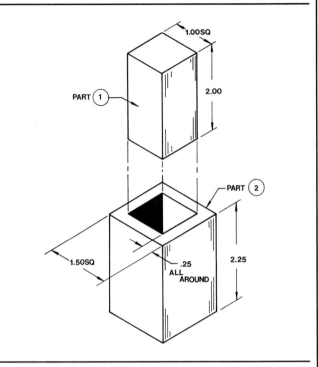

Figure 8-138

EX8-55

Dimension and tolerance parts 1 and 2 of Figure 8-139 so that part 1 always fits into part 2 with a minimum clearance of 0.03 mm. The tolerance for part 1's diameter is 0.05 mm. Take into account the fact that the interface is long relative to the diameters.

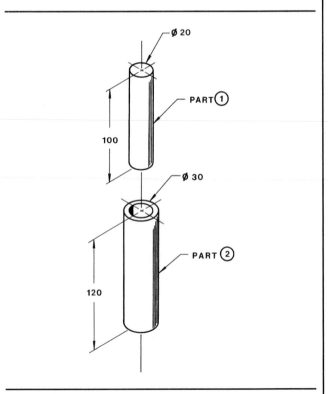

Figure 8-139

EX8-56 INCHES

Assume that there are two copies of the part in Figure 8-140 and that these parts are to be joined together using six fasteners in the floating condition. Draw front and top views of the object, including dimensions and tolerances. Add the following tolerances and specifications to the drawing, then draw front and top views of a shaft that can be used to join the two objects. The shaft should be able to fit into any of the six holes.

A. Surface 1 is datum –A–.
B. Surface 2 is round within .003.
C. Specify the positional tolerance for the six holes applied at MMC.
D. The clearance between the shafts and the holes is to be .001 minimum and .003 maximum.

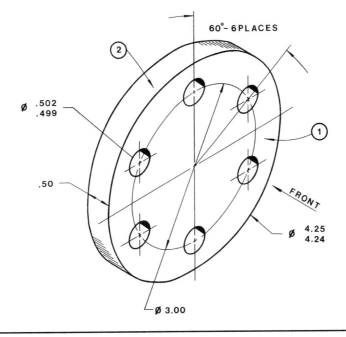

Figure 8-140

EX8-57 MILLIMETERS

The assembly shown is made from parts defined in Section 5-7.

1. Draw an exploded assembly drawing.
2. Draw a BOM.

3. Use the drawing layout mode and draw orthographic views of each part. Include dimensions and geometric tolerances. The pegs should have a minimum clearance of 0.02. Select appropriate tolerances and define them for each hole using positional tolerance.

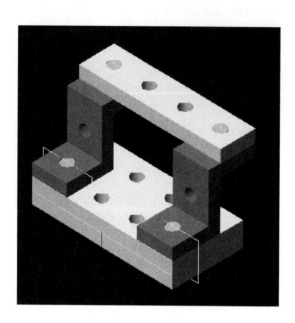

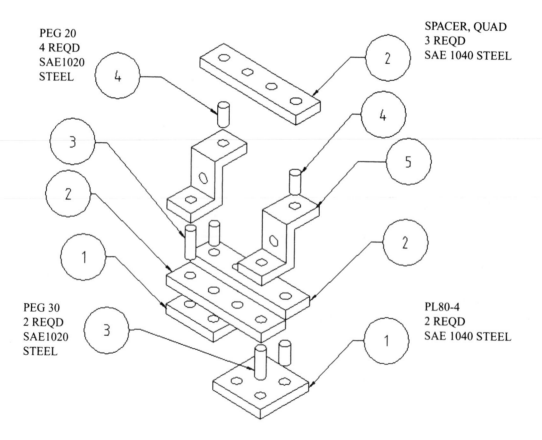

PEG 20
4 REQD
SAE1020
STEEL

SPACER, QUAD
3 REQD
SAE 1040 STEEL

PEG 30
2 REQD
SAE1020
STEEL

PL80-4
2 REQD
SAE 1040 STEEL

CHAPTER 9

Bearings and Shafts

9-1 INTRODUCTION

This chapter discusses bearings, shafts, and support structures. There are many different types of bearings. This chapter concentrates on just two types: plain cylindrical (sleeve) and ball. The chapter reviews manufacturers' specifications, and it explains how to use bearing tolerances to interface with shafts and support plates, and how to size bearings based on given force and wear values. The chapter also explains how to work with manufacturers' catalogs and web sites.

9-2 SLEEVE BEARINGS

Sleeve bearings are hollow cylinders made from either a soft material such as bronze or a material with a very low coefficient of friction such as Teflon. Figure 9-1 shows two plain cylindrical bearings drawn using Inventor. One of the bearings shown in Figure 9–1 is a flanged bearing. Flanges are helpful for when mounting bearings in support plates.

Typically rotating shafts are inserted into bearings, and the bearings in turn are mounted into a supporting structure. Shafts rotate within sleeve bearings, so the clearance between the shaft and the bearing is critical.

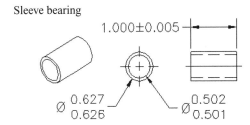

Sleeve bearing

Flanged sleeve bearing

Figure 9-1

CLEARANCES—Sleeve Bearings

Nominal Ø	Below 600 RPM		Above 600 RPM	
	Min	Max	Min	Max
.250	.0004	.0015	.0007	.0024
.500	.0005	.0018	.0009	.0030
.750	.0006	.0022	.0012	.0035
1.000	.0007	.0025	.0014	.0040
1.250	.0010	.0027	.0016	.0046
1.500	.0012	.0030	.0018	.0050
1.750	.0013	.0033	.0020	.0053
2.000	.0014	.0036	.0023	.0056

Figure 9-2

Sleeve bearing and shaft clearance

The hole in the sleeve bearing, called a *bore,* shown in Figure 9-1 has a limit tolerance of 0.502 and 0.501 for a total tolerance of 0.001. The tolerances for the shaft must allow a clearance between the bearing and the shaft so that the shaft can rotate freely, but the clearance must not be so great as to allow the rotating shaft to create excessive vibrations. Figure 9-2 shows a listing of sleeve bearing clearances as related to rotation speed and bearing diameters. If the sleeve bearing presented in Figure 9-1 had a shaft inserted that was rotating at under 600 RPM, what would be the shaft tolerances?

The tolerance range as specified in Figure 9-2 for a nominal bearing bore of .5000 is a minimum clearance of .0005 and a maximum clearance of .0018. The minimum clearance occurs when the bore diameter is at its minimum size and the shaft diameter is at its maximum size; therefore, given the minimum bore diameter of .5010 and the minimum clearance of .0005, the following equation may be completed to determine the maximum shaft diameter:

```
  .5010   Bore diameter—min
−.0005    Clearance—min
  .5005   Shaft diameter—max
```

The maximum shaft diameter is determined by subtracting the minimum clearance from the minimum bore diameter.

The minimum shaft diameter is determined by subtracting the maximum clearance from the maximum bore diameter:

```
  .5020   Bore diameter—max
−.0018    Clearance—max
  .5002   Shaft diameter—min
```

The limit tolerances for the shaft are therefore Ø.5005/.5002, for a total tolerance of .0003.

Standard shafts

Many manufacturers make shafts with tolerances designed to match their gear and bearing products. For example, W.M. Berg, Inc. makes a shaft with a nominal .500 diameter. The actual shaft dimensions are Ø.4997, +.0000, −.0002.

Figure 9-3 shows Mechanical desktop drawings of two W. M. Berg parts. The bore diameter of the bearing is Ø.5020/.5010, and the diameter of the shaft is Ø.4997, +.0000, −.0002. This means that the maximum clearance is

```
  .5020   Bore diameter—max
−.4995    Shaft diameter—min
  .0025   Clearance—max
```

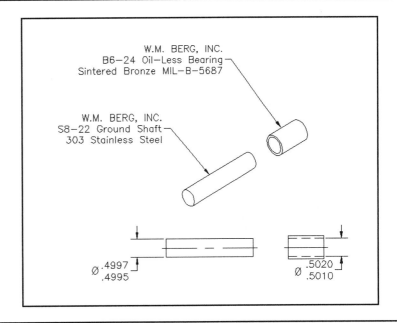

Figure 9-3

and the minimum clearance is:

> .5010 Bore diameter—min
> −.4997 Shaft diameter—max
> .0013 Clearance—min

The maximum and minimum values are within the range specified in Figure 9-2 for RPMs greater than 600.

Pressure velocity (*PV*) factor

A bearing's *PV* factor is a measure of a bearing's capability given a load and an operating speed. The *PV* factor is defined as follows:

$$PV = \frac{.262Fn}{L}$$

$$PV = \frac{F}{A}$$

where

> A = bearing ID × bearing length (sq in.)
> V = .262 × RPM × Øshaft (ft/min)
> F = load (lb)
> L = bearing length (in.)
> n = speed (RPM)

Figure 9-4 lists *PV* factors from Boston Gear. The values are typical of those presented by bearing manufacturers and will be used throughout the book.

9-3 SAMPLE PROBLEM SP9-1

A sleeve bearing made from Nyloil has a .500-in. inside diameter and is a 1.000 in. long. It is operating at 200 RPM under a load of 300 lb. Is the bearing acceptable for this application?

$$PV = \frac{.262(300)(200)}{1.000} = 15,720$$

PV Factors

Material	PV_{max}	P_{max}	V_{max}
BOST-BRONZ	50,000	2,000	1,200
Nyloil	16,000	2,000	400
UHMW-PE with internal wear strip	4,000	1,400	100

Figure 9-4 Courtesy of Boston Gear

Wear Rate Factor

Material	K Value
Delrin	50×10^{-10}
Teflon-filled nylon	13×10^{-10}
Nylon	12×10^{-10}

Figure 9-5

The calculated PV value of 15,720 indicates that the bearing is acceptable as it is within the specified 16,000 limit.

It is good practice to check the individual P and V values to ensure they do not exceed their individual limits.

$$V = .262(300)(.5) = 26.2 \text{ ft/min}$$

The Vmax limit value is 400 ft/min, so the calculated V is well within this limit.

$$P = \frac{PV}{V} = \frac{15,720}{26.2} = 600 \text{ lb/sq in. (psi)}$$

The P_{max} limit value is 2000, so the 600 value is within the designated limit.

The bearing is acceptable for the specified application.

Lubrication

Lubrication is critical when designing with bearings. Many designs encase bearings with a reservoir of oil. Others specify frequent lubrication. As a shaft turns within a sleeve bearing it theoretically rides on a cushion of oil, meaning there is very little friction and very little wear. This is called *full film* lubrication. There is wear during the initial startup and if, for any reason, the lubrication is removed.

Wear

For nonmetallic, nonlubricated bearings wear is determined by

$$t = K(PVT)$$

where

$V = .262 \times RPM \times DIA \text{ (ft/min)}$

$t = \text{wear (in.)}$

$K = \text{wear rate factor (See Figure 9-5.)}$

$T = \text{running time (hours)}$

9-4 SAMPLE PROBLEM SP9-2

A sleeve bearing made from nylon is operating at 400 RPM with a load of 200 lb. The bearing has a bore of .500 in. and is 1.000 in. long. It has been determined that the maximum allowable wear is 0.05 in. What is the estimated running time?

$$T = \frac{t}{KPV} = \frac{0.5}{12 \times 10^{-10}(400)(52.4)} = 1987 \text{ hours}$$

$$V = .262(\emptyset \text{shaft})(RPM) = 262(.5)(400)$$
$$= 52.4 \text{ ft/min}$$

$$P = \frac{F}{A} = \frac{200}{(.500)(1.000)} = 400 \text{ psi}$$

The bearing can operate for approximately 1987 hours before the wear limit is reached.

Selecting sleeve bearings on the web

Many major bearing manufacturers' now operate web sites that allow a designer to select sleeve bearings. Figure 9-6 shows a page from the Boston Gear web site (www.bostgear.com). Bearing size requirements may be entered in the text boxes at the top of the page. Figure 9-7 shows the same page with size specifications entered. Bearing B810-8 matches the entered specifications. Clicking on the Spec button leads to a detailed definition of the selected bearing. See Figure 9-8. Note that the maximum PV, P, and V values are listed. If the appropriate software is available, a drawing of the bearing can be downloaded.

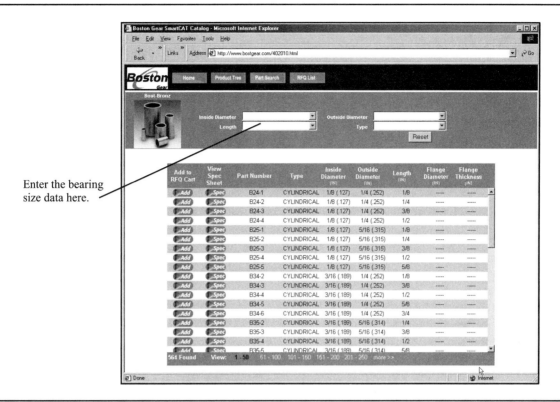

Enter the bearing size data here.

Figure 9-6   Courtesy of Boston Gear

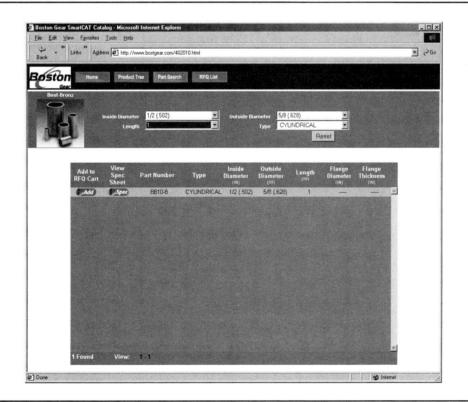

Figure 9-7   Courtesy of Boston Gear

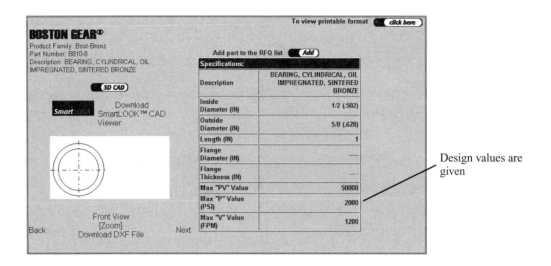

Figure 9-8   Courtesy of Boston Gear

A page from the W.M. Berg web site

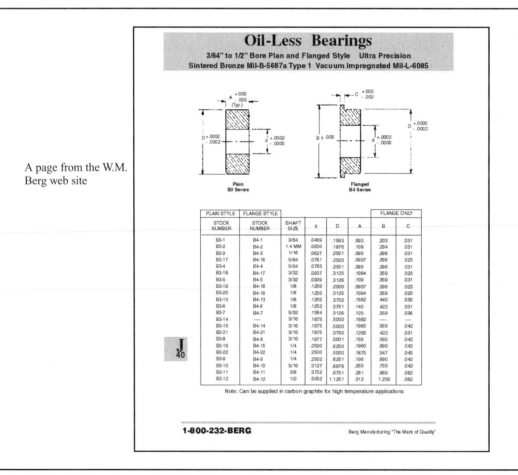

Figure 9-9   Courtesy of W.M. Berg.

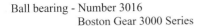

Ball bearing - Number 3016
Boston Gear 3000 Series

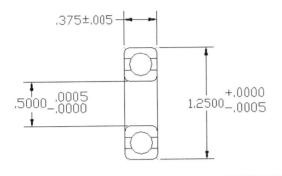

.375±.005

$.5000^{+.0005}_{-.0000}$

$1.2500^{+.0000}_{-.0005}$

Figure 9-10

Nominal Size Range	Limits of Interference	Class LN1	
		Standard Limits	
Over To		Hole H6	Shaft n5
0.40–0.71	0 / 0.8	+0.4 / 0	+0.8 / +0.4
O.D. 1.19–1.97	0 / 1.1	+0.6 / 0	+1.1 / +0.6

Figure 9-11

Figure 9-9 shows another sleeve bearing web page. This page is from W.M. Berg, Inc. (www.wmberg. com). The page lists a variety of available bearings along with their sizes and tolerances.

9-5 BALL BEARINGS

Ball bearings are capable of operating at higher speeds and carrying greater loads than sleeve bearings. They are also more expensive and require more area for installation.

Shafts are mounted into ball bearings using a light interference fit, and the bearing itself is mounted into its support structure, called a *housing,* also using a light interference fit. Heavy fits, that is, tolerances that create large interferences, could damage the bearings.

Class LN1 fits

Figure 9-10 shows the tolerances for a typical ball bearing. Determine the shaft and housing dimensions and tolerances using a Class LN1 fit for both.

Figure 9-11 shows the tolerance specifications for a Class LN1 fit. Additional sizes can be found in the complete table in the appendix.

The hole tolerance specified for Class LN1 is 0.0004. This differs from the bearing bore tolerance of .0005 by .0001. This type of variance is common when dealing with exisiting parts. It is important to maintain the limits of the tolerance. The lower tolerance limit is 0, so the lower limit of the shaft is defined as .5005 to match the bore limit. The upper limit of tolerance is +0.8, so the upper limit of the shaft is .5008. See Figure 9-12.

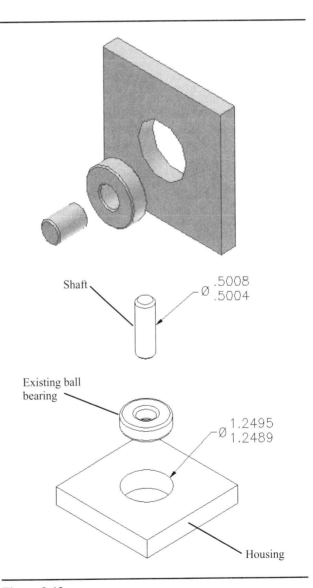

Shaft — $\varnothing {.5008 \atop .5004}$

Existing ball bearing — $\varnothing {1.2495 \atop 1.2489}$

Housing

Figure 9-12

| Bearing Bore | | Stationary Shaft Diameter | | Bearing O.D. | | Stationary Shaft Diameter | |
Max	Min	Max	Min	Max	Min	Max	Min
.2500	.2495	.2495	.2490	.6875	.6870	.6880	.6875
.3125	.3120	.3120	.3115	.8750	.8745	.8755	.8750
.3750	.3745	.3745	.3740	.9063	.9058	.9068	.9063
.4375	.4370	.4370	.4365	1.1250	1.1245	1.1255	1.1250
.5000	.4995	.4995	.4990	1.3750	1.3745	1.3755	1.3750
.6250	.6245	.6245	.6240	1.625	1.6245	1.6258	1.6250
.7500	7495	.7495	.7490	1.7500	1.7495	1.7508	1.7500
.8750	.8745	.8745	.8740	2.0000	1.994	2.0010	2.0000
1.0000	.9995	.9995	.9990				

Some possible tolerances between bore and shaft, and between bearing O.D. and housing

Figure 9-13 Courtesy of Boston Gear

The O.D. of the bearing is fitted into a hole in the support housing. For purposes of fit calculations the bearing becomes the shaft. A Class LN1 fit calls for a minimum clearance of 0, so the largest hole diameter is set equal to the smallest bearing O.D.: 1.2495. The maximum limit of tolerance is 1.1000, so the smallest hole diameter is 1.2489.

Manufacturers' specifications

Bearing manufacturers usually recommend tolerances for use with their bearings. They often also manufacture shafts designed to match their bearing requirements. Figure 9-13 shows a listing of tolerances recommended by Boston Gear. The maximum interference between the bearing's bore and shaft for a .5000 nominal bore is .0010, or .0002 greater than that specified for an LN1 tolerance. The maximum interference between the bearing's 1.1250 nominal O.D. and a housing is .0010, or .0001 less than is specified for an LN1 fit. These differences are, except for extreme cases, negligible; however, when using products from a manufacturer it is usually best to consider matching products from the same manufacturer.

9-6 SHAFTS

Shafts are made from hard materials with a smooth surface finish. Figure 9-14 shows a shaft manufacturered by W.M. Berg. The surface finish is specified as 12. Surface

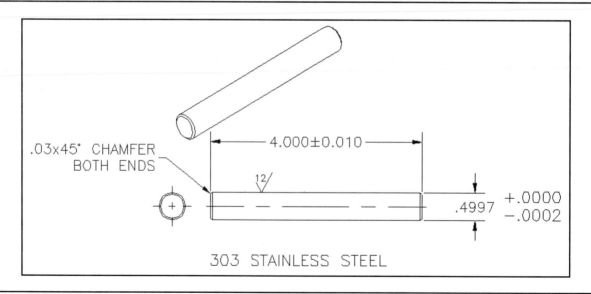

Figure 9-14 Courtesy of W. M. Berg.

finishes were defined in Section 8-24. The metric equivalent to a surface finish of 12 μin. is approximately 0.3 μm.

9-7 SHEAR AND MOMENT DIAGRAMS

Shear and moment diagrams are graphic representations of the shear and moment forces acting on a shaft. In this section, only simply supported shafts—shafts supported at both ends—with point loads will be considered. The values derived from the diagrams are used to calculate a shaft's minimum diameter, deflection, and critical speed.

Figure 9-15 shows a beam with a 32.5-lb load. The supporting forces at each end are labeled R1 and R2. The left end of the shaft is labeled A. The shaft is in a *static* condition, that is, it is not moving. All the forces are balanced. The forces R1 and R2 support the 32.5- lb load, but R1 does not equal R2. The forces R1 and R2 are determined by first calculating the moments about A, the left end of the shaft. A *moment* is a force times a distance. Again, because the shaft is in a static condition, the sum of the moments is equal to 0.

$$\sum M_A = 0$$

Considering the 32.5-lb force and the supporting force R2,

$$\sum M_A = 0 = 2.63(32.5) - 3.88R2$$

The 32.5-lb force and the force R2 act in opposite directions, so they are assigned opposite signs.

Solution of the equation yields R2 = 22.0 lb.

The supporting forces R1 and R2 must equal the 32.5-lb load:

R1 + R2 = 32.5

R1 + 22.0 = 32.5

so

R1 = 10.5 lb

To draw a shear diagram

A shear diagram shows the pound forces acting on a shaft. See Figure 9-16. The 10.5-lb force at R1 is represented by a horizontal line drawn above the shaft. A scale of 10 lb = .25 in. was selected. Any scale may be used as long as it is applied consistently and presents an acceptable graphic representation of the forces involved.

The horizontal line representing R1 continues across the shaft to the 32.5-lb load. The 32.5-lb force acts downward to a distance of 22.0 lb. A horizontal line representing the 22.0 lb is drawn below the shaft.

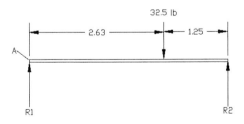

A simply supported beam.

Figure 9-15

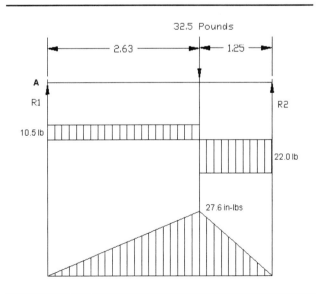

Figure 9-16

To calculate the maximum moment

Moments are measured in inch-pounds. The moments in a shaft are calculated by starting at the left end and working across the shaft to the right. The moment at the extreme left end of the shaft is

$$M_0 = (10.5)(0) = 0$$

The subscript 0 indicates the distance the moment was taken from the left end of the shaft. Forces to the right of the selected point are not considered.

At a point 2.63 in. from the left end of the shaft,

$$M_{2.63} = (10.5)(2.63) - 32.5(0)$$

$$= 27.6 \text{ in.-lb}$$

The moment at the extreme right end of the shaft is

$$M_{3.8} = (10.5)(3.88) - (32.5)(1.25) \approx 0$$

The results are not exactly equal to 0 because of round-off error.

The maximum moment of 27.6 in.-lb is used to create the moment diagram.

For two loads

Figure 9-17 shows a simply supported shaft that has two point loads. The values for the shear and moment diagrams are determined as follows. Point A is the left end of the shaft.

$$\Sigma M_A = 0 = 3(22) + 7(37) - 12R2$$

$$R2 = 27 \text{ lb}$$

The supporting forces must equal the load.

$$R1 + R2 = 22 + 37$$

$$R1 + 27 = 59$$

$$R1 = 32 \text{ lb}$$

The shear diagram starts from the left end of the shaft with the horizontal line representing 32 lb drawn above the shaft line. The horizontal line to the right of the 22 lb load represents 10 lb, and the horizontal line to the right of the 37-lb load represents –27 lb which is exactly equal to R2.

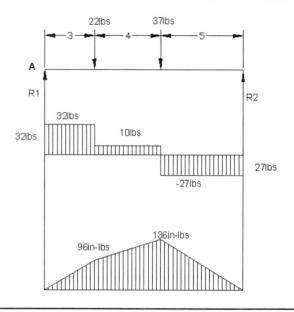

Figure 9-17

The moments are calculated starting at the left end of the shaft as follows:

$$M_0 = 0(32) = 0$$

$$M_3 = 3(32) - 0(22) = 96 \text{ in.-lb}$$

$$M_7 = 7(32) - 4(22) - 0(37) = 136 \text{ in.-lb}$$

$$M_{12} = 12(32) - 9(22) - 5(37) = 0$$

Figure 9-17 shows the completed shear and bending moment diagram.

Metric values

Metric units specify shear loads in newtons (N) or kilonewtons (kN) and moments in newton-meters (N-m) or kilonewton-meters (kN-m). Centimeters or millimeters may also be used depending on the size of the values.

Figure 9-18 shows a shear and moment diagram calculated using kN and kN-mm values. The calculations are as follows:

$$\Sigma M_A = 0 = 40(230) - 66R2$$

$$R2 = 139.4 \text{ kN}$$

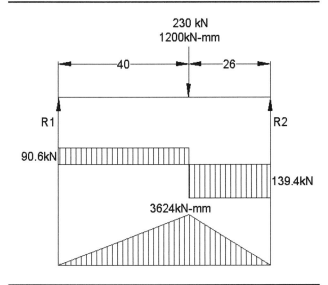

Figure 9-18

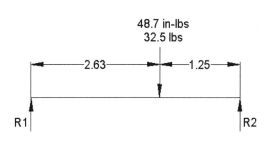

Figure 9-19

To find R1:

$$R1 + R2 = 230.0 \text{ KN}$$
$$R1 + 139.4 = 230.0 \text{ KN}$$
$$R1 = 90.6 \text{ KN}$$

The maximum moment calculation is

$$M_{40} = 40(90.6)$$
$$= 3624 \text{ KN-mm}$$

9-8 MINIMUM SHAFT DIAMETERS

The minimum diameter of a shaft is determined by the forces applied to the shaft. There are two principal forces that affect shaft diameter size: moment and torsional. Figure 9-19 shows a simply supported shaft with a single point load and a torsional load. This is the same shaft that was presented in Figure 9-15 but with the addition of a torsional load.

Maximum torque

The maximum torque acting on a shaft is a combination of the bending moment and the torsional load. These forces are assumed to be acting at 90° to each other, so the Pythagorean theorem can be used to determine the resulting maximum torque (T_{max}):

$$T_{max} = \sqrt{M^2 + T^2}$$
$$= \sqrt{(27.6^2) + (48.7^2)}$$
$$= 55.98 \text{ in.-lb}$$

The 27.6-in.-lb value came from the shear and moment diagram shown in Figure 9-16.

The maximum torque value can be used to determine the minimum shaft diameter using the following equation:

$$d = \sqrt[3]{\frac{5.1 T_{max}}{T_{allowable}}}$$

where

$$d = \text{minimum shaft diameter (in.)}$$
$$T_{max} = \text{maximum torque (in.-lb)}$$
$$T_{allowable} = \text{allowable torsional shearing stress (psi)}$$

For purposes of this book and its examples and exercise problems, the allowable torsional shearing stress for steel is 6000 psi, and 12,000 psi for 303 stainless steel.

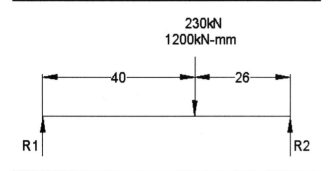

230kN
1200kN-mm

←—40—→|←—26—→

R1   R2

Figure 9-20

If the shaft shown in Figure Figure 9-19 is made from steel with an allowable torsional shearing stress of 6000 psi, the diameter calculation is as follows:

$$d = \sqrt[3]{\frac{5.1\,(55.98)}{6000}}$$
$$= .363 \text{ in.}$$

Standard sizes

Shafts are generally manufactured in .125-in. increments, so the closest standard shaft diameter for the example problem would be specified as .375 in. nominal. Shafts can be manufactured to any size specification, but standard sizes are preferred, as they can more easily be matched to standard bore diameters for bearings, gears, and pulleys.

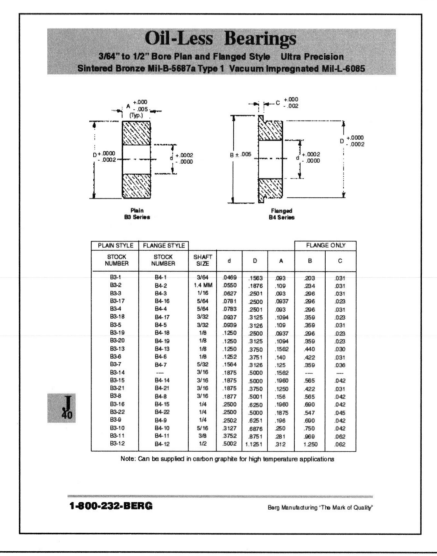

Figure 9-21 A sample page from W.M. Berg, Inc. Courtesy of W.M. Berg.

Metric values

The preceding equation can also be applied to metric values. Allowable torsional shearing stress is measured in newtons per square millimeter, and the moment is measured in newton-millimeters. The allowable stress for power-transmitting shafts is generally 28 N/mm^2.

Figure 9-20 shows the same shaft that was presented in Figure 9-18.

The maximum torque is

$$M_{40} = 40(90.6) = 3624 \text{ kN-mm}$$

The minimum shaft diameter is

$$d = \sqrt[3]{\frac{5.1(3{,}817{,}500)}{413{,}700}}$$
$$= 3.61 \text{ mm.}$$

Figure 9-21 shows a page from the W.M. Berg catalog as found on the web at www.wmberg.com

9-9 SHOCK FACTORS

Shock factors are used to account for sudden loads or impacts. Figure 9-22 shows a listing of shock factors. The shaft loads are multiplied by the shock factors to determine the torque values used for calculating the maximum torque. The K_m factors are applied to moments; the K_t factors are applied to torsional loads.

If the shaft loading for Figure 9-19 was further subjected to minor shocks and sudden loads for both moment and torsional loads, the resulting torque would be as follows:

$$T_{max} = \sqrt{1.7(M)^2 + 1.3(T)^2}$$
$$= \sqrt{1.7(27.6^2) + 1.3(48.7^2)}$$
$$= 66.16 \text{ in.-lb}$$

Shock Factors

Type of Load	K_m	K_τ
Gradually applied load	1.0	1.0
Minor shocks, sudden loads	1.7	1.3
Heavy shocks, sudden heavy loads	2.5	2.3

Figure 9-22

Use of this value to calculate the minimum shaft diameter yields:

$$d = \sqrt[3]{\frac{5.1(66.16)}{6000}}$$
$$= .38 \text{ in.}$$

The shock values thus increase the minimum shaft diameter from .36 in to .38 in.

9-10 SHAFT DEFLECTION

Figure 9-23 shows a listing of formulas used to calculate shaft deflection

where:

y = shaft deflection (in.)

F = load (lb)

L = shaft length

E = modulus of elasticity (psi)

I = moment of inertia (in.)

Formulas for shaft deflection (y)

Single load centered on shaft

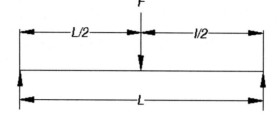

$$y = \frac{FL^3}{48EI}$$

Single load not centered on shaft

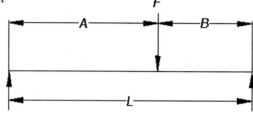

$$y = \frac{FA^2B^2}{3EIL}$$

Two loads randomly located on the shaft

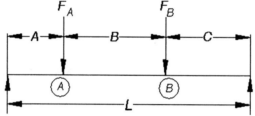

Deflection at Ⓐ caused by the load at Ⓐ

$$y = \frac{F_A C^2 (B+C)^2}{3EIL}$$

Deflection at Ⓑ caused by the load at Ⓑ

$$y = \frac{F_B C^2 (A+B)^2}{3EIL}$$

Deflection at Ⓐ caused by the load at Ⓑ

$$y = \frac{F_B AC}{6EIL} (L^2 - C^2 - A^2)$$

Deflection at Ⓑ caused by the load at Ⓐ

$$y = \frac{F_A AC}{6EIL} (L^2 - C^2 - A^2)$$

Figure 9-23

Moment of inertia

A *moment of inertia* is a measure of a shape's ability to resist any change to the shape. The value of the moment of inertia depends on the shape. Figure 9-24 shows four common shaft shapes and the formulas used to calculate their moments of inertia.

Modulus of elasticity (Young's modulus)

The *modulus of elasticity* is a ratio of unit stress to unit strain of a material or a measure of how much a material will stretch under a load. For all steels used in this book the modulus of elasticity is 30,000,000 psi. A modulus of elasticity for different materials may be found in reference books or on the web.

Formulas for Moments of Inertia

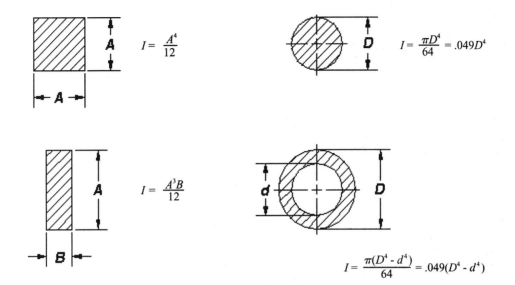

Figure 9-24

9-11 SAMPLE PROBLEM SP9-1

Figure 9-25 shows a shaft. The shaft is simply supported at the endpoints R1 and R2 and is made from SAE 1040 steel. Determine the deflection.

The modulus of elasticity for SAE 1040 steel is 30,000,000 psi.

1. Calculate the *I* value.

$$I = \frac{\pi D^4}{64} = \frac{(3.14)(.25)^4}{64} = .0002 \text{ in}^4.$$

2. Calculate the deflection.

$$y = \frac{FA^2 B^2}{3EIL}$$

$$= \frac{(35)(2.25)(6.25)}{3(30,000,000)(.0002)(4)}$$

The deflection value of .0068 in. may seem small, but many of the tolerances for bearings and gears are also in the .0005 range. If the shaft deflection becomes too large, gears may not mesh properly, or pulley belts may start to slip. Include the deflection value in any tolerance calculation.

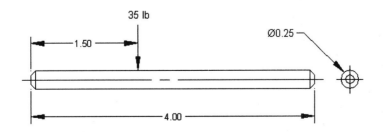

Figure 9-25

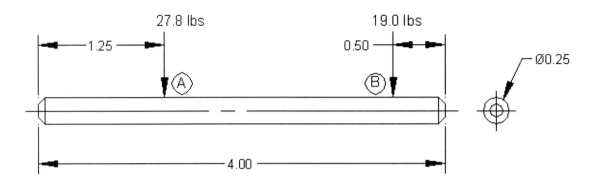

Figure 9-26

9-12 SAMPLE PROBLEM SP9-2

Figure 9-26 shows a simply supported shaft with two loads. The load at point A is 27.8 lb, and the load at point B is 19.0 lb. The total deflection caused by two loads is found by first determining the deflection due to the load acting directly on a point, then adding the deflection caused by the other load. The shaft is solid and is made from SAE 1020 steel. Calculations are rounded to four places.

Fixed values

The modulus of elasticity for the shaft is 30,000,000 psi and the moment of inertia is .0002 in. The inertia calculation is the same as was presented in Section 9-10.

The EIL denominator of the following formulas is a constant:

$$EIL = (30,000,000)(.0002)(4) = 24,000$$

The deflection at point A caused by the load at A

$$y = \frac{F_A A^2 (B + C)^2}{3EIL}$$

$$= \frac{(27.8)(1.25)^2(2.75)^2}{3(24,000)}$$

$$= .0046 \text{ in.}$$

The 2.75 value for B is the distance from the load to the right end of the shaft.

The deflection at point A caused by the load at point B

$$y = \frac{F_B AC}{6EIL}(L^2 - C^2 - A^2)$$

$$= \frac{(19.0)(.50)(1.25)}{6(24,000)}(4.00^2 - .50^2 - 1.25^2)$$

$$= .0012 \text{ in.}$$

Total deflection at A

$$.0046 + .0012 = .0058 \text{ in.}$$

The deflection at point B caused by the load at B

$$y = \frac{F_B C^2 (A + B)^2}{3EIL}$$

$$= \frac{(19.0)(3.50)^2(.50)^2}{3(24,000)}$$

$$= .0008 \text{ in.}$$

The deflection at point B caused by the load at point A

$$y = \frac{F_A AC}{6EIL}(L^2 - C^2 - A^2)$$

$$= \frac{(27.8)(1.25)(.50)}{6(24,000)}(4^2 - .5^2 - 1.25^2)$$

$$= .0002 \text{ in.}$$

Total deflection at B

.0008 + .0002 = .0010 in.

9-13 CRITICAL SPEED

As the rotating speed of a shaft increases, its vibration increases. At some speed the shaft will become dynamically unstable, resulting in possible damage. This speed is called the shaft's *critical speed.* The critical speed for a simply supported shaft with one load is calculated using the formula

$$N_c = 187.7\sqrt{\frac{1}{y}}$$

where

N_C = Shaft's critical speed (RPM)
y = shaft's static deflection (in.)

The deflection for the shaft presented in Figure 9-25 is .0068 in. See Section 9-10. The shaft's critical speed is therefore

$$N_c = 187.7\sqrt{\frac{1}{y}} = 187.7\sqrt{\frac{1}{.0068}} = 2276 \text{ RPM}$$

For shafts supporting two concentrated loads

Section 9-11 shows how to calculate the deflection for a shaft with two loads. The deflection values derived can be used to determine the shaft's critical speed as follows:

$$N_c = 187.7\sqrt{\frac{F_A y_A + F_B y_B}{F_A y_A^2 + F_B y_B^2}}$$

$$= 187.7\sqrt{\frac{(27.8)(.0058) + (19.0)(.0010)}{(27.8)(.0058)^2 + (19.0)(.0010)^2}}$$

$$= 2579 \text{ RPM}$$

9-14 SAMPLE PROBLEM SP9-3

Figure 9-27 shows a shaft supported by two sleeve bearings that are in turn mounted into two housings. The shaft carries a static load of 40 lb and a torsional load of ¼ horsepower (HP) rotating at 100 RPM. The physical properties of the shaft are listed in Figure 9-27.

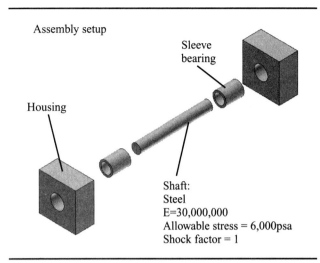

Assembly setup

Sleeve bearing

Housing

Shaft:
Steel
E=30,000,000
Allowable stress = 6,000psa
Shock factor = 1

Figure 9-27

Problem

1. Select the appropriate bearings to support the specified loading. Use the Boston Gear web site for the selection.
2. Specify the dimensions and tolerances for the housings.
3. Calculate the shaft deflection.

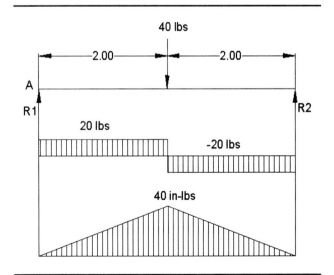

Figure 9-28

Loads

Figure 9–28 shows the shear and bending diagrams for the shaft. There is only one static load located at the center of the shaft, so each bearing supports a 20-lb load. The maximum moment is 40 in.-lb.

The horsepower can be converted into torque as follows:

$$T = \frac{63,025 \text{ HP}}{N} = \frac{63.025(.25)}{100} = 157.5 \text{ in.--lb}$$

The maximum load is calculated as follows:

$$T_{max} = \sqrt{M^2 + T^2} = \sqrt{(40)^2 + (157.5)^2}$$
$$= 162.5 \text{ in.--lb}$$

Shaft diameter

The shaft diameter is determined as follows:

$$d = \sqrt[3]{\frac{5.1 T_{max}}{T_{allowable}}} = \sqrt[3]{\frac{5.1(162.5)}{6000}}$$
$$= .52 \text{ in.}$$

The next largest standard bearing bore diameter is estimated to be .5625 in. The nominal diameter of the shaft is therefore determined to be .5625 in.

Shaft inertia

The shaft inertia value is determined as follows:

$$I = \frac{\pi D^4}{64} = \frac{(.5625)^4}{64} = \frac{(.5625)}{64} = .0049 \text{ in.}^4$$

Shaft deflection

The shaft deflection is determined as follows:

$$y = \frac{FL^3}{48EI} = \frac{(40)(4)^4}{48(30,000,000)(.100)}$$
$$= .00001 \text{ in.}$$

Shaft's critical speed

The shaft's critical speed is determined as follows:

$$N_c = 187.7\sqrt{\frac{1}{y}} = 59,355 \text{ RPM}$$

The shaft is operating at 100 RPM, which is well below its critical speed.

Selecting a bearing

There is now enough information to select a bearing. After a bearing is selected the *PV* value for the shaft conditions will be calculated. If it is within the bearing's *PV* value, the bearing is acceptable. If not, another bearing must be selected.

1. Go to a manufacturer's web site or catalog and search for a bearing with a nominal bore diameter of .5625.

In this example the Boston Gear web site was selected (www.bostgear.com).

2. Select the Products option, then bearings, then plain sleeve.

Searching a web site or a catalog will vary from manufacturer to manufacturer. In this example the Boston Gear web page allows the specification of the bearing's inside diameter at the top of the page.

Other information can also be added to help select a bearing. A length of 1.00 in. was selected, as a bearing's length should be equal to or about twice the bearing's outside diameter. The outside diameter was selected as ¾ or .753, and the bearing type as cylindrical. The site recommended bearing number B912-8. See Figure 9-29.

3. Click the Spec Sheet button to find out more about the bearing.

See Figure 9-30. The spec sheet lists the bearing's *PV*, Max *P*, and Max *V* values. The shaft's operational data can now be used to calculate the required *P*, Max *P*, and Max *V* values.

Enter the bearing parameters here.

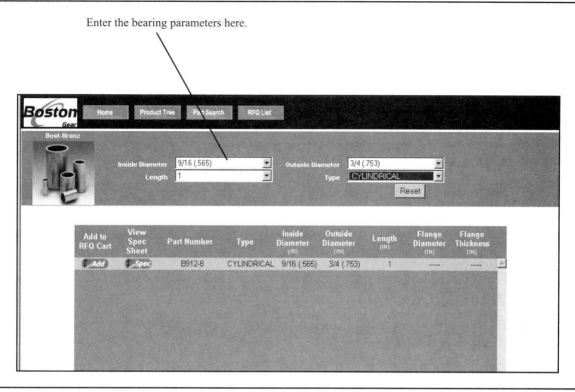

Figure 9-29 Courtesy of Boston Gear

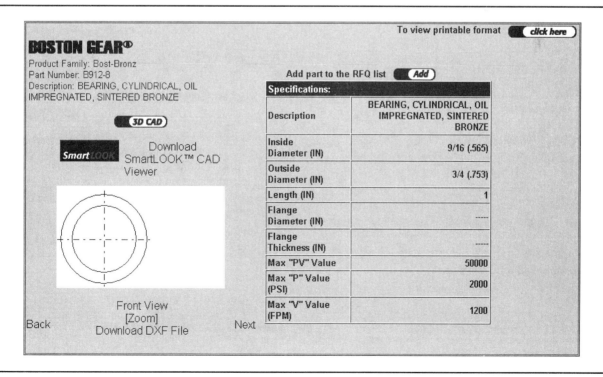

Figure 9-30 Courtesy of Boston Gear

PV value

The *PV* requirements for the shaft are calculated as follows:

$$PV = \frac{.262FN}{L} = \frac{.262(20)(100)}{1.00} = 524$$

The calculated value of 524 is well below the Max *PV* value of 50,000, so the bearing is acceptable. The Max *P* and Max *V* values are checked as follows. The diameter specified by Boston Gear is $\varnothing.565$.

See Figure 9-30.

$$V_{max} = .262DN = .262(.565)(100) = 14.8$$

$$P_{max} = \frac{(PV)}{V} = \frac{524}{14.8} = 35.4$$

Both values are within the bearing's specified limits.

Shaft/bore tolerances

The listed tolerance for the the shaft bore is $+0.0000/-0.0010$. According to Boston Gear the clearance range for *Bost-Bronz* bronze sleeve bearings is 0.0009 minimum and 0.0018 maximum. These values are slightly more conservative than the 0.0005 minimum and 0.0020 maximum listed in Figure 9-2.

The calculations for the shaft tolerances are as follows:

.5640	Hole—min
−.5631	Shaft—max
.0009	Clearance—min

.5650	Hole—max
−.5632	Shaft—min
.0018	Clearance—max

based on Boston Gear clearance values.

These calculations generate a shaft tolerance of .5631 to .5632, or a total variation of .0001. This a very tight tolerance. The conditions of this sample problem are relatively slow speed and moderate loads, so a looser tolerance would be acceptable. The Boston Gear clearance data are referred to in the catalog as a *guide,* so the values may be varied.

Using the clearance values presented in Figure 9-2, the following shaft tolerance is calculated:

.5640	Hole—min
−.5635	Shaft—max
.0005	Clearance—min

.5650	Hole—max
−.5630	Shaft—min
.0020	Clearance—max

The shaft tolerance is now .0005 with dimensions of .5635 and .5630.

Bearing O.D./housing tolerance

The Boston Gear engineering data suggest an interference range of .001 to .003. The operating conditions for this problem are moderate, so a Class LN2 fit is selected. An LN2 fit has an interference range of 0.0013.

The Boston Gear specifications for the outside diameter of the bearing are $\varnothing.753$ with a tolerance of $+.00005$ and $+.0015$. The values for a Class LN2 fit are in the appendix. The calculations are as follows. For these calculations the bearing O.D. values are the shaft values.

.7545	Shaft—max
−.7532	Hole—min
.0013	Interference—max

.7535	Shaft—min
−.7535	Hole—max
.0000	Interference—min

The hole in the housing is therefore $\varnothing.7535-.7532$. See Figure 9-31.

Interference effects on bearing

As a sleeve bearing is pressed into the housing the inside diameter becomes slightly smaller. In general, the shrinkage is about .75 of the press fit allowance.

In this example the maximum press fit allowance is 0.0013, so the maximum amount the I.D. can shrink by $.75(.0013) = .000975$, or .001. The shaft dimensions may be modified by subtracting the shrinkage value. The modified shaft dimensions are $\varnothing.5625$ and $\varnothing.5620$.

Figure 9-31 shows the final dimensions for the housing and shaft.

The completed assembly

Boston Gear Sleeve Bearing B912-8

Shaft

Presentation Drawing

SLEEVE BEARING

HOUSING

SHAFT

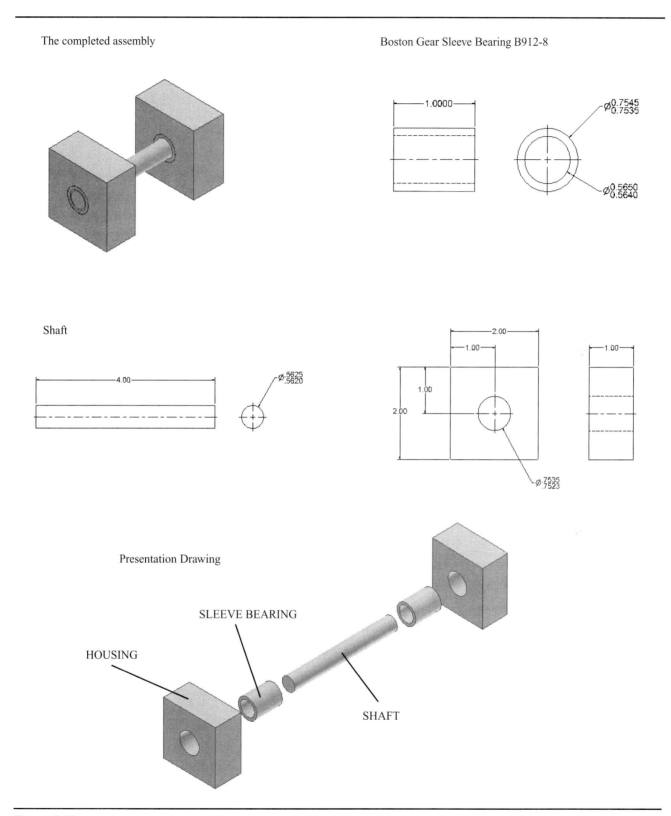

Figure 9-31

9-15 EXERCISE PROBLEMS

EX9-1

A sleeve bearing has an inside diameter of .2500, +.0000, −.0005. It is supporting a shaft that is rotating at 250 RPM. According to the values presented in Figure 9-2, what are appropriate maximum and minimum shaft diameters?

EX9-2

A sleeve bearing has an inside diameter of .3750, +.0000, −.0005. It is supporting a shaft that is rotating at 800 RPM. According to the values presented in Figure 9-2, what are appropriate maximum and minimum shaft diameters?

EX9-3

A sleeve bearing has an inside diameter of .7500, +.0000, −.0005. It is supporting a shaft that is rotating at 500 RPM. According to the values presented in Figure 9-2, what are appropriate maximum and minimum shaft diameters?

EX9-4

The outside diameter of a ball bearing is Ø1.0000, +.0000, −.0005. The bearing is to be mounted in a housing using an LN1 interference fit. What are the appropriate maximum and minimum diameters for the hole in the housing?

EX9-5

The outside diameter of a ball bearing is Ø1.5000, +.0000, −.0005. The bearing is to be mounted in a housing using an LN2 interference fit. What are the appropriate maximum and minimum diameters for the hole in the housing?

EX9-6

The outside diameter of a ball bearing is Ø0.7500, +.0000, −.0005. The bearing is to be mounted in a housing using an LN1 interference fit. What are the appropriate maximum and minimum diameters for the hole in the housing?

EX9-7

A bearing has an bore of .2500 and a length of .7500. It is operating at 125 RPM and a load of 75 lb. Is the bearing acceptable according to the *PV* factors shown in Figure 9-4?

EX9-8

A bearing has a bore of .5000 and a length of 1.0000. It is operating at 300 RPM and a load of 125 lb. Is the bearing acceptable according to the *PV* factors shown in Figure 9-4?

EX9-9

A bearing has a bore of .5625 and a length of 1.0000. It is operating at 800 RPM and a load of 200 lb. Is the bearing acceptable according to the *PV* factors shown in Figure 9-4?

EX9-10

Determine the required shaft diameter and critical speed for the following conditions.

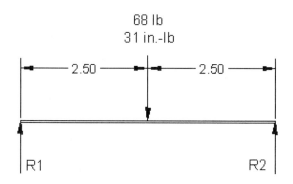

EX9-11

Determine the required shaft diameter and critical speed for the following conditions. Calculate for a minor shock.

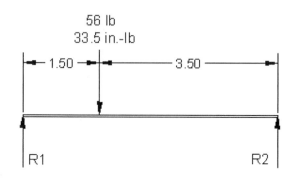

EX9-12

Determine the required shaft diameter and critical speed for the following conditions.

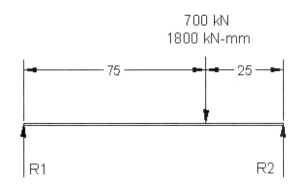

EX9-13

Determine the required shaft diameter and critical speed for the following conditions. Calculate for a heavy shock.

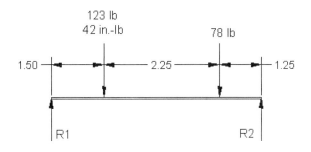

EX9-14

Determine the required shaft diameter and critical speed for the following conditions.

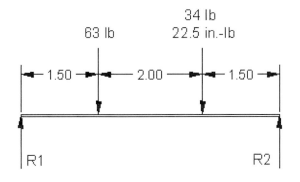

After completing the calculations for EX9-15 through EX9-20, create a drawing similar to that shown here to present your answers.

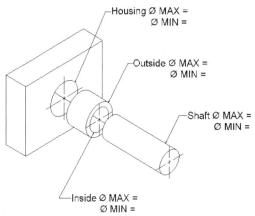

Selected Bearing =

EX9-15

Given the following loading,

1. Determine the appropriate shaft diameter.
2. Select appropriate sleeve bearings.
3. Specify the dimensions for the bearing O.D.
4. Specify the dimensions for the housing.

Speed = 100 RPM.

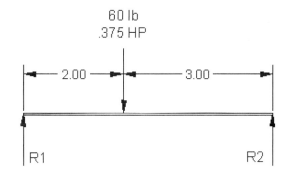

EX9-16

Given the following loading,

1. Determine the appropriate shaft diameter.
2. Select appropriate ball bearings.
3. Specify the dimensions for the bearing O.D.
4. Specify the dimensions for the housing.

Speed = 150 RPM.

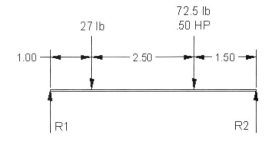

EX9-17

Given the following loading,

1. Determine the appropriate shaft diameter.
2. Select appropriate ball bearings.
3. Specify the dimensions for the bearing O.D.
4. Specify the dimensions for the housing.

   Speed = 300 RPM.

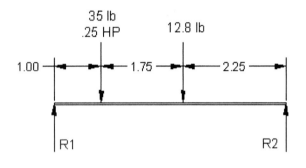

EX9-19

Given the following loading,

1. Determine the appropriate shaft diameter.
2. Select appropriate ball bearings.
3. Specify the dimensions for the bearing O.D.
4. Specify the dimensions for the housing.

   Speed = 400 RPM.

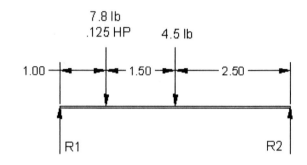

EX9-18

Given the following loading,

1. Determine the appropriate shaft diameter.
2. Select appropriate ball bearings.
3. Specify the dimensions for the bearing O.D.
4. Specify the dimensions for the housing.

   Speed = 200 RPM.

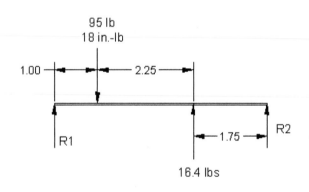

EX9-20

Given the following loading,

1. Determine the appropriate shaft diameter.
2. Select appropriate ball bearings.
3. Specify the dimensions for the bearing O.D.
4. Specify the dimensions for the housing.

   Speed = 400 RPM.

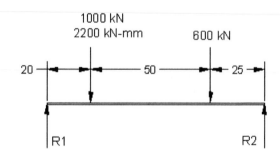

C H A P T E R 10

Gears

10-1 INTRODUCTION

This chapter explains how to design using gears. It discusses gear terminology, gear ratios, gear trains, and forces in gears. The design approach is to select gears from manufacturer's catalogs to satisfy specific design requirements.

There are several types of gears; among them are spur, bevel, worm, helical, and rack. Figure 10-1 shows representations of the different types of gears. Note how the different types of gears are used to change the direction of motion and forces.

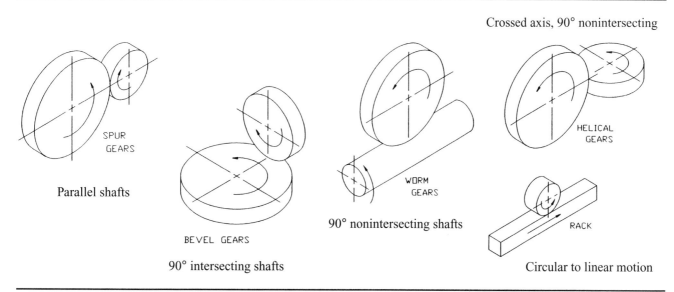

Parallel shafts

SPUR GEARS

BEVEL GEARS

90° intersecting shafts

WORM GEARS

90° nonintersecting shafts

Crossed axis, 90° nonintersecting

HELICAL GEARS

RACK

Circular to linear motion

Figure 10-1

10-2 GEAR TERMINOLOGY

Pitch diameter (D): The diameter used to define the spacing of gears. Ideally, gears are exactly tangent to each other along their pitch diameters.

Diametral pitch (P): The number of teeth per inch. Meshing gears must have the same diametral pitch. Manufacturers' gear charts list gears with the same diametral pitch.

Module (M): The pitch diameter divided by the number of teeth. The metric equivalent of diametral pitch.

Number of teeth (N): The number of teeth of a gear.

Circular pitch (CP): The circular distance from a fixed point on one tooth to the same position on the next tooth as measured along the pitch circle. The circumference of the pitch circle divided by the number of teeth.

Preferred pitches: The standard sizes available from gear manufacturers. Whenever possible, use preferred gear sizes.

Center distance (CD): The distance between the center points of two meshing gears.

Backlash: The difference between a tooth width and the engaging space on a meshing gear.

Addendum (a): The height of a tooth above the pitch diameter.

Dedendum (d): The depth of a tooth below the pitch diameter.

Whole depth: The total depth of a tooth. The addendum plus the dedendum.

Working depth: The depth of engagement of one gear into another. Equal to the sum of the two gear's adendeums.

Circular thickness: The distance across a tooth as measured along the pitch circle.

Face width (F): The distance from front to back along a tooth as measured perpendicular to the pitch circle.

Outside diameter: The largest diameter of the gear. It equals the pitch diameter plus the addendum.

Root diameter: The diameter of the base of the teeth. The pitch diameter minus the dedendum.

Clearance: The distance between the addendum of the meshing gear and the dedendum of the mating gear.

Pressure angle: The angle between the line of action and a line tangent to the pitch circle. Most gears have pressure angles of either 14.5° or 20°.

See Figure 10-2. In general, 20° gears are for heavy loads, but 14.5° gears are more widely used.

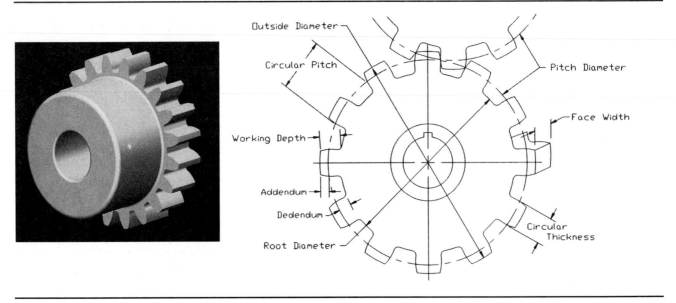

Figure 10-2   Courtesy of Boston Gear

Diametral pitch (P)	$P = \dfrac{N}{D}$
Pitch diameter (D)	$D = \dfrac{N}{P}$
Number of teeth (N)	$N = DP$
Addendum (a)	$a = \dfrac{1}{P}$

Metric

Module (M)	$M = \dfrac{D}{N}$

Figure 10-3

10-3 GEAR FORMULAS

Figure 10-3 shows a chart of formulas commonly associated with gears. The formulas are for spur gears.

10-4 GEAR RATIOS

The speed ratio between two meshing gears, 1 and 2, is defined using angular velocity as

$$\frac{\omega_2}{\omega_1}$$

which is equivalent to

$$\frac{N_2}{N_1}$$

The speed ratio between two gears is directly proportional to the number of teeth on the two gears. Figure 10-4 shows two meshing gears tangent at their pitch diameters represented by two cylinders. Gear 1 has 20 teeth, and gear 2 has 60 teeth. The speed ratio between the gears is 3:1.

$$\frac{60}{20} = \frac{3}{1}$$

If the smaller gear has a speed of 1750 RPM, then the larger gear has a speed of

$$\frac{1750}{3} = 583.3 \, \text{RPM}$$

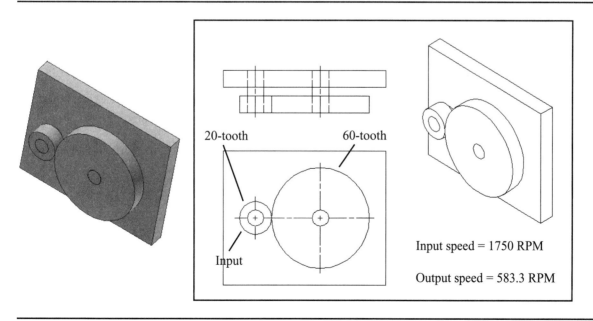

20-tooth   60-tooth

Input

Input speed = 1750 RPM

Output speed = 583.3 RPM

Figure 10-4

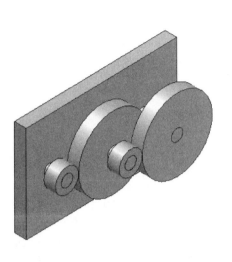
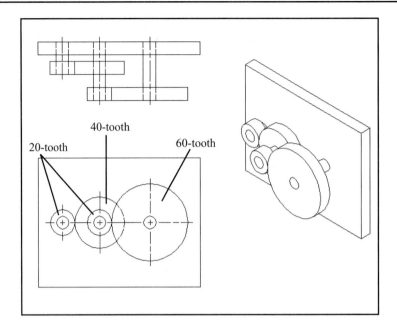

Figure 10-5

10-3 GEAR TRAINS

When more than two gears are used in a design the combination is called a *gear train*. Figure 10-5 shows a gear train that contains four gears: two 20-tooth gears, one 40-tooth, and one 60-tooth. The speed ratio between the input RPM and output RPM is determined by multiplying the individual gear ratios together. Observe that the 40-tooth gear and one of the 20-tooth gears are mounted on the same shaft.

There is no speed ratio between these two gears, as they have the same angular velocity. The speed ratio is

$$\left(\frac{40}{20}\right)\left(\frac{60}{20}\right) = \frac{6}{1}$$

For an input speed of 1750 RPM, the output speed would be

$$1750\left(\frac{1}{6}\right) = 292 \text{ RPM}$$

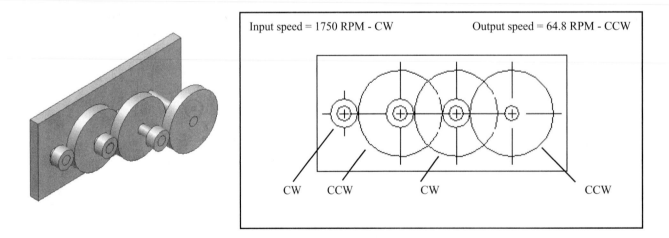

Figure 10-6

Figure 10-6 shows another gear train that includes six gears: three 20-tooth gears and three 60-tooth gears. The speed ratio between input and output speeds is

$$\left(\frac{60}{20}\right)\left(\frac{60}{20}\right)\left(\frac{60}{20}\right) = \frac{27}{1}$$

For an input speed of 1750 RPM, the output speed would be

$$\frac{1750}{27} = 64.8 \text{ RPM}$$

Gear direction

Meshing gears always rotate in opposite directions. If gear 1 in Figure 10-6 were to rotate clockwise (CW), then gear 2 would rotate counterclockwise (CCW). Gear 3 would also rotate CCW, driving gear 4 in a CW direction. Gear 5 would rotate in the CW direction and drive gear 6 in the CCW direction. A gear called an *idler* may be added to a gear train for the sole purpose of changing the direction of the final rotation. Idler gears are usually identical with one of the gears they are meshing with so as not to affect the final speed ratio. See Figure 10-7.

10-6 DESIGNING GEAR SPEED RATIOS

A typical speed ratio design problem starts with a given input speed from a motor and requires a desired output speed. The problem is to select gears that generate the required output speed from the given input speed.

> **NOTE:**
> The range limit for a single set of spur gears is 6:1. Ratios larger than 6:1 are not recommended.

Sample problem

Design a gear train that will reduce input speed by 100:1. Select gears from the chart shown in Figure 10-8.

If three identical gears are used, the speed ratio will be 4.64:1.

$$\sqrt[3]{100} = 4.64$$

The available gears cannot produce this ratio. The closest this set of gears can produce is a 70-tooth gear

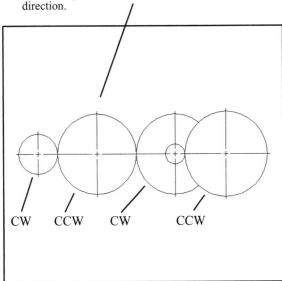

An idler gear. An idler does not change the speed ratio between input and output, but does affect the output direction.

CW   CCW   CW   CCW

Figure 10-7

Gear Specifications	
Pressure angle = 14.5°	
Face width = 1.000 in.	
Material = SAE 1040	
Diametral pitch = 10	
N	**Pitch diameter (in.)**
12	1.20
16	1.60
18	1.80
20	2.00
24	2.40
30	3.00
36	3.60
40	4.00
48	4.80
50	5.00
60	6.00
70	7.00
80	8.00
90	9.00
100	10.00

Figure 10-8

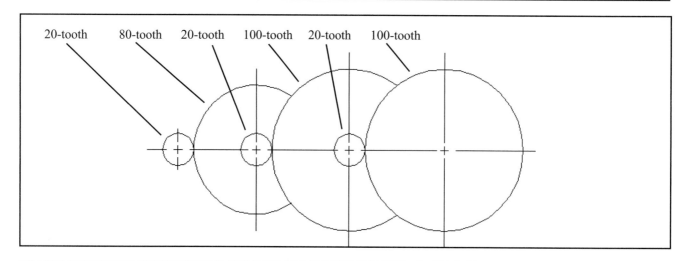

Figure 10-9

meshed with a 16-tooth gear, which yields a ratio of 4.375. Rounding off the ratio to 5:1 or 4:1 and trying several combinations leads to the following ratios. Gears that produce ratios of 4:1 and 5:1 are available from the given gear selection.

$$\left(\frac{1}{5}\right)\left(\frac{1}{5}\right)\left(\frac{1}{4}\right) = \frac{1}{100}$$

Figure 10-9 shows a possible solution that generates a speed reduction of 100:1.

10-7 FORCES IN GEARS

Gears provide an excellent way to transfer forces and create mechanical advantages. Forces are transferred between gears along the pitch circle. Figure 10-10 shows the forces acting between two meshing gears. The force acting between two gears is divided into two components, the *tangential force* and the *radial force*. The angle is defined as the *pressure angle* and is either 14.5° or 20°. The tangential force is the force transferred between the two gears.

Figure 10-11 shows two meshed gears: a 20-tooth and a 60-tooth selected from the chart in Figure 10-8. The diameter of the 20-tooth gear is 2.00 in. and the diameter of the 60-tooth gear is 6.00 in. The 20-tooth gear is driven by a motor that generates 0.25 HP at 1750 RPM.

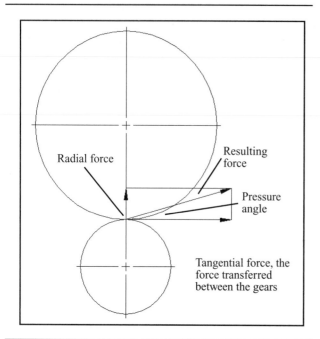

Figure 10-10

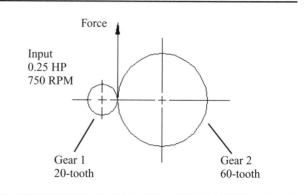

Figure 10-11

Horsepower can be related to torque using the formula

$$HP = \frac{T\omega}{63,000}$$

where

HP = horsepower transmitted
ω = speed of the gear (RPM)
T = torque on the gear (in.-lb)

Torque can also be defined as

$$T = Fr$$

where

F = tangential force acting along the pitch circle
r = radius of the gear

Substituting Fr for T and solving for F yields

$$F = \frac{(HP)(63,000)}{r\omega}$$

For the given problem

$$F = \frac{(0.25)(63,000)}{(1)(1,750)} = 9 \text{ lb}$$

The 9 lb is transferred between the gears. The 9-lb force is a tangential force and has a radial component. See Figure 10-10.

The physics

Three important concepts must be understood. The tangential force, linear speed, and horsepower transferred between the two gears are equal, based on the conservation of energy theorem. Both gears have the same tangential force, linear speed, and horsepower.

Force transfer

The force transferred between the two gears is the same force. The torque of the two gears is different, but the force is equal. If the transfer force is the same, then

$$F = \frac{T_1}{r_1} \quad \text{and} \quad F = \frac{T_2}{r_2}$$

Therefore,

$$\frac{T_1}{r_1} = \frac{T_2}{r_2}$$

or

$$T_1 r_2 = T_2 r_1$$

The radius of the smaller gear equals 1, and the radius of the larger gear equals 3, so the torque must also differ by a factor of 3. The torque on the smaller gear is

$$T_1 = \frac{(63,000)(HP)}{\omega_1}$$

$$= \frac{(63,000)(0.25)}{1750} = 9 \text{ in.-lb}$$

so the torque on the larger gear is

$$T_2 = \frac{T_1 r_2}{r_2} = \frac{(9)(3)}{1} = 27 \text{ in.-lb}$$

Speed transfer

The linear speed between the two gears is equal. The rotary speed is different, but the linear speed is the same. From the information presented in Section 10-3, the speed of the larger gear is

$$\omega_2 = \omega_1 \frac{20}{60} = 1750\frac{1}{3} = 583.3 \text{ RPM}$$

Horsepower transfer

If the torque and speed values derived for the larger gear are applied to the formula for horsepower, the resulting horsepower value will be the same as the input horsepower value:

$$HP = \frac{T_2 \omega_2}{63,000} = \frac{(27)(583.3)}{63,000} = 0.25 \text{ HP}$$

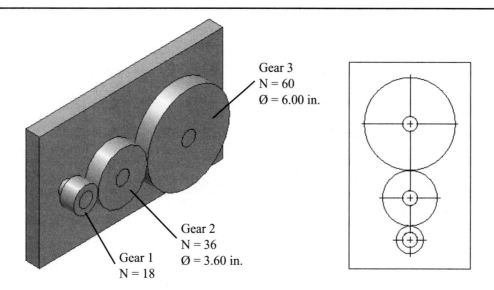

Figure 10-12

10-8 SAMPLE PROBLEM SP10-1

Figure 10-12 shows three gears and their individual specifications. The three gears form a gear train. The input is 0.375 HP at 1750 RPM. What is the speed and force in gear 3?

Gear 1 to gear 2

The torque in gear 1 is

$$T = \frac{(63,000)(HP)}{\omega_1} = \frac{(63,000)(0.375)}{1750}$$
$$= 13.5 \text{ in.-lb}$$

Therefore the torque in gear 2 is

$$T_2 = \frac{T_1 r_2}{r_1} = \frac{(13.5)(1.80)}{0.90} = 27.0 \text{ in.-lb}$$

The speed of gear 2 is

$$\omega_2 = \omega_1 \frac{1.80}{3.60} = 1750(0.50) = 875 \text{ RPM}$$

Gear 2 to gear 3

The torque in gear 3 is

$$T_3 = \frac{T_2 r_3}{r_2} = \frac{(27)(3.00)}{1.80} = 45 \text{ in.-lb}$$

The speed of gear 3 is

$$\omega_3 = \omega_2 \frac{36}{60} = 875(0.60) = 525 \text{ RPM}$$

Checking the answers

The conservation of energy theorem requires that the horsepower be the same in all gears in a train. If the results are correct, gear 3 should be operating at .375 HP, the same as the initial input.

$$HP = \frac{T_3 \omega_3}{63,000} = \frac{(45)(525)}{63,000} = 0.375 \text{ HP}$$

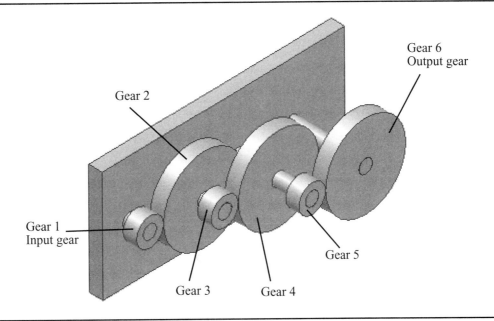

Figure 10-13

10-9 SAMPLE PROBLEM SP10-2

Figure 10-13 shows a gear train that includes three 20-tooth gears and three 60-tooth gears, labeled gears 1, 2, 3, 4, 5, and 6 as shown. The initial input is 0.25 HP at 1750 RPM. The diameter of the 20-tooth gear is 2.00 in., and the diameter of the 60-tooth gear is 6.00 in. What is the force in gear 6?

The initial torque value

$$T = \frac{(63,000)(HP)}{\omega_1} = \frac{(63,000)(.25)}{1750} = 9 \text{ in.-lb}$$

Gear 1 to gear 2

$$T_2 = \frac{T_1 r_2}{r_1} = \frac{(9)(3)}{1} = 27 \text{ in.-lb}$$

Gear 2 to gear 3

Gears 2 and 3 are mounted on the same shaft, so they have the same rotation speed and the same torque:

$$T_2 = T_3 = 27 \text{ in.-lb}$$

Gear 3 to gear 4

$$T_4 = \frac{T_3 r_4}{r_3} = \frac{(27)(3)}{1} = 81 \text{ in.-lb}$$

Gear 4 to gear 5

Gears 4 and 5 are on the same shaft, so

$$T_4 = T_5 = 81 \text{ in.-lb}$$

Gear 5 to gear 6

$$T_6 = \frac{T_5 r_6}{r_5} = \frac{(81)(3)}{1} = 243 \text{ in.-lb}$$

If the horsepower formula is applied to the condition of gear 6, it will be determined that gear 6 is operating at 0.25 HP, or the same force as the initial input value:

$$HP = \frac{T_6 \omega_6}{63,000} = \frac{(243)(64.8)}{63,000} = 0.25 \text{ HP}$$

See Figure 10-14.

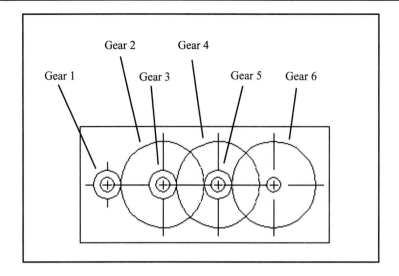

Figure 10-14

10-10 FORCES IN GEAR TRAINS

Once the forces in a gear train have been determined, gears must be selected that can withstand the forces. The Lewis equation as modified by Barth can be used to determine the bending strength of a gear tooth. If the tooth is strong enough for the given loads, the gear may be used. If the tooth is not strong enough, another gear must be selected. The Lewis formula is generally limited to pitch circle velocities of 1500 (FPM ft/min).

The Lewis equation with Barth's modification is

$$F_b = \frac{\sigma A Y}{P} = \left(\frac{600}{600 + V}\right)$$

where

F_b = bending load (lb)

σ = allowable stress on the tooth (psi) (See Figure 10-15.)

A = the gear's face width (in.)

Y = the Lewis tooth factor (See Figure 10-16.)

P = diametral pitch

V = pitch velocity (FPM)

and

$$V = 0.262(D)(\omega)$$

where

D = pitch diameter

Allowable Tooth Stress

Material	psi
Plastic	5,000
Bronze	10,000
Cast iron	12,000
SAE1020	20,000
SAE1040	25,000

These values are for use only in the examples and exercises of this book. Refer to gear manufacturers' catalogs for actual values.

Figure 10-15

10-11 SAMPLE PROBLEM SP10-3

A gear has the following specifications:

Pressure angle = 14.5°
Face width = 1.000 in.
Material = SAE1040 steel—untreated
Diametral pitch = 10
Pitch diameter = 4.00 in.
Number of teeth = 40

The gear is operating at 1750 RPM. What is the allowable bending load on the tooth?

$$F_b = \frac{(25,000)(1.000)(0.336)}{10}\left(\frac{600}{600 + 1834}\right)$$

$$= 207 \text{ lb}$$

where

$$V = 0.262(4)(1750) = 1834 \text{ FPM}$$

10-12 SAMPLE PROBLEM SP10-4

The loads for a gear train are known. The problem is then to select gears that can withstand the required loads. This may be done by rewriting the Lewis equation and solving for A, the face width. Select a gear and calculate the required face width. If the gear has a face width greater than the calculated value, it is acceptable.

Lewis Gear Tooth Form Factor

N	14.5°	20°
10	0.176	0.201
11	0.192	0.226
12	0.210	0.245
14	0.236	0.276
16	0.255	0.295
18	0.270	0.308
20	0.283	0.320
22	0.292	0.330
24	0.302	0.337
26	0.308	0.344
28	0.314	0.352
30	0.318	0.358
32	0.322	0.364
34	0.325	0.370
36	0.329	0.377
38	0.332	0.383
40	0.336	0.389
45	0.340	0.399
50	0.346	0.408
55	0.352	0.415
60	0.355	0.421
65	0.358	0.425
70	0.360	0.429
75	0.361	0.433
80	0.363	0.436
90	0.366	0.442
100	0.368	0.446
150	0.375	0.458
200	0.378	0.463
Rack	0.390	0.484

Figure 10-16

A gear is subject to the following conditions:

Torque = 315 in.-lb
Speed = 600 RPM
Diametral pitch = 10
Material = SAE 1020 (20,000 psi)

Pitch diameter (D) = 2.00 in.
Pressure angle = 20°
Number of teeth = 20
Lewis factor = 0.320
The face width for the gear is 1.250 in.

The Lewis equation:

$$F_b = \frac{\sigma A Y}{P}\left(\frac{600}{600 + V}\right)$$

Rewriting the Lewis equation to solve for A:

$$A = \frac{F_b P}{\sigma Y}\left(\frac{600 + V}{600}\right)$$

Calculating V:

$$V = 0.262(D)(\omega)$$
$$= 0.262(2)(600) = 314.5 \text{ FPM}$$

Using the given torque values and solving for F_b:

$$F_b = \frac{T}{r} = \frac{315}{1} = 315 \text{ lb}$$

Solving for A:

$$A = \frac{(315)(10)}{(20,000)(0.320)}\left(\frac{600 + 314.4}{600}\right)$$
$$= .748 \text{ in.}$$

The selected gear has a face width of 1.25, so it is acceptable for this application.

10-13 SAFETY FACTORS

A *safety factor* is used to assure that the load and force values used to calculate the strength of a design exceed the expected operational values. A safety factor of 2 means that the design is built twice as strong as is needed. Saftey factors may be applied to gears by increasing the bearing load requirements.

Service Factors

| | Hours per Day | |
Load	8–10	24
Steady	1.00	0.80
Light shock	0.80	0.65
Medium shock	0.65	0.55
Heavy shock	0.55	0.50

Figure 10-17

In sample problem SP10-4 presented in the previous section the face width requirement of a gear tooth was calculated based on a bearing load of 315 lb. If the design was to have a safety factor of 2, then the bearing load would be $(2)(315) = 630$ lb. This would change the minimum acceptable tooth width from 0.748 in. to 1.486 in. or twice as wide as the original load calculations required.

10-14 SERVICE FACTORS

A *service factor* is a factor used to take into account the action of gears rubbing against each other. A gear box that operates 8 hours a day under a steady load will experience less stress than the same gear box operating 24 hours a day under heavy shocks. See Figure 10-17.

Service factors are applied to the Lewis equation by multiplying the service factor by the allowable stress of the tooth material. If in the preceding sample problem SP10-4, the additional parameter of 24-hour operation under light shock were added to the problem, the allowable stress of 20,000 psi would be modified by the service factor 0.65. The allowable stress would then be

$$\sigma(0.65) = (20,000)(0.65) = 13,000 \text{ psi}$$

Lubrication Factors

Type of Lubrication	Lubrication Factor
Submerged in oil	1.00
Oil drip	0.80
Grease	0.65
Intermittent lubrication	0.50

Figure 10-18

If the 13,000 psi value was then used to calculate *A*, the required tooth width would be 1.15 in. If the gear was used 8 hours a day under a steady load, the original calculated value for the tooth width of 0.748 in. would be acceptable.

10-15 LUBRICATION FACTORS

A *lubrication factor* takes into account the friction generated as two teeth mesh against each other. See Figure 10-18. The lubrication factor is applied to the Lewis equation by multipling the allowable stress by the lubrication factor. If the gears in the preceding sample problem SP10-4 were lubricated using only an oil drip, the allowable stress value would be

$$(20{,}000)(0.80) = 16{,}000 \text{ psi}$$

10-16 SAMPLE PROBLEM SP10-5

It is possible for a gear to be subject to a safety factor, a service factor, and a lubrication factor. If the parameters of sample problem SP10-4 were modified to include a safety factor of 1.5, a service factor of 0.80, and a lubrication factor of 0.65, the resulting tooth width calculation would be modified as follows:

safety factor

$$F_b = (315)(1.5) = 472.5$$

service factor lubrication factor

$$\sigma = (20{,}000)(0.80)(0.65) = 10{,}400$$

Solving for *A*:

$$A = \frac{(472.5)(10)}{(10{,}400)(0.320)}\left(\frac{600 + 314.4}{600}\right) = 2.156 \text{ in.}$$

The tooth face width value of 2.156 in. is considerably greater than the original 0.748 in. value, showing the dramatic effects saftey, service, and lubrication factors can have on gear selection.

10-17 MANUFACTURERS' CATALOGS

Gear manufacturers list gears in their catalogs not only by size but also by horsepower and torque ratings. Figure 10-19 shows a page from a Boston Gear catalog (14 Hayward Street, Quincy, MA 02171, Tel: 617-328-3300).

The data include a torque rating and a horsepower rating for given speeds. The catalog can be perused to locate the closest acceptable gear for a given torque and speed requirement. A cross-reference leads to a listing of gears that satisfy the requirements.

Note in Figure 10-19 that a service factor of 1.0 is specified along with the gears' diametral pitch, material, and pressure angle.

SPUR GEARS

APPROXIMATE HORSEPOWER AND TORQUE* RATINGS
FOR CLASS I SERVICE (Service Factor = 1.0)

10 DIAMETRAL PITCH STEEL **20° PRESSURE ANGLE** **1-1/4" FACE** REFERENCE PAGE 32.

No. Teeth	25 RPM		50 RPM		100 RPM		200 RPM		300 RPM		600 RPM		900 RPM		1200 RPM		1800 RPM		3600 RPM	
	H.P.	Torque	H.P.	Torque	H.P.	Torque	H.P.	Torque	H.P.	Torque	H.P.	Torque	H.P.	Torque	H.P.	Torque	H.P.	Torque	H.P.	Torque
12	.14	363	.28	358	.55	349	1.06	333	1.51	318	2.66	280	3.57	250	4.30	226	5.40	189	7.27	127
14	.19	477	.37	469	.72	456	1.37	431	1.95	409	3.37	354	4.46	312	5.31	279	6.58	230	8.63	151
15	.21	533	.42	525	.81	509	1.52	479	2.16	453	3.70	389	4.87	341	5.78	303	7.10	249	9.22	161
16	.23	580	.45	571	.88	552	1.64	518	2.32	488	3.96	416	5.18	363	6.12	321	7.47	262	9.60	168
18	.27	679	.53	667	1.02	642	1.90	599	2.67	561	4.48	471	5.79	406	6.79	356	8.19	287	10.33	181
20	.31	784	.61	768	1.17	737	2.16	682	3.02	635	5.00	526	6.41	449	7.45	391	8.90	311	11.04	193
24	.39	983	.76	958	1.45	913	2.65	834	3.65	767	5.89	619	7.41	519	8.50	447	9.98	349		
25	.41	1032	.80	1005	1.52	956	2.76	870	3.80	799	6.10	641	7.64	535	8.74	459	10.21	358		
28	.47	1195	.92	1161	1.74	1097	3.14	990	4.29	901	6.76	710	8.37	586	9.50	499	10.99	385		
30	.52	1300	1.00	1260	1.88	1187	3.38	1064	4.59	964	7.16	752	8.80	616	9.94	522	11.42	400		
35	.64	1615	1.24	1558	2.31	1454	4.08	1284	5.47	1150	8.33	875	10.08	706	11.27	592				
40	.75	1896	1.44	1820	2.67	1685	4.65	1467	6.18	1299	9.20	966	10.99	770	12.17	639				
45	.87	2190	1.66	2092	3.05	1920	5.23	1649	6.88	1445	10.04	1054	11.85	830						
48	.92	2328	1.76	2218	3.21	2026	5.48	1727	7.16	1504	10.33	1085	12.12	849						
50	.96	2420	1.83	2301	3.32	2095	5.64	1777	7.34	1542	10.52	1105	12.29	861						

10 DIAMETRAL PITCH CAST IRON **20° PRESSURE ANGLE** **1-1/4" FACE** REFERENCE PAGE 32.

No. Teeth	25 RPM		50 RPM		100 RPM		200 RPM		300 RPM		600 RPM		900 RPM		1200 RPM		1800 RPM		3600 RPM	
	H.P.	Torque	H.P.	Torque	H.P.	Torque	H.P.	Torque	H.P.	Torque	H.P.	Torque	H.P.	Torque	H.P.	Torque	H.P.	Torque	H.P.	Torque
55	.65	1638	1.23	1550	2.22	1400	3.72	1173	4.80	1009	6.77	711	7.84	549						
60	.71	1778	1.33	1675	2.38	1501	3.94	1243	5.05	1061	7.01	737	8.06	564						
70	.84	2114	1.57	1974	2.77	1743	4.48	1413	5.65	1187	7.65	803								
80	.98	2462	1.81	2279	3.15	1984	5.00	1576	6.22	1307	8.23	865								
90	1.09	2742	2.00	2517	3.43	2162	5.35	1686	6.58	1382	8.54	897								
100	1.20	3016	2.18	2746	3.70	2329	5.67	1786	6.89	1448	8.80	924								
120	1.45	3650	2.59	3271	4.30	2709	6.40	2016	7.64	1605	9.48	996								
140	1.66	4177	2.93	3688	4.74	2989	6.88	2167	8.09	1699										
160	1.91	4814	3.32	4191	5.28	3329	7.49	2359	8.69	1826										
200	2.30	5802	3.90	4920	5.99	3773	8.17	2573												

8 DIAMETRAL PITCH STEEL **20° PRESSURE ANGLE** **1-1/2" FACE** REFERENCE PAGE 33.

No. Teeth	25 RPM		50 RPM		100 RPM		200 RPM		300 RPM		600 RPM		900 RPM		1200 RPM		1800 RPM		3600 RPM	
	H.P.	Torque	H.P.	Torque	H.P.	Torque	H.P.	Torque	H.P.	Torque	H.P.	Torque	H.P.	Torque	H.P.	Torque	H.P.	Torque	H.P.	Torque
12	.27	678	.53	667	1.03	647	1.93	609	2.74	576	4.71	495	6.19	434	7.35	386	9.03	316	11.72	205
14	.35	890	.69	874	1.34	843	2.50	787	3.51	738	5.92	622	7.68	537	9.01	473	10.91	382	13.81	242
15	.39	996	.77	976	1.49	939	2.77	873	3.88	816	6.49	681	8.35	585	9.76	513	11.73	411	14.70	257
16	.43	1084	.84	1061	1.62	1018	2.99	943	4.18	877	6.92	727	8.85	620	10.30	541	12.30	431		
18	.50	1268	.98	1238	1.88	1183	3.44	1086	4.77	1003	7.78	817	9.84	689	11.35	596	13.40	469		
20	.58	1462	1.13	1424	2.15	1354	3.91	1233	5.39	1131	8.64	908	10.82	758	12.38	650	14.47	507		
22	.66	1651	1.27	1604	2.41	1518	4.35	1371	5.95	1250	9.41	989	11.67	817	13.27	698	15.36	538		
24	.73	1831	1.41	1775	2.65	1672	4.75	1498	6.46	1357	10.08	1059	12.39	868	14.00	735	16.08	563		
28	.88	2224	1.70	2145	3.18	2003	5.61	1768	7.54	1583	11.47	1204	13.88	972	15.51	815				
32	1.06	2664	2.03	2557	3.76	2367	6.54	2060	8.68	1824	12.92	1358	15.44	1081	17.10	898				
36	1.22	3082	2.34	2944	4.29	2703	7.37	2321	9.68	2034	14.13	1484	16.68	1168						

Ratings are based on strength calculation. Basic static strength rating, or for hand operation of above gears is approximately 3 times the 100 RPM rating.

NOTE: Ratings to right of heavy line are not recommended, as pitch line velocity exceeds 1000 feet per minute.
They should be used for interpolation purposes only.

*Torque Ratings (Lb. Ins.).

BOSTON GEAR®

Figure 10-19   Courtesy of Boston Gear

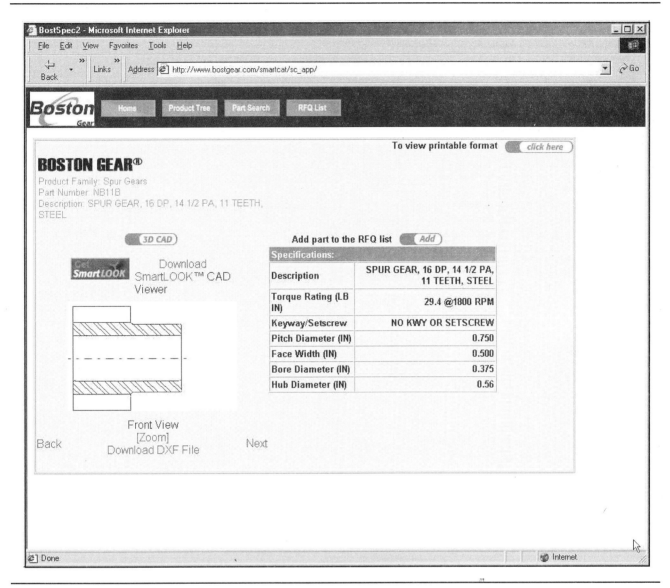

Figure 10-20 Courtesy of Boston Gear

10-18 WEB SITES

Figure 10-20 shows a page from the Boston Gear web site (www.bostgear.com). It is typical of the information available from gear manufacturers. The gears are rated by torque and speed so that evaluations can be based on initial conditions.

Note that the web site also includes access to drawing files that can be added directly to drawings. The .dxf file is compatible with AutoCAD, Mechanical Desktop, and Inventor. Gear drawings can be downloaded, then modified as needed. The .dxf files are two-dimensional drawings. Three-dimensional drawings are available as well.

10-19 FORCES IN NONMETALLIC GEARS

The Lewis equation may be applied to nonmetallic gears if modified as shown:

$$F_b = \frac{\sigma A Y}{P}\left(\frac{150}{200 + V} + 0.25\right)$$

According to Figure 10-15, the allowable stress for plastic or Delrin gears is approximately

$$\sigma = 5000 \text{ psi}$$

10-20 SAMPLE PROBLEM SP10-6

A student working on a design project is considering a gear with the following specifications:

Pressure angle = 20°
Face width = .125 in.
Material = molded Delrin with brass insert
Diametral pitch = 48
Pitch diameter = .417 in.
Number of teeth = 20

The gear will be subject to a load of 5.3 lb at 4765 RPM. Is the gear acceptable for the application?

The Lewis equation is used to calculate the minimum acceptable tooth width (face width). If the results are larger than 0.125 in., the gear is not acceptable.

$$A = \frac{F_b}{\sigma Y \left(\dfrac{150}{200 + V} + 0.25 \right)}$$

$$= \frac{(5.3)(48)}{(5000)(0.320) \left(\dfrac{150}{200 + 521} + 0.25 \right)}$$

$$= .347 \text{ in.}$$

The gear is not acceptable. It is too small to carry the designated load.

10-21 METRIC SPUR GEARS

Calculations for metric spur gears are also made using the Lewis equation. The diametral pitch is replaced by the term *module* (*m*).

$$F_b = \frac{\sigma A Y}{m} \left(\frac{600}{600 + V} \right)$$

where

F_b = bending load (Mpa)

σ = allowable stress on the tooth (Mpa)

A = gear face width (mm)

Y = Lewis tooth factor (See Figure 10-21.)

m = module

V = pitch velocity (RPM)

Lewis Factor—metric

Gear parameters:
Pressure angle = 20°.
Module equals the dedendum.

Number of Teeth	Y
12	.335
14	.360
16	.379
18	.395
20	.408
22	.419
24	.428
26	.436
28	.443
30	.449
34	.459
38	.467
45	.478
50	.484
60	.494
75	.503
100	.513
150	.523
Rack	.479

Figure 10-21

A gear module is defined as the millimeters of pitch diameter per tooth:

$$m = \frac{D}{N}$$

where

D = pitch diameter
N = number of teeth

The conversion factor for pounds per square inch (psi) to pascals (Pa) is

1 psi = 6895 Pa

Also,

1 MPa = 1,000,000 Pa

The gear shown in Figure 10-22 is made from plastic. The allowable stress for plastic is 5000 psi.

$$(5000)(6895) = 34,475,000 \text{ Pa or } 34.475 \text{ Mpa}$$

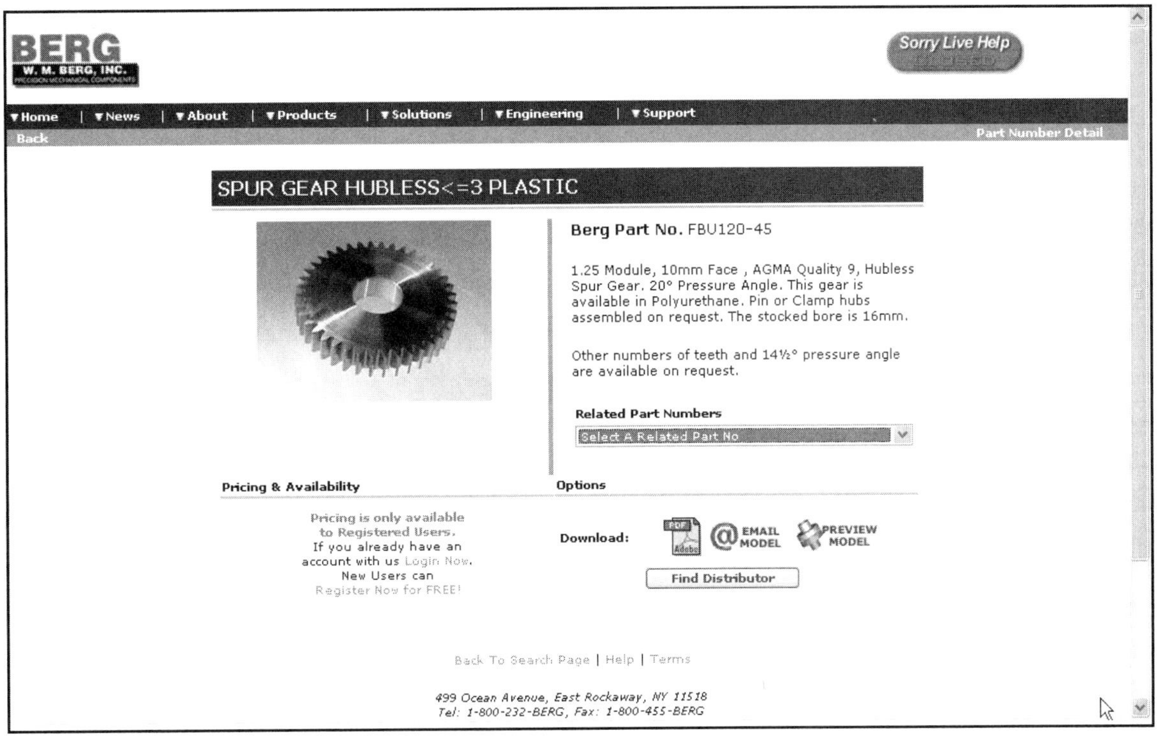

Figure 10-22 Courtesy of W.M. Berg

10-22 SAMPLE PROBLEM SP10-7

What is the allowable bending load for the W. M. Berg gear FBU120-45 shown in Figure 10-22 if $V = 850$?

$$F_b = \frac{\sigma AY}{m}\left(\frac{600}{600 + V}\right)$$

$$= \frac{(34.48)(10)(.478)}{1.25}\left(\frac{600}{600 + 850}\right)$$

$$= 54.56 \text{ Mpa}$$

10-23 BEVEL GEARS

Bevel gears are gears that operate at 90° to each other. As a general rule, speed ratios of greater than 4:1 are not acceptable.

The Lewis equation may be applied to bevel gears with the following modifications:

$$F_b = \frac{\sigma AY}{P}\left(\frac{600}{600 + V}\right)(.75)$$

Figure 10-23 shows the Lewis factor values for bevel gears.

10-24 WORM GEARS

The speed ratio between a worm gear and a spur gear is equal to the number of teeth on the spur gear and the number of starts on the worm gear. If the worm gear has one tooth in contact with the spur gear, the worm gear will make a complete turn for every tooth on the spur gear. If a worm gear is meshed with a 60-tooth spur gear, the speed ratio will be 60:1. Speed ratios greater than 75:1 are not generally acceptable.

10-25 MATCHING GEARS AND SHAFTS

The hole in the center of a gear is called the *bore*. Bore tolerances vary from manufacturer to manufacturer. For exercises and examples in this book bore tolerances will be either $\pm.0005$ in. or ± 0.002 mm.

Lewis Factors for Bevel Gears

No. of teeth Pinion	Ratios 1		1.5		2		3		4	
	Pin	Gear	Pin	Gear	Pin	Gear	Pin	Gear	Pin	Gear
12					.345	.283	.355	.302	.358	.305
14			.349	.292	.367	.301	.377	.317	.380	.323
16		.333	.367	.311	.386	.320	.396	.333	.402	.339
18		.342	.383	.328	.402	.336	.415	.346	.427	.364
20		.352	.402	.339	.418	.349	.427	.355	.456	.386
24		.371	.424	.364	.443	.368	.471	.377	.377	.506
28		.386	.446	.383	.462	.386	.509	.396	.543	.421
32		.399	.462	.396	.487	.402	.540	.412		
36		.408	.477	.408	.518	.415	.569	.424		
40		.418			.542	.424	.594	.434		

Figure 10-23   Courtesy of Boston Gear

From Chapter 9 the designated shaft tolerance used in this book is $+.0000/-.0004$ in. or $+0.000/-0.010$ mm. Therefore, the maximum clearance between a gear's bore and a shaft is

+.0005	Gear bore—max
−.0004	Shaft—min
Δ = .0009	

Considering that the height of a tooth with a diametral pitch of 16 is about .135 in., this is a very exact tolerance.

10-26 CENTER DISTANCE AND BACKLASH

Center distance is the distance between the centers of two meshing gears. See Figure 10-24. The center distance is calculated using the gears' pitch diameters, as shown by the formula in Figure 10-24. This is an ideal distance and does not include any tolerances.

Backlash

Backlash is the difference by which the amount of excess tooth space exceeds the thickness of the meshing tooth. As center distance increases, backlash increases. If backlash becomes too large, the gear's teeth can become damaged.

$$CD = \frac{\text{Pitch } \varnothing_1 + \text{Pitch } \varnothing_2}{2}$$

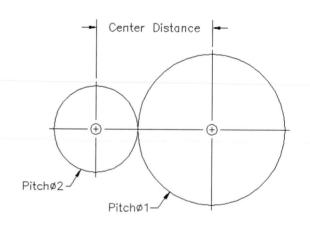

Figure 10-24

All meshing gears include some backlash. If the center distance between the gears is ideal, based on the preceding formula, there is some backlash between the gears. The amount of backlash varies according to the size, type, and quality of the gear. The quality of gears is rated by the American Gear Manu-

Backlash - Boston Gear Inc.

Diametral Pitch	Backlash (in.)
3	.013
4	.010
5	.008
6	.007
7	.006
8–9	.005
10–13	.004
14–32	.003
33–36	.0025

Values are for perfect center distance condition.

Figure 10-25 Courtesy of Boston Gear

facturer's Society (AGMS) by the letters A, B, C, D, and E, with A being the lowest quality level and E the highest. For purposes of this book, the lesser quality types will be used. These gears are acceptable for all but the most exacting applications.

Gear manufacturers specify the backlash for their gears. Figure 10-25 shows the backlash specifications for gears manufactured by Boston Gear.

Boston Gear recommends that the increase in the center distance be equal to half the backlash specified for ideal center distance conditions. Thus, for two meshing gears of 10 diametral pitch, the average allowable backlash would be half of .004 or .002 in.

The center distance tolerance should not exceed an additional 50% allowance. A tolerance of +.001/−2.000 generates a total tolerance of .001, so it would be within the specified .002 limit. See Figure 10-26.

W.M. Berg, Inc. specifies backlash in mating gears at .001 to .002 for Class C Quality 10 gears with 16 to 48 diametral pitch. Again, a tolerance of ±.0005 meets these requirements.

Backlash and center distance tolerances are only one of several tolerance that affect the accuracy of gears' interfaces. The following sample problem will consider all the tolerances involved.

10-27 SAMPLE PROBLEM SP10-8

This problem will be solved using gears from Boston Gear Inc. There are many other gear manufacturers whose products would fit the typical data presented here.

The +0.001/-0.000 tolerance adds a total tolerance of 0.001, so is acceptable.

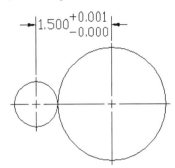

The diametral pitch for both gears is 10.

Figure 10-26

PROBLEM: A student is working on a design project. The project will be powered by a 0.10 HP motor operating at 1800 RPM. The project requires a speed output of between 200 and 230 RPM.

The required final gear ratio is determined by dividing the input RPM by the output RPM.

$$\frac{1800}{200} = 9$$

$$\frac{1800}{230} = 7.8$$

The final gear ratio must be between 7.8 and 9. The recommended highest gear ratio for spur gears is 6:1, so a gear train will be required.

Several different combinations of gears will satisfy the requirements. Gear sets with tooth values of 20/40 and 20/80 will generate a final gear ration of 8:1; however, two sets of 20/60 will generate a final gear ratio of 9:1 and require only two different gears. Figure 10-27 shows the preliminary gear train.

Boston Gear specifies a horespower rating for its gears. The company's web site lists a set of steel gears with 24 diametral pitch, 14.5° pressure angle, and 0.25 face width that will satisfy the project requirements. The 20-tooth gear is rated at .42 HP at 1800 RPM, and .19 HP at 600 RPM; and the 60-tooth gear is rated at .53 HP at 600 RPM, and .24 HP at 200 RPM. Both values are well above the project's 0.10 HP requirement. Figure 10-28 shows the specification sheets for the two gears.

After the gears have been selected, the bearings are selected. The 20-tooth gear has a nominal bore diameter of

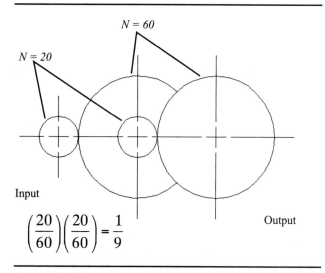

$$\left(\frac{20}{60}\right)\left(\frac{20}{60}\right) = \frac{1}{9}$$

Figure 10-27

0.312, and the 60-tooth, 0.375. See Figure 10-28. Both gears are equipped with setscrews.

Bearings—PV calculations

The gears' bore diameters are used to intially select sleeve bearings. If the project exceeds the capabilities of the sleeve bearing, the use of ball bearings will be investigated.

A Boston Gear Bost-Bronz oil-impregnated, sintered sleeve is selected. Figure 10-29 shows two pages from Boston Gear's web page.

The sleeve bearing requirements were entered in the text boxes. A length of .75 in. and an outside diameter of .502 in. were selected. Sleeve bearing B58-6 was recommended to meet these requirements.

Figure 10-29 shows the bearing's specification sheet. The maximum values for PV, V, and P are listed.

The maximum PV value occurs on the bearing supporting the output gear shaft. The bore of the gear is $\varnothing.3750$, so it will be assumed that the mating shaft diameter is very

Catalog Number
H2420

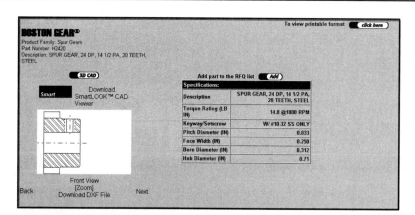

Catalog Number
H2460

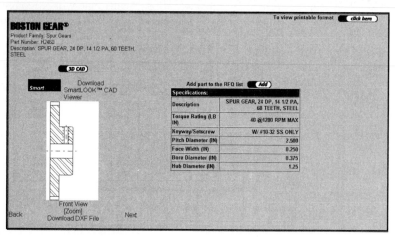

Figure 10-28   From www.bostgear.com; Courtesy of Boston Gear

nearly the same. The gear is turning at 200 RPM. The maximum velocity is

$$V = .262 \ shaft(\varnothing)(N) = .262(.375)(200)$$
$$= 19.65 \ ft/min$$

The maximum force (F) is

$$F = \left(\frac{33,000(HP)}{V}\right) = \frac{33,000(.1)}{19.65} = 167.9 \ lb$$

Therefore, the maximum PV value is

$$PV = \frac{.262(F)(N)}{L} = \frac{.262(168.4)(200)}{.75} = 11,731$$

which is well below the specified 50,000 PV limit for the bearing.

A check of the system's input maximum V and P values also indicates that they are below the bearing's specified limits; therefore, the recommended bearing is acceptable.

$$V_{max} = .262(shaft\ \varnothing)(N)$$
$$= .262(.312)(1800) = 147$$

$$P_{max} = \frac{PV}{V} = \frac{11,765}{147} = 80$$

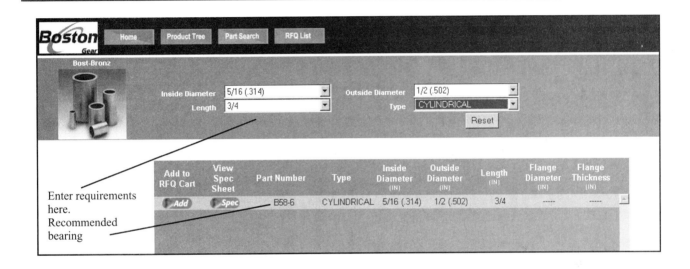

Enter requirements here.
Recommended bearing

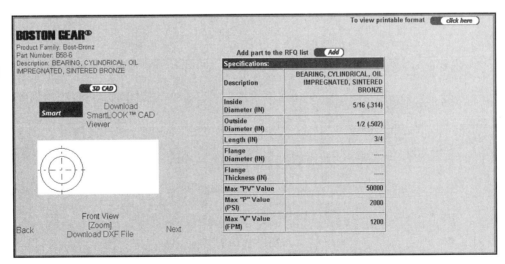

Specification sheet of recommended bearing.

Figure 10-29 Courtesy of Boston Gear

The selected sleeve bearing is well within the bearing's specified design limits, so a second bearing for the larger gear bore can be selected from the same bearing lisiting. In this example a B68-8 was selected.

Shaft

Three different shafts are required for the gear train: one shaft for the 20-tooth input gear, one for the 60-tooth output gear, and one shaft that can accommodate both gears.

Figure 10-30 shows an intial layout for a shaft that holds two gears and is mounted into two different sleeve bearings. The layout is used to size the shaft in terms of both the two required diameters and the overall length.

The right side of the shaft supports the 20-tooth H2420 gear and is in turn supported by the B58-6 sleeve bearing. The inside diameter of the bearing is Ø.3140 with a tolerance of +.0000/−.0010.

Boston Gear suggests a clearance tolerance of .0004 minimum and .0014 maximum. The shaft is operating at 600 RPM, so the data in Figure 9-2 suggest a tolerance of .0004 minimum and .0015 maximum. The data in Figure 9-2 also suggest a tolerance of .0007 minimum and .0024 maximum for speeds above 600 RPM. The shaft tolerance for the higher RPM values is calculated as follows:

```
 .3130—Hole—min
−.3126—Shaft—max
 .0004—Clearance—min
```

```
 .3140—Hole—max
−.3116—Shaft—min
 .0024—Clearance—max
```

These calculations define the shaft dimensions as .3126 maximum and .3116 minimum, for a total tolerance of .0010.

The bore of the 20-tooth gear is .3125±.0005. The minimum bore diameter is .3120, or .0006 smaller than the maximum shaft diameter as defined for the bearing interface. *This tolerance interference is not acceptable* because in the worst case the gear will not fit onto the shaft. Its bore could be too small. A new bearing must be selected.

When selecting parts for a design application it is not unusual to have to make several iterations. Parts are selected and analyzed. If problems occur, other parts are selected until an acceptable solution is reached.

Boston Gear sleeve bearing B47-4 is now selected. This bearing has an inside diameter of .2520 with a tolerance of +.0000, −.0010. Figure 10-31 shows the bearing's specification sheet.

The maximum *PV, P,* and *V* values are calculated again, and the B47-4 bearing is within the specified limits. The shaft tolerances are calculated as follows:

```
 .2510   Hole—min
−.2506   Shaft—max
 .0004   Clearance—min
```

```
 .2520   Hole—max
−.2496   Shaft—min
 .0024   Clearance—max
```

Initial layout for a shaft that holds two gears

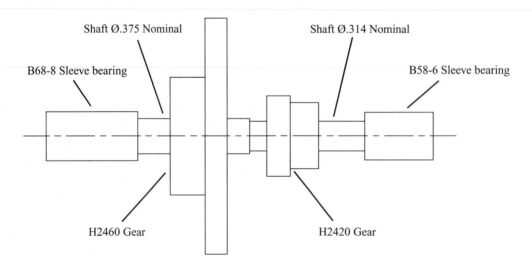

Shaft Ø.375 Nominal

Shaft Ø.314 Nominal

B68-8 Sleeve bearing

B58-6 Sleeve bearing

H2460 Gear

H2420 Gear

Figure 10-30

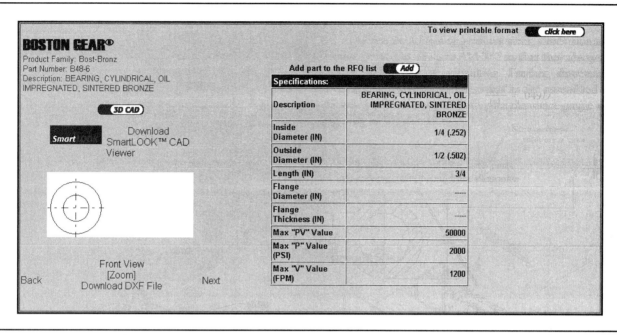

Figure 10-31   Courtesy of Boston Gear

The shaft tolerances are ∅.2506/.2496 for the portion of the shaft that fits into the bearing.

The portion of the shaft that supports the gear can now be toleranced. It is desirable to have the gear fitted with a clearance fit so that it can easily be removed if necessary. The gear includes a #10-32 setscrew that when tightened will facilitate the force transfer between the shaft and the gear. A minimum clearance of .0005 is selected, with a total shaft tolerance of .0010. The shaft dimension calculations are as follows:

.3120	Gear bore—min
−.0005	Specified minimum clearance
.3115	Shaft—maximum diameter
−.0010	Specified shaft tolerance
.3105	Shaft—minimum diameter

Therefore, the maximum clearance will be .0025. This value is acceptable for the applications of this problem. If the design situation determined that this value was too large, then the tolerances would have to be recalculated.

The 60-tooth gear bore is .3750±.0005 using the same shaft specifications as were used for the 20-tooth gear, the shaft tolerances are as follows:

.3745	Gear bore—min
−.0005	Specified minimum clearance
.3740	Shaft—maximum diameter
−.0010	Specified shaft tolerance
.3730	Shaft—minimum diameter

The shaft tolerances are ∅.3740/.3730 for the portion of the shaft supporting the 60-tooth gear.

It is good design practice, whenever possible, to minimize the number of different parts used in an assembly. For this reason it was decided to use the same sleeve bearing at both ends of the shaft. Figure 10-32 shows the resulting shaft. The shaft is designated as Shaft-1.

The surface finish designations indicated by the check marks with numbers are discussed in Section 8-25. The numbers 12 and 32 are in general compliance with AGMA Quality 10 standards and can be used for exercise problems throughout the book.

Figure 10-37 shows the input shaft, and Figure 10-38 shows the output shaft. The extra length of 1.00 in. is for coupling to the drive motor and to have the output shaft extend beyond the housing.

Housing

The bearings must now be fitted into a support housing. The outside diameter of an R48-6 Boston Gear sleeve bearing is ∅.502 with a tolerance of +.0000,−.0010. The recommended press fit allowance for the bearing in a steel housing is approximately .0007 minimum and .0022 maximum. See Figure 10-34. The hole in the housing is calculated as follows. In this example the shaft dimension is the outside diameter of the bearing.

.5020   Shaft—max
<u>.4998</u>   Hole—min
.0022   Interference—max

.5010   Shaft—min
<u>.5003</u>   Hole—max
.0007   Interference—min

The housing hole dimensions are .5003/.4998. Because the system was designed so that all three shafts are supported by the same bearing, all the housing holes are the same size.

Press fit shrinkage

As a sleeve bearing is pressed into a housing the bearing inside diameter is reduced. The general rule is that the shrinkage is about .75 of the press fit allowance. In this example the average press fit allowance is .0013, so the shrinkage is as follows:

$$(.0013)(.75) = .000975 \approx .001$$

The .001 shrinkage value is accounted for by subtracting .001 from the ends of the shafts that fit into the bearings. The revised shaft tolerance is Ø.2406/.2396. Figure 10-32 shows a dimensioned drawing for the shaft that supports two gears.

Center distance

The center distance between meshing gears is calculated as follows:

$$CD = \frac{PD1 + PD2}{2}$$

where PD is the pitch diameter of the gear. For the two gears selected for this example the center distance is 1.6665 calculated as follows:

$$CD = \frac{.833 + 2.500}{2} = 1.6665$$

The center distance must include a tolerance, which will affect the gear backlash. The recommended backlash for a 24 diametral pitch gear is .003. It is also recommended that the tolerance of the gear's center distance not exceed half the average backlash, or in this example, .0015. In this example it was decided to assign a positional tolerance of .0010 about the holes' center point at maximum material condition. See Section 8-42 for an explanation of positional tolerance. Figure 10-33 shows an inital drawing for the support housing plate. No thickness or mounting holes are included.

The gears are drawn using *phantom lines,* which are used to show objects that are not part of the the object being defined but whose size and shape are relative to the part being defined.

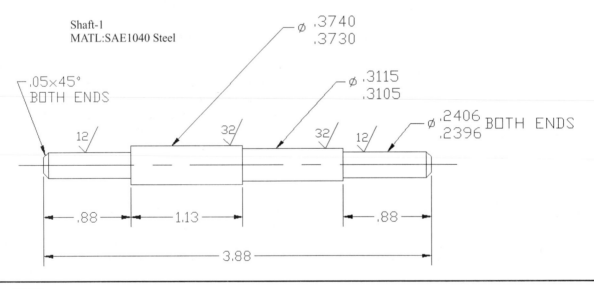

Figure 10-32

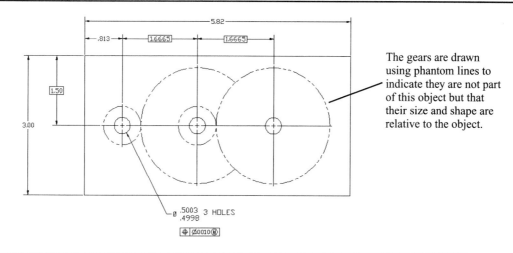

The gears are drawn using phantom lines to indicate they are not part of this object but that their size and shape are relative to the object.

Figure 10-33

Tolerances

A review of the various tolerances that affect the gear meshing is shown in Figure 10-34. Consider the *worst-case* conditions. The interference fit will not contribute to the overall clearance of the system. As the shaft spins within the bearing it will tend to center itself and ride on the lubricating film, lessening its effects on the total clearance value. There will be maximum clearance when the system is at rest. As power is applied the system will center itself, but during the initial activation there could be a maximum clearance of as much as .0050. This value is .0020 more than the recommended backlash tolerance of .0030. Once the system is up to speed, the worst-case clearance would be be .0030, or the same as the recommended backlash allowance. The startup time is very fast, and the overall loads on the system are well within the design limits of both the gears and bearings, so the tolerances as stated are acceptable for the application.

Gear box

The shafts are to be supported at both ends by plates that will be part of a housing or gear box. The support plates are in turn aligned and held in place by other plates. Figure 10-35 shows a dimensioned and toleranced support plate for the gears based on the derived center distances and gear sizes. The box is held together by 18¼-20 UNC × 1.00 flat head screws. Each countersunk hole has a .266 clearance hole, and each blind threaded hole is set for a .50 thread depth.

Figure 10-36 shows a dimensioned and tolerance base plate that will support two support plates. A second base plate could be added to the top of the support plates, and then two side plate created to enclose the gears. For this example only two support plates and one base plate will be used so that the final gear assembly will be more visible.

Final shaft drawings

Figures 10-37 and 10-38 show detail drawings of the other two shafts: Shaft-2 and Shaft-3. The third shaft was detailed in Figure 10-32. Shaft-2 and Shaft-3 are longer than Shaft-1 so that they extend through and beyond the support plates. This allows them to be coupled to either the drive motor or the output device.

Tolerance Chart			
	Max	Min	Total Tolerance
Gear Bore/Shaft	.0025	.0005	.0020 Clearance
Shaft/Bearing	.0024	.0004	.0020 Clearance
Bearing/Housing	.0022	.0007	.0015 Interference
Center Distance	.0000	.0010	.0010 Clearance

Figure 10-34

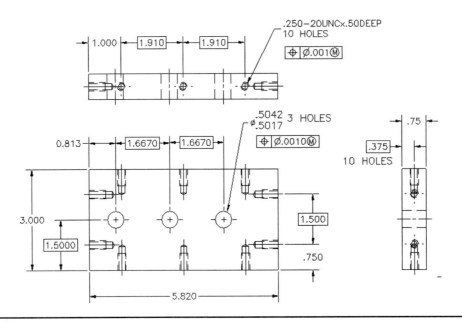

Figure 10-35

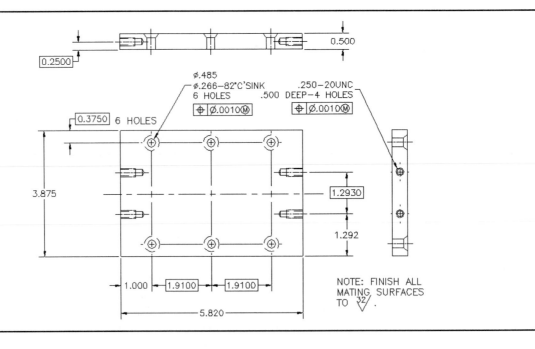

Figure 10-36

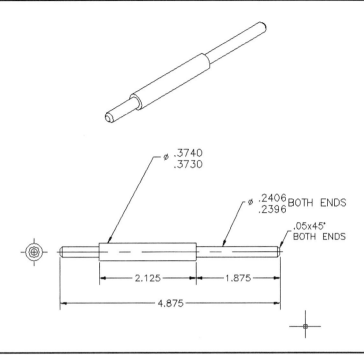

Figure 10-37

Final assembly drawing

Figure 10-39 shows an exploded scene and an exploded isometric drawing of the gear box assembly. Only one fastener has been included.

Parts list

Figure 10-40 shows the assembly's part list (BOM). The headings have been modified from the default Inventor format. A column for part numbers has been added, and the Standard column has been deleted. Parts lists are discussed in more detail in Section 5-12.

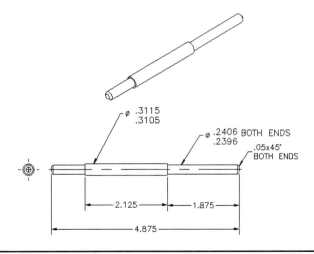

Figure 10-38

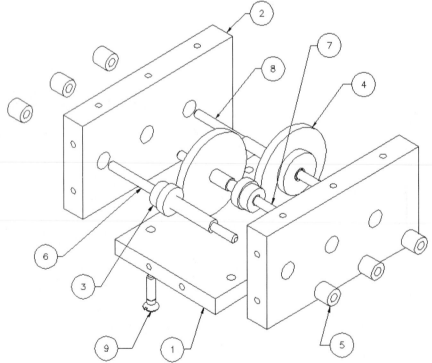

Figure 10-39

	Item	Qty	Name	Material	Part Number	Vendor
+	1	1	PLATEBA	SAE1020	BU-1	
+	2	2	PLATESUP	SAE1020	BU-2	
+	3	2	GEARH2420	Steel	H2420	Boston Gear
+	4	2	GEARH2460	Steel	H2460	Boston Gear
+	5	6	BEARB47-4	BOST-BRONZ	B47-4	Boston Gear
+	6	1	SHAFT-3A	SAE1040	AM312-1	
+	7	1	SHAFT-1A	SAE1040	AM312-2	
+	8	1	SHAFT-2A	SAE1040	AM312-3	
	9	1	SLOTTED FLAT COUNTERSUNK	Steel	.250-20UNCx1.000	

Figure 10-40

The Boston Gear part numbers were used for Boston Gear parts. All other part numbers were created for this exercise. Standard screw notation is used to define the flat head screws.

The names of the parts are the file names created when the individual models were created. These cannot be changed.

Figure 10-41 shows the assembled gear box. The different shades of gray were obtained by using different colors for the parts.

10-28 GEAR ASSEMBLY ANIMATION

Gear assemblies can be *animated,* that is, made to rotate on screen. A gear ratio for the rotation may also be defined, enabling the gears to rotate at different velocities.

Gear rotation is done manually, unlike assembly drawing that can be animated using the Drive Constraint tool, as was explained in Section 5-16. A point is located

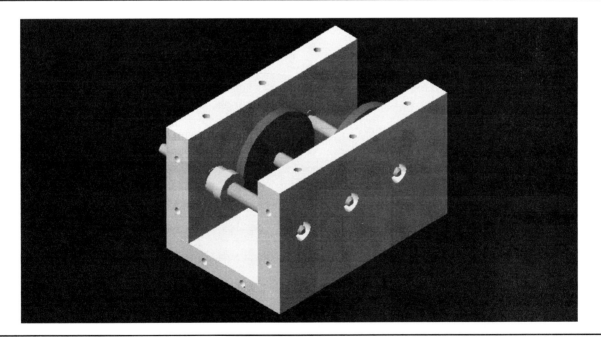

Figure 10-41

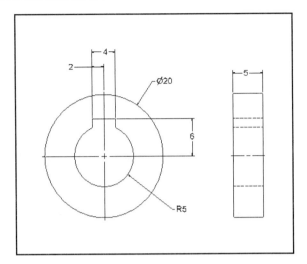
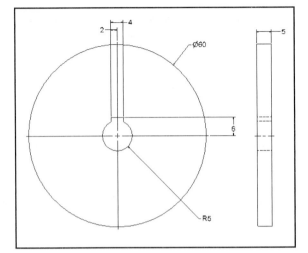
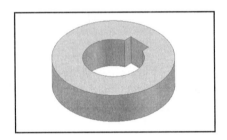
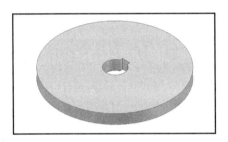

Figure 10-42

near the outside edge of the gear, and the cursor is moved in a rotating pattern, resulting in an equivalent rotation of the gears.

Figure 10-42 shows two cylinders that represent two gears. The smaller gear has a diameter of 20 mm, and the larger gear a diameter of 60 mm. Their rotation ratio is 3:1. Both have keyways dimensioned as shown. Tooth representations could be added to the cylinders. Both gears were saved as Standard (mm).ipt files.

To animate two gears

1. Create a Standard (mm).iam file and use the Place Component tool to place copies of the two gears on the screen.

   See Figure 10-43.

2. Use the Place Constraint tool and Tangent option to align the edges of the gears. Use the Flush tool to constraint the top surfaces of both gears to each other.

3. Right-click the Ø20G-key gear in the browser box and remove the Grounded constraint.

   The pushpin will disappear from the gear callout.

4. Create a work axis for each gear through the bore.
5. Ground the two work axes.
6. Click the Place Constraint tool, and select the Motion option and the Reverse option.
7. Set the ratio for 1/3, then click the Ø60 gear first, then the Ø20. Click the Apply button.
8. Locate the cursor on the outside edge of the top surface of the Ø60 gear and slowly move the cursor in a rotary motion.

   The gears will move in opposite directions at a 3:1 ratio.

10-29 SIMULATED GEAR DRAWINGS

Inventor can be used to draw estimated gear shapes. These drawing are not true gear drawings but may be used in presentation drawings to create a more realistic picture.

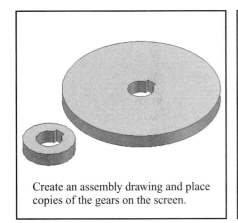

Create an assembly drawing and place copies of the gears on the screen.

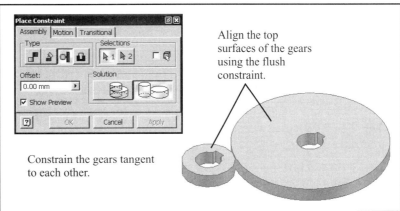

Align the top surfaces of the gears using the flush constraint.

Constrain the gears tangent to each other.

Remove the Grounded constraint from the first gear entered.

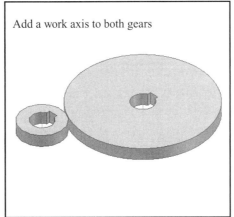

Add a work axis to both gears

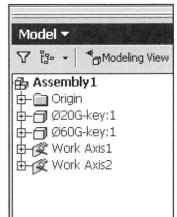

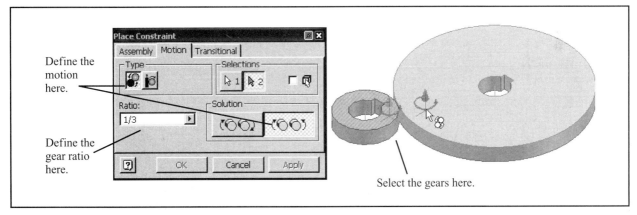

Define the motion here.

Define the gear ratio here.

Select the gears here.

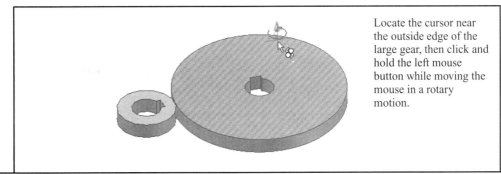

Locate the cursor near the outside edge of the large gear, then click and hold the left mouse button while moving the mouse in a rotary motion.

Figure 10-43

To draw an estimated gear

1. Start a new drawing using the Standard.ipt format.

   In this example the metric option was selected.

2. Draw a Ø30 circle.
3. Draw a tooth using the dimensions shown in Figure 10-44.

   The dimensions used were calculated to create 30 teeth around the Ø30 circle.

4. Use the Circular Pattern tool and array 30 teeth around the Ø30 circle.
5. Access the Part Features panel and extrude the circle and all 30 teeth a distance of 1.50 mm.

6. Create a new sketch plane and locate a hole center point at the center of the gear.
7. Add a Ø5 hole to the center of the gear.

Figure 10-44 shows two of the estimated gears meshed together.

Another estimation

Gears also may be estimated using a triangular shape. The gear shown in Figure 10-45 was created by first drawing a Ø1.50 in. circle, adding a triangular shape, then using the Circular Pattern tool to add enough teeth to make the gear look realistic. In this example 60 triangles were added. A Ø.125 hole was placed in the center of the gear.

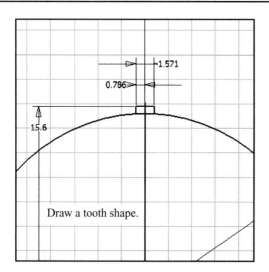

Draw a tooth shape.

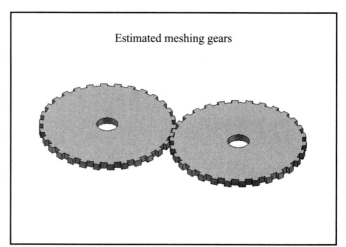

Estimated meshing gears

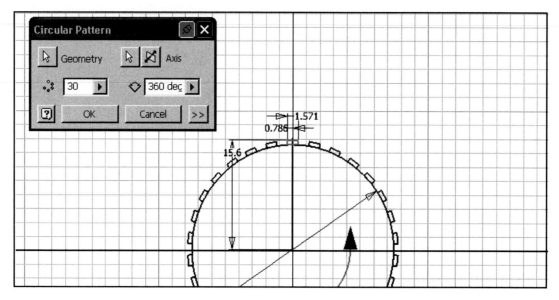

Figure 10-44

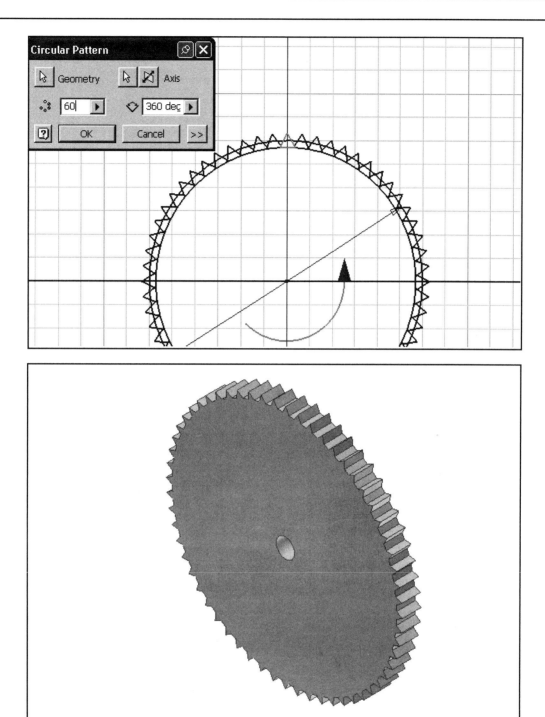

Figure 10-45

10-30 EXERCISE PROBLEMS

EX10-1

Given the gear setup shown in Figure 10-46, what is the output speed if the input speed is

A. 1750 RPM
B. 1600 RPM
C. 2400 RPM

and if the gears have the following tooth specifications:

D. Gear 1 = 20, Gear 2 = 40,
   Gear 3 = 20, Gear 4 = 50
E. Gear 1 = 20, Gear 2 = 36,
   Gear 3 = 24, Gear 4 = 60
F. Gear 1 = 20, Gear 2 = 38,
   Gear 3 = 22, Gear 4 = 52

EX10-2

If an input motor generates .375 HP at 1750 RPM, and the gear setup is as shown for EX10-1 E, what is the output torque on gear 4?

EX10-3

Given the gear setup shown in Figure 10-47, what is the output speed if the input speed is

A. 1650 RPM
B. 1000 RPM
C. 3600 RPM

and if the gears have the following tooth specifications:

D. Gear 1 = 20, Gear 2 = 40
   Gear 3 = 20, Gear 4 = 50
   Gear 5 = 20, Gear 6 = 40
E. Gear 1 = 20, Gear 2 = 34
   Gear 3 = 22, Gear 4 = 50
   Gear 5 = 24, Gear 6 = 64
F. Gear 1 = 20, Gear 2 = 50
   Gear 3 = 22, Gear 4 = 58
   Gear 5 = 20, Gear 6 = 68

EX10-4

If an input motor generates .500 HP at 1750 RPM, and the gear setup is as shown for EX10-3 F, what is the output torque on gear 6?

EX10-5

If an input motor generates .125 HP at 3600 RPM, and the gear setup is as shown for EX10-3 D, what is the output torque on gear 6?

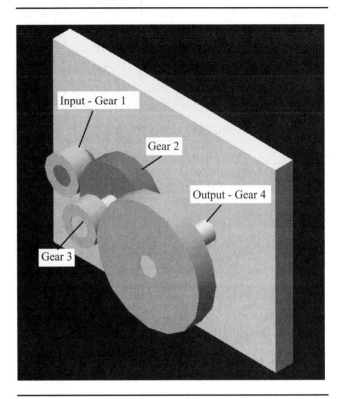

Figure 10-46

EX10-6

If an input motor generates .750 HP at 1250 RPM, and the gear setup is as shown for EX10-3 E, what is the output torque on gear 6?

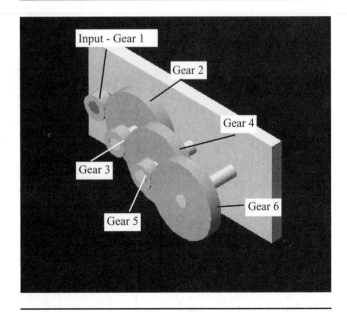

Figure 10-47

EX10-7

Design a gear train that reduces an input speed of 1800 RPM to an approximate output speed as listed. No two meshing gears may have a ratio greater than 4:1.

Present your solution as cylindrical models such as shown in Figure 10-5. Label each gear.

A. 400 RPM
B. 300 RPM
C. 200 RPM
D. 150 RPM
E. Animate the gear train.

EX10-8

Go to a gear manufacturer's web site and select gears that will satisfy the ratios called for in EX10-7. List the selected gears by manufacturer's part number. Create a drawing based on the pitch diameters specified by the manufacturers. Two possible web sites are

A. bostgear.com
B. wmberg.com

EX10-9

A. A gear has the following specifications:

Pressure angle = 14.5°
Face width = .750 in.
Material = SAE 1040 steel—untreated
Diametral pitch = 8
Number of teeth = 56
Pitch diameter = 7.000 in.

If the gear is operating at 1750 RPM, what is the allowable bending load on the tooth?

B. What is the allowable bending if the gear is subject to a safety factor of 1.25, is operated under a light shock load for 24 hours a day, and is lubricated using an oil submerging condition?

EX10-10

A. A gear has the following specifications:

Pressure angle = 14.5
Face width = 1.250 in.
Material = cast iron
Diametral pitch = 48
Number of teeth = 96
Pitch diameter = 2.000 in.

If the gear is operating at 1750 RPM, what is the allowable bending load on the tooth?

B. What is the allowable bending if the gear is subject to a safety factor of 2, is operated under a heavy shock load for 24 hours a day, and is lubricated using an oil drip method?
C. Specify a gear by manufacturer and part number that will satisfy the part A requirements.

EX10-11

A. A gear has the following specifications:

Pressure angle = 20
Face width = .125 in.
Material = bronze
Diametral pitch = 24
Number of teeth = 20
Pitch diameter = 0.833 in.

If the gear is operating at 3600 RPM, what is the allowable bending load on the tooth?

B. What is the allowable bending if the gear is subject to a safety factor of 2, is operated under a light shock load for 8 hours a day, and is lubricated using an oil drip method?

EX10-12

A. A gear has the following specifications:

Pressure angle = 20
Face width = .250 in.
Material = plastic
Diametral pitch = 10
Number of teeth = 60
Pitch diameter = 6.000 in.

If the gear is operating at 1250 RPM, what is the allowable bending load on the tooth?

B. What is the allowable bending if the gear is subject to a safety factor of 1.5?

EX10-13

A gear has the following requirements:

Pressure angle = 14.5°
Allowable bending = 100 lb
Material = SAE 1040 steel—untreated
Diametral pitch = 16
Number of teeth = 24
Pitch diameter = 1.000 in.

If the gear is operating at 1750 RPM, what is the required face width?

EX10-14

A gear has the following specifications:

Pressure angle = 20°
Torque = 75 in.-lb
Material = cast iron
Diametral pitch = 48
Number of teeth = 54
Pitch diameter = 1.125 in.

If the gear is operating at 1750 RPM, what is the required face width?

EX10-15

A gear has the following requirements:

Pressure angle = 14.5°
HP = .500
Material = SAE 1040 steel—untreated
Diametral pitch = 32
Number of teeth = 96
Pitch diameter = 3.000 in.

If the gear is operating at 1750 RPM, what is the required face width?

EX10-16

A gear has the following requirements:

Pressure angle = 14.5°
Allowable bending = 25 lb
Material = plastic
Diametral pitch = 1.000 in.
Number of teeth = 48

If the gear is operating at 1750 RPM, what is the required face width?

EX10-17

Two gears with the following specifications are meshed together. Given the backlash requirements stated in Figure 10-22, what is the appropriate center distance and tolerance?

Gear 1:
Diametral pitch = 24
Number of teeth = 16
Pitch diameter = .667
Bore = .3125±.0005
Gear 2:
Diametral pitch = 24
Number of teeth = 48
Pitch diameter = 2.000
Bore = .3750±.0005

EX10-18

Two gears with the following specifications are meshed together. Given the backlash requirements stated in Figure 10-22, what is the appropriate center distance and tolerance?

Gear 1:
Diametral pitch = 20
Number of teeth = 18
Pitch diameter = .900
Bore = .3750±.0005
Gear 2:
Diametral pitch = 20
Number of teeth = 50
Pitch diameter = 2.500
Bore = .3750±.0005

EX10-19

Two gears with the following specifications are meshed together. Given the backlash requirements stated in Figure 10-22, what is the appropriate center distance and tolerance?

Gear 1:
Diametral pitch = 32
Number of teeth = 16
Pitch diameter = .500
Bore = .1875±.0005
Gear 2:
Diametral pitch = 32
Number of teeth = 72
Pitch diameter = 3.000
Bore = .3125±.0005

EX10-20

Design a gear box that reduces an input speed of 1750 RPM to approximately 275 RPM. No gear ratio may exceed 5:1. The input horsepower is .25 HP. Select the gears from a manufacturer's catalog or from a catalog on the web.

A. Gears
B. Sleeve bearings
C. Design and dimension appropriate shafts. Include all tolerances.
D. Specify the center distances with tolerances.
E. Design two support plates and a base plate as shown in Figure 10-36. Include all dimensions and tolerances.
F. Prepare models then detail drawings of all shafts and plates.
G. Prepare an exploded scene of the gear box and an exploded isometric assembly drawing with assembly numbers.
H. Prepare a parts list for the gear box.
I. Animate the gear train.

EX10-21

Design a gear box that reduces an input speed of 1750 RPM to approximately 100 RPM. No gear ratio may exceed 5:1. The input horsepower is .50 HP. Select the gears from a manufacturer's catalog or from a catalog on the web.

A. Gears
B. Sleeve bearings
C. Design and dimension appropriate shafts. Include all tolerances.
D. Specify the center distances with tolerances.
E. Design two support plates and a base plate as shown in Figure 10-36. Include all dimensions and tolerances.
F. Prepare models then detail drawings of all shafts and plates.
G. Prepare an exploded scene of the gear box and an exploded isometric assembly drawing with assembly numbers.
H. Prepare a parts list for the gear box.
I. Animate the gear train.

EX10-22

Design a gear box that reduces an input speed of 2400 RPM to approximately 350 RPM. No gear ratio may exceed 5:1. The input horsepower is .75 HP. Select the gears from a manufacturer's catalog or from a catalog on the web.

A. Gears
B. Sleeve bearings
C. Design and dimension appropriate shafts. Include all tolerances.
D. Specify the center distances with tolerances.
E. Design two support plates and a base plate as shown in Figure 10-36. Include all dimensions and tolerances.
F. Prepare models then detail drawings of all shafts and plates.
G. Prepare an exploded scene of the gear box and an exploded isometric assembly drawing with assembly numbers.
H. Prepare a parts list for the gear box.
I. Animate the gear train.

EX10-23

Design a gear box that reduces an input speed of 1750 RPM to approximately 50 RPM. No gear ratio may exceed 5:1. The input horsepower is .375 HP. Select the gears from a manufacturer's catalog or from a catalog on the web.

A. Gears
B. Sleeve bearings

C. Design and dimension appropriate shafts. Include all tolerances.
D. Specify the center distances with tolerances.
E. Design two support plates and a base plate as shown in Figure 10-36. Include all dimensions and tolerances.
F. Prepare models then detail drawings of all shafts and plates.
G. Prepare an exploded scene of the gear box and an exploded isometric assembly drawing with assembly numbers.
H. Prepare a parts list for the gear box.
I. Animate the gear train.

EX10-24

Design a gear box that reduces an input speed of 1750 RPM to approximately 125 RPM. No gear ratio may exceed 5:1. The input horsepower is .50 HP. Select the gears from a manufacturer's catalog or from a catalog on the web.

A. Gears
B. Sleeve bearings
C. Design and dimension appropriate shafts. Include all tolerances.
D. Specify the center distances with tolerances.
E. Design two support plates and a base plate as shown in Figure 10-36. Include all dimensions and tolerances.
F. Prepare models then detail drawings of all shafts and plates.
G. Prepare an exploded scene of the gear box and an exploded isometric assembly drawing with assembly numbers.
H. Prepare a parts list for the gear box.
I. Animate the gear train.

EX10-25 DRAWING PROJECTS THAT CAN BE BUILT

The gear assembly in Figure 10-48 is made from Etech parts that are available at many toy stores or on the web (one possible site is www.constructiontoys.com).

A. Measure and draw the parts as needed, then create the following assembly.
B. What is the gear train's ratio?
C. If the input speed is 6000 RPM, what is the output speed of the last gear?
D. Add M4 setscrews.
E. Add collars as needed to hold the shafts in place.
F. Draw a parts list. (All parts are made from mild steel.)

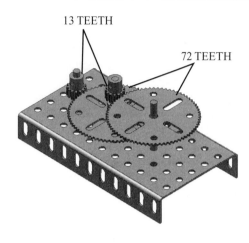
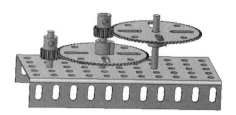

13 TEETH

72 TEETH

Figure 10-48

EX10-26 DRAWING PROJECTS THAT CAN BE BUILT

The gear assembly in Figure 10-49 is made from Etech parts that are available at many toy stores or on the web (one possible site is www.constructiontoys.com).

A. Measure and draw the parts as needed, then create the following assembly.
B. What is the gear train's ratio?
C. If the input speed is 9000 RPM, what is the output speed of the last gear?
D. Add M4 setscrews and M4 screws and nuts as needed to complete the assembly.
E. Add collars as needed to hold the shafts in place.
F. Draw a parts list. (All parts are made from mild steel.)

EX10-27 CLASS ASSIGNMENT

Select a group of gears from a manufacturer's catalog. The gears must all have the same pitch and be made of the same material. Assign each member of the class a different gear, and have each member prepare an accurate drawing of the assigned gear. Save the gear drawings on a disk to create a common gear library. The gear drawings in the library may then be used to create gear trains.

EX10-28

Given an input motor speed of 7200 RPM, design and draw a gear train similar to that shown in EX10-25 that reduces the output speed to approximately 43 RPM.

EX10-29

Given an input motor speed of 10,000 RPM, design and draw a gear train similar to that shown in EX10-26 that reduces the output speed to approximately 6 RPM.

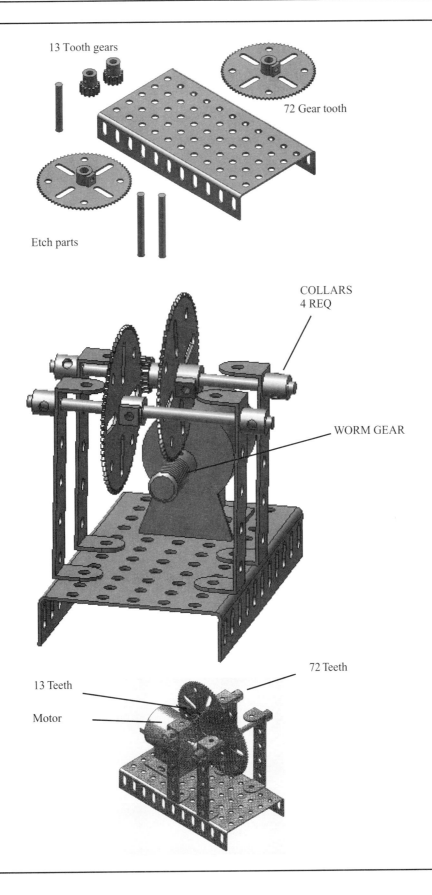

13 Tooth gears

72 Gear tooth

Etch parts

COLLARS
4 REQ

WORM GEAR

13 Teeth

72 Teeth

Motor

Figure 10-49

Cams, Springs, and Keys

11-1 INTRODUCTION

This chapter explains how to design cams and to select keys. It also explains how to match cams to shafts using standard tolerances.

Cams are eccentric objects that convert rotary motion into linear motion. Cams are fitted onto rotating shafts, and lift and lower followers as they rotate.

Figure 11-1 shows a cam drawn using Inventor Release 8. The cam bore includes a keyway.

11-2 DISPLACEMENT DIAGRAMS

Displacement diagrams are used to define the motion of a cam using a linear diagram. The distances are then transferred to a base circle to create the required cam shape.

Figure 11-2 shows a displacement diagram and a cam shape generated from the information on the diagram. The displacement diagram shown is drawn using only straight lines, which is called *uniform motion*. This diagram results in points of discontinuity that can result in erratic follower motion. Several different shapes can be used to smooth out these areas and create smoother follower motion.

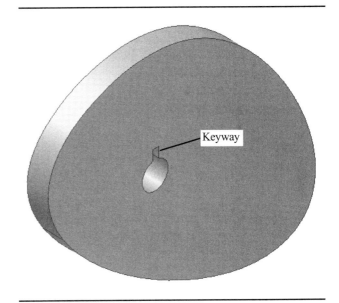

Figure 11-1

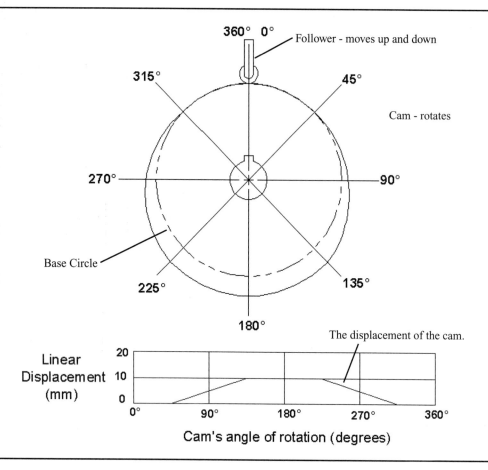

Figure 11-2

Modified uniform motion

Figure 11-3 shows a displacement diagram that lifts the follower using modified uniform motion. Modified uniform motion is created from uniform motion by adding an arc at the beginning and end of the uniform straight line. The radius of the arc is equal to half the total displacement.

In the example shown circles were used to create the required arcs. The circles were drawn, then lines were drawn tangent to the circles. The Trim command was used to remove the excess lines.

Uniformly accelerated and retarded motion

Figure 11-4 shows an example of uniformly accelerated and retarded motion. The defining construction line drawn to the left of the diagram is divided into 18 equal parts. The angle of the line and the dividing spacing is optional. The line is then marked off in proportional segments of 1, 3, 5, and the proportioned segments are projected onto the displacement diagram. There is one space between points

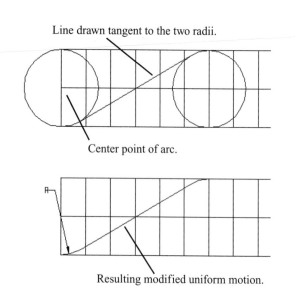

Figure 11-3

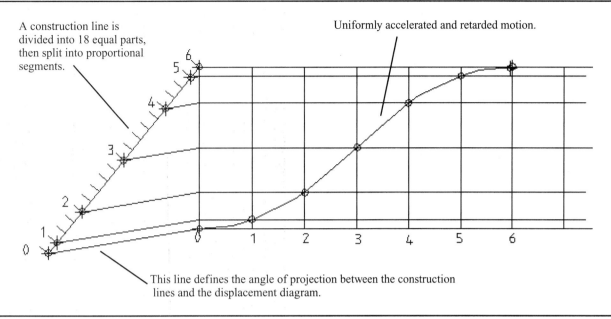

A construction line is divided into 18 equal parts, then split into proportional segments.

Uniformly accelerated and retarded motion.

This line defines the angle of projection between the construction lines and the displacement diagram.

Figure 11-4

0 and 1 on the construction line, 3 spaces between points 1 and 2, and 5 spaces between 2 and 3. The top portion of the line is symmetrical to the bottom portion. The angle of projection is determined by drawing a line from the end of the construction line to the lower left corner of the diplacement diagram.

Harmonic motion

Figure 11-5 shows an example of simple harmonic motion. A semicircle is drawn with its center point on the left edge of the centerline of the displacement diagram. The semicircle is divided into six segments. Six equally spaced vertical lines are added to the displacement diagram. More than six segments and lines could be used provided the number of segments equaled the number of added parallel lines.

The intersections of the ray lines with the edge of the semicircle are projected into the displacement diagram as shown to define the simple harmonic motion.

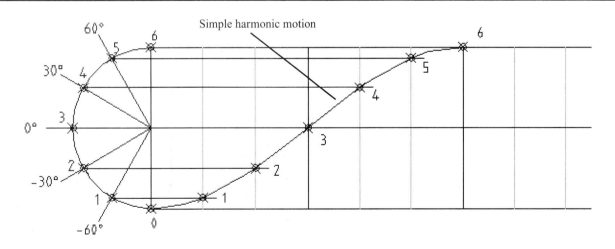

Simple harmonic motion

Figure 11-5

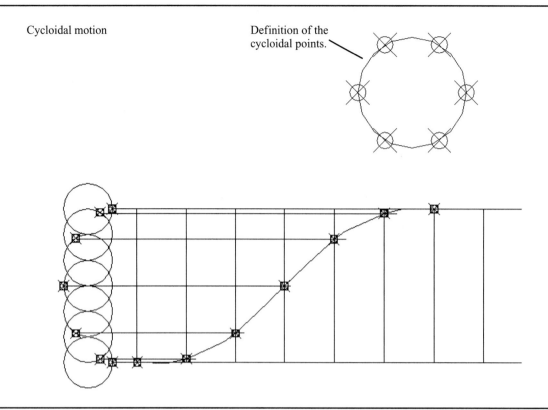

Figure 11-6

Cycloidal motion

Figure 11-6 shows an example of cycloidal motion. Cycloidal motion is defined by the rotation of a point around a circle. In the example shown, six points on the circle are used. Six equal distances are marked off on the displacement diagram (seven lines). The center points of seven circles are drawn equally spaced along the vertical axis of the displacement diagram, and the six points are defined equally spaced on the circumference of the circle. Points 1 and 7 are the same point. The radius of the circles equals ⅙ of the total vertical displacement. Lines are then projected into the displacement diagram intersecting the six equally spaced vertical lines, defining the cycloidal motion. Horizontal to the right equals 0°.

11-3 CREATING A CAM FROM A DISPLACEMENT DIAGRAM

Once a displacement diagram has been defined the vertical distances are used to define the shape of the cam. Figure 11-7 shows a displacement diagram that rises 10 mm using simple harmonic motion and descends 10 mm using modified linear motion. The dwell sections of the cam are as indicated.

Base circle

A cam starts with a base circle. The base circle may be of any diameter but should be large enough so that the change in diameter created by the cam's change of shape does not cause problems with the speed of the follower. Figure 11-7 shows a base circle above the displacment diagram. The diameter of the base circle is 100 mm.

The base circle is divided into sectors using rays. Rays are added as needed. Dwell areas have a constant radius, so only the begining and end points need be defined. The first 90° of the displacement diagram uses simple harmonic motion, so the cam shape is constantly changing. More rays are needed to help define the cam shape.

To add rays

Figure 11-8 shows how to add rays using the Circular Pattern tool.

1. Draw a vertical line from the circle's center point to its edge.
2. Click the Circular Pattern tool located on the 2D Sketch Panel bar.

The Circular Pattern dialog box will appear. See Figure 11-8.

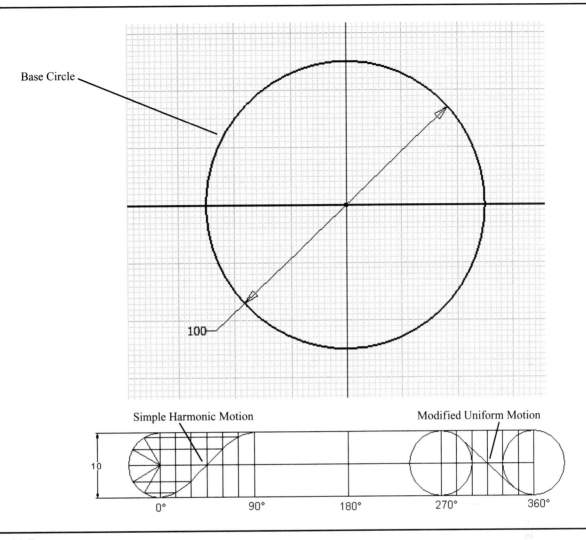

Figure 11-7

3. Select the line as the Geometry, click the arrow in the Axis area, and click the circle's center point. Set the number of rays to 7 (6 sectors defined by 7 rays) and the degrees to 90.
4. Click the Axis box to shift the rays to the right of the vertical line.
5. Add 7 more rays to the lower left quadrant using the same procedure.

To determine the displacement distances

Figure 11-9 shows an enlarged view of the first 90° of the displacement diagram. It was drawn using the commands on the 2D Sketch Panel. The General Dimension tool was used to determine the distance at every 15° along the displacement diagram for the first 90°.

To transfer the displacement distance to the cam

The distances are transferred to the cam by drawing concentric circles about the base circle offset at the distances derived in Figure 11-9. The ray lines are then extended to the concentric circles, and the resulting intersections are used to create a spline that defines the edge of the cam. See Figure 11-10. The values used in Figure 11-10 are diameters, twice the distances defined in Figure 11-9.

Figure 11-11 shows the completed cam sketch and a 3D model of the final cam.

To add a bore

The bore of a cam is the center hole. In this example a Ø16 bore was added to the center of the cam.

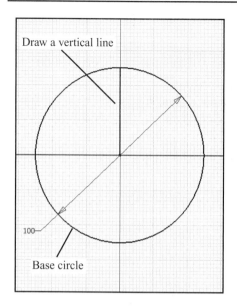

Draw a vertical line

Base circle

100

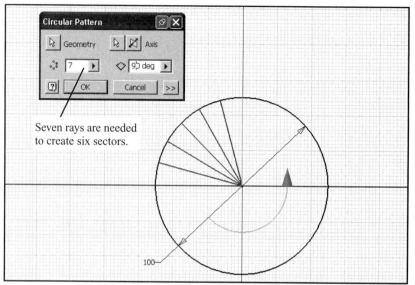

Seven rays are needed to create six sectors.

100

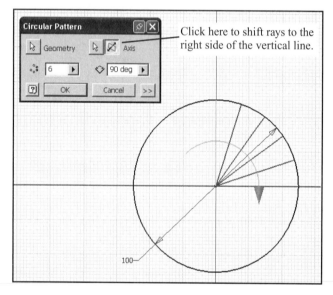

Click here to shift rays to the right side of the vertical line.

100

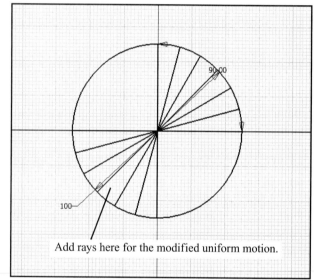

90.00

100

Add rays here for the modified uniform motion.

Figure 11-8

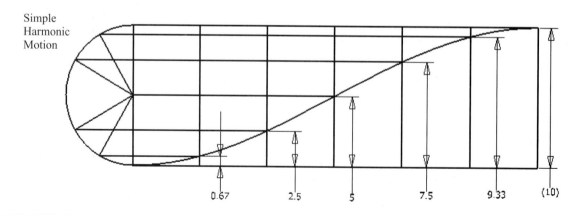

Simple Harmonic Motion

0.67   2.5   5   7.5   9.33   (10)

Figure 11-9

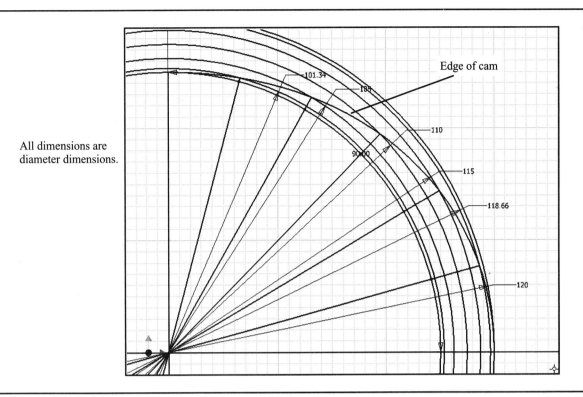

All dimensions are
diameter dimensions.

Figure 11-10

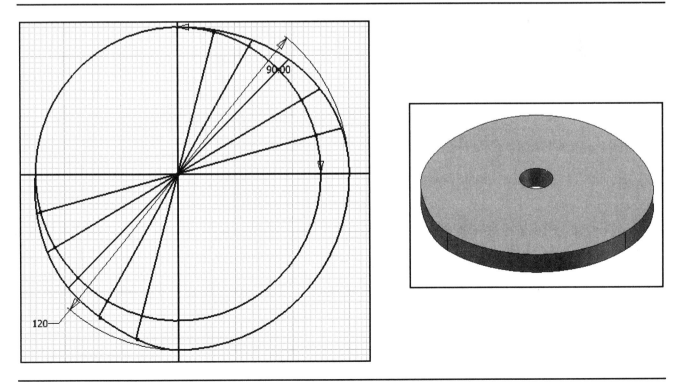

Figure 11-11

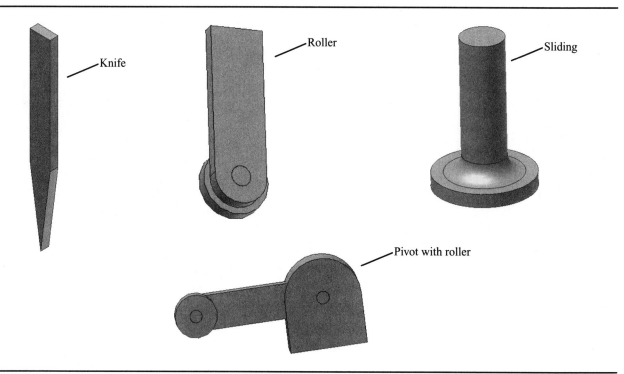

Figure 11-12

1. Click on the cam surface and select the New Sketch option.
2. Create a hole center point, finish the sketch, and add a hole.

Figure 11-11 shows the cam with a bore.

11-4 FOLLOWERS

There are many different types of cam followers. Figure 11-12 shows four different types. As a cam turns, the follower is pushed up and down. A spring is often used to force the follower to stay in contact with the cam surface.

11-5 SPRINGS

Springs are drawn in Inventor with the Coil tool. The procedure is as follows.

1. Sketch a circle and a line as shown in Figure 11-13.

In the example presented the circle is Ø5.00 mm and the line is drawn 15 mm from the circle's center point.

2. Click the Finish the Sketch option to access the Part Features panel, which contains the Coil command.

3. Click the Coil tool.

The Coil dialog box will appear. The circle will automatically be selected as the profile.

4. Select the line as the axis.

The direction of the coils is controlled using the Rotation box.

5. Click the Coil Size tab on the Coil dialog box and set the Pitch for 10 and the Revolution for 6.
6. Click OK.

Figure 11-13 shows the resulting spring. Figure 11-13 also shows a spring based on the same Ø5.00 profile and axis line. The pitch was changed to 6, and the Coil Ends tab was used to define the Start transition angle as 90°. Note the differences between the two springs.

11-6 DESIGNING SPRINGS

Spring design is a function of the spring's material and load requirements. A common spring material is music wire ASTM A228. For simplicity, all springs in this book will be made from round music wire.

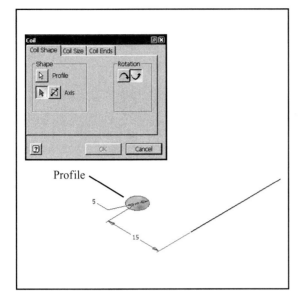

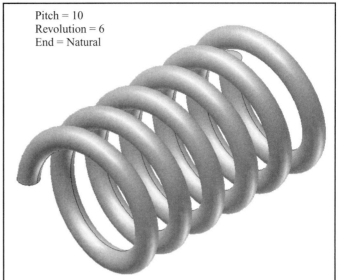

Pitch = 10
Revolution = 6
End = Natural

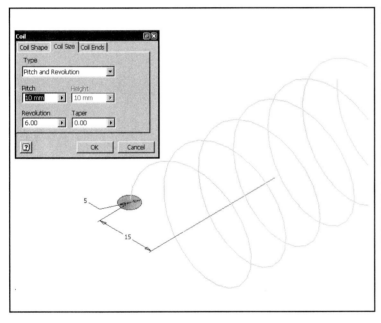

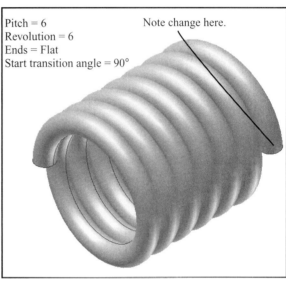

Pitch = 6
Revolution = 6
Ends = Flat
Start transition angle = 90°

Note change here.

Figure 11-13

The deflection of a spring (y) is defined as

$$y = \frac{8PND^3}{Gd^3}$$

where

y = deflection (in.)

P = load (lb)

N = number of active coils

D = mean coil diameter (in.)

G = modulus of elasticity (psi)

d = diameter of the wire (in.)

Figure 11-14 defines the D and d terms.

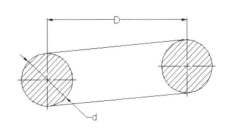

Figure 11-14

Spring Formulas

Number of active coils

$$N = \frac{Gd^4y}{8PD^3}$$

Load

$$P = \frac{Dd^4y}{8ND}$$

Torsional stress

$$\sigma_\tau = \frac{Gdy}{\pi ND^2}$$

Wire diameter

$$d = \frac{\pi \sigma_\tau ND^2}{Gy}$$

where　　σ_τ = torsional stress (psi)

Figure 11-15

Figure 11-15 shows other spring formulas, including two that include torsional stress. Remember that torsional stress is not the same as allowable stress.

Active coils

The spring formulas presented refer to *active coils*. The number of active coils is different from the number of total coils. The number of active coils in a spring depends on how the ends of the spring are manufactured. Figure 11-16 shows four ways that spring ends can be manufactured and the formulas used to calculate the number of active coils in each.

11-7 SAMPLE PROBLEM SP11-1

A group of students are working on a design project that requires them to design a spring. The spring must fit around a ∅.375-in. cam follower and will be subjected to a load of 1.25 lb and a deflection of 0.63 in. It has been decided to make the spring from ASTM A228 music wire that has a modulus of elasticity of 11,500,000 psi, or 200 GPa.

The formula

$$N = \frac{Gd^4y}{8PD^3}$$

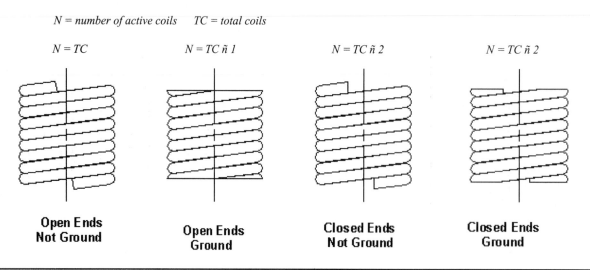

N = number of active coils　　TC = total coils

$N = TC$　　　　$N = TC ñ 1$　　　　$N = TC ñ 2$　　　　$N = TC ñ 2$

Open Ends Not Ground　　**Open Ends Ground**　　**Closed Ends Not Ground**　　**Closed Ends Ground**

Figure 11-16

requires values for D and d, which are not known at this time. It is not unusual for there to be more than one unknown. The procedure is to estimate the values for D and d and calculate the number of coils required. If the number is not acceptable, new D and d values are entered.

If $D = 0.50$ in. and $d = 0.10$ in.:

$$N = \frac{Gd^4 y}{8PD^3} = \frac{(11{,}500{,}000)(0.1)^4(0.63)}{8(1.25)(0.50)^3}$$
$$= 579.6$$

This is deemed to be too many coils.
If $D = 0.50$ and $d = 0.05$:

$$N = \frac{Gd^4 y}{8PD^3} = \frac{(11{,}500{,}000)(0.05)^4(0.63)}{8(1.25)(0.50)^3}$$
$$= 36.23$$

This number of coils is deemed acceptable.

11-8 SPRINGS ON THE WEB

There are several different web sites that deal with springs. Spring manufacturers or distributors often include web pages that allow you to enter a spring's specifications, and the web site will locate a matching spring. Figure 11-17 shows a spring page from the GlobalSpec web page.

Other web sites calculate the required spring from given inputs. The web site www.eFunda.com includes many calculator pages such as the one shown in Figure 11-18. The values used are metric values. Individual values for Young's modulus of elasticity, Poisson's ratio, and density of materials can be found for different materials by clicking on the appropriate headings.

Hooke's law

Hooke's law relates the force applied to a spring to the amount of displacement:

$$P = ky$$

where:

P = force

y = displacement

k = spring constant

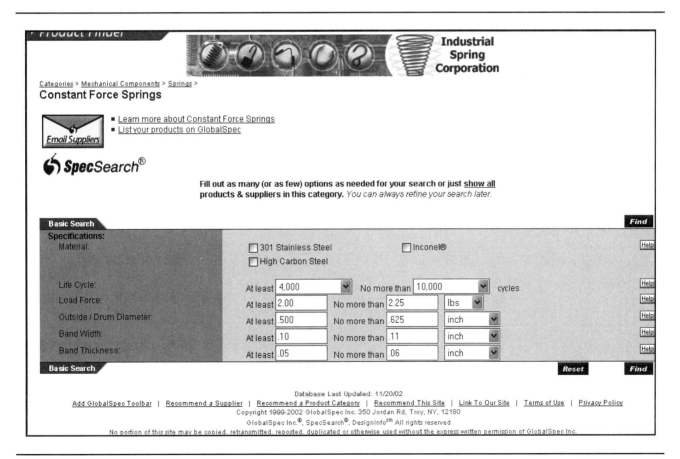

Figure 11-17 Courtesy of www.globalspec.com

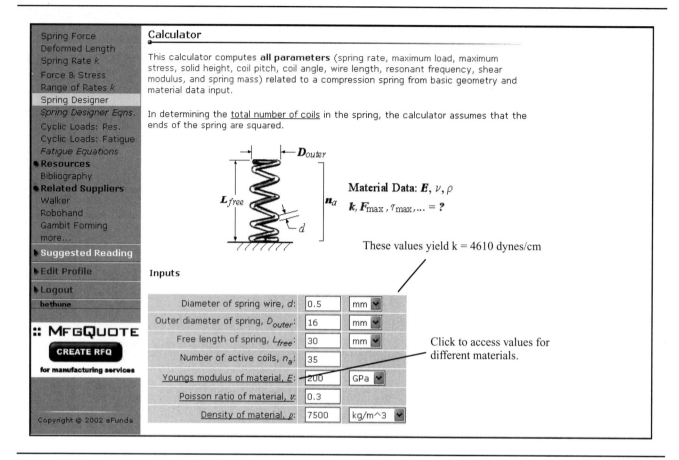

Figure 11-18 Courtesy of www.eFunda.com

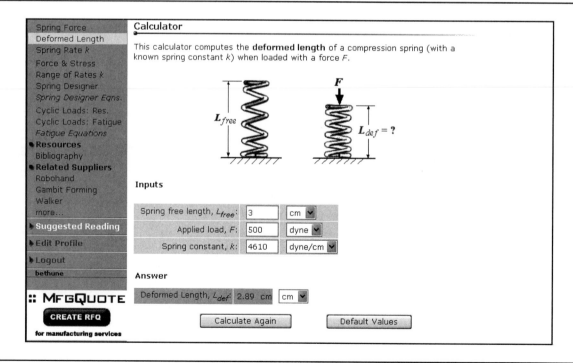

Figure 11-19 Courtesy of www.eFunda.com

The *k* value will vary according to the spring's material, wire diameter, number of active coils, outer diameter, and free length. The spring defined in Figure 11-18 will have a *k* value of 4610 dynes/cm. Figure 11-19 shows another calculation page from eFunda.com, where the *k* value is entered along with a free-length value of 30 mm and a force of 500 dynes. The resulting deformation length is 2.89 cm, or 28.9 mm. The deformation is 0.11 cm.

Calculator pages on web sites allow you quickly to try many different values when designing springs.

Standard wire sizes for springs

Figures 11-20 and 11-21 list standard wire sizes. It is always best to design using standard sizes because they are more readily available from distributors and are more cost efficient than nonstandard sizes.

11-9 KEYS AND KEYWAYS

Cams are mounted onto rotating shafts, and power is transmitted to the cam using keys. Setscrews and pins can also be used.

Types of keys

There are five different types of keys: square, flat, gib, Pratt and Whitney, and Woodruff. See Figure 11-22. This text discusses only square keys, as they are acceptable for the projects outlined in Chapter 13.

Square keys are rectangular prisms that fit into square grooves cut in the shaft and cam called *keyways.* The cross section of a square key is a square. Nominally the keyseats are cut to a depth of half the height of the key.

Keys may be manufacturerd to any size, but standard sizes are available. Figure 11-23 is an abbreviated list of nominal shaft diameter and the recommended key sizes.

Figure 11-24 lists recommended shaft and cam dimensions. Note how the keyways are dimensioned. This is the dimension format recommended by American National Standards Dimensioning and Tolerancing Y14 as published by the the American Society of Mechanical Engineers. All shafts and cams should be dimensioned using this format. Chapters 7and 8 also adhere to ANSI Y14.

To draw a keyway in a shaft

Figure 11-25 shows a Ø1.000 shaft.

1. Click on the front circular surface of the shaft and create a new sketch plane.
2. Draw a vertical line starting at the shaft's center point that can be used with the General Dimension tool to center and size the keyway.

Figures 11-23 and 11-24 indicate that a Ø1.000 nominal shaft should use a ¼ (0.250)-in. square key. The recommended

Millimeters	
Preferred	Second Preference
.10	
.12	.14
.16	.18
.20	.22
.25	.28
.30	.35
.40	.45
.50	.55
.60	.65
.80	.90
1.00	1.10
1.20	1.40
1.60	1.80
2.00	2.20
2.50	2.80
3.00	3.20
3.50	3.80

Figure 11-20

Decimal Inches			
Gauge	Thickness	Gauge	Thickness
1	.010	19	.043
2	.011	20	.045
3	.012	21	.047
4	.013	22	.049
5	.014	23	.051
6	.016	24	.055
7	.018	25	.059
8	.020	26	.063
9	.022	27	.067
10	.024	28	.071
11	.026	29	.075
12	.029	30	.080
13	.031	31	.085
14	.033	32	.090
15	.035	33	.095
16	.037	34	.100
17	.039	35	.106
18	.041	36	.112

Figure 11-21

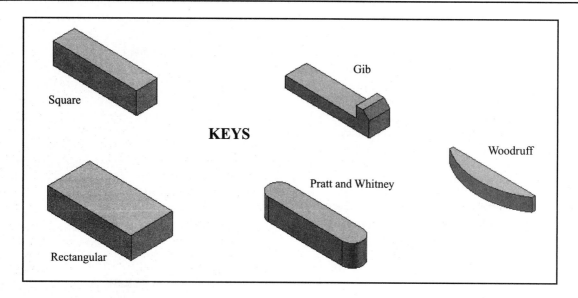

Figure 11-22

distance from the bottom of the shaft to the botttom of the keyway is .859.

3. Draw a rectangle on the front surface of the shaft and use the General Dimension tool to size and center the rectangle.
4. Delete the vertical line, and access the Part Features panel.
5. Extrude the rectangle 0.75 into the shaft and use the Cut option to create the keyway.

Nominal Shaft Ø	Square Key Size
Over - to	
$\frac{5}{16}$ (.3125) - $\frac{7}{16}$ (.4375)	$\frac{3}{32}$ (.0938)
$\frac{7}{16}$ (.4375) - $\frac{9}{16}$ (.5625)	$\frac{1}{8}$ (.1250)
$\frac{9}{16}$ (.5625) - $\frac{7}{8}$ (.8750)	$\frac{3}{16}$ (.1875)
$\frac{7}{8}$ (.8750) - $1\frac{1}{4}$ (1.2500)	$\frac{1}{4}$ (.2500)

Figure 11-23   Courtesy of *Machinery's Handbook*, Industrial Press, New York.

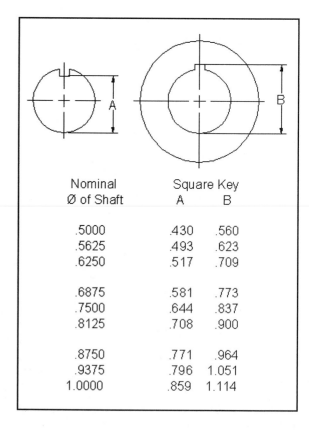

Nominal Ø of Shaft	Square Key A	B
.5000	.430	.560
.5625	.493	.623
.6250	.517	.709
.6875	.581	.773
.7500	.644	.837
.8125	.708	.900
.8750	.771	.964
.9375	.796	1.051
1.0000	.859	1.114

Figure 11-24   Courtesy of *Machinery's Handbook*, Industrial Press, New York.

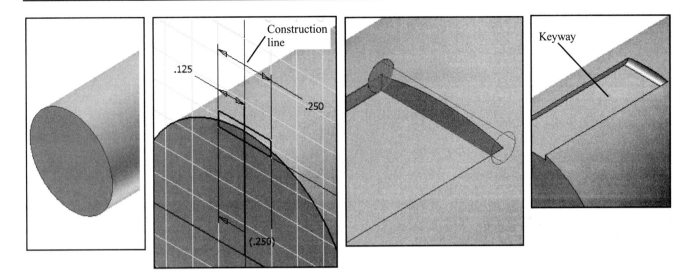

Figure 11-25

6. Enlarge the view of the shaft and create a new sketch plane along the keyway surface as shown.
7. Draw a circle with its radius equal to the depth of the keyway and its center point on the back top edge of the keyway.
8. Use the Extrude tool to cut the quarter-round segment at the back of the keyway.

See Figure 11-25.

To draw a keyway in a cam

Figure 11-26 shows a cam with a Ø1.000 bore.

1. Click on the front circular surface of the cam and create a new sketch plane.

2. Draw a vertical line starting at the cam's center point that can be used with the General Dimension tool to center and size the keyway.

Figures 11-23 and 11-24 indicate that a Ø1.000 nominal bore should use a ¼ (0.250)-in. square key. The recommended distance from the bottom of the bore to the top of the keyway is 1.114.

3. Draw a rectangle on the front surface of the shaft and use the General Dimension tool to size and center the rectangle.
4. Delete the vertical line, and access the Part Features panel.
5. Extrude the rectangle through the cam and use the Cut option to create the keyway.

See Figure 11-26.

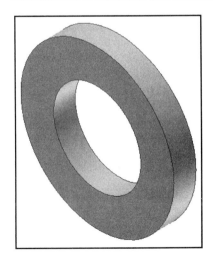
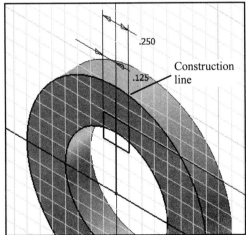
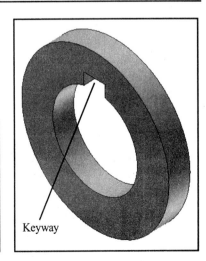

Figure 11-26

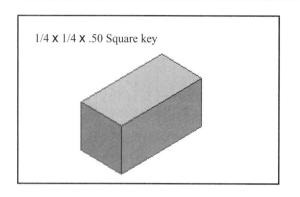

1/4 X 1/4 X .50 Square key

Figure 11-27

To draw a square key

Keys may be cut to any length. In this example a ¼ × ¼ square key will be made 0.50 in. long.

1. Draw a .250 × .250 rectangle.
2. Extrude the rectangle a distance of 0.50.
3. Save the key.

See Figure 11-27.

To draw a cam, shaft, and key assembly and a presentation drawing

1. Create a new assembly drawing using Standard .iam.

See Figure 11-28.

2. Assemble the components using the Constraint tool. See Chapter 5.

3. Create a presentation drawing using Standard .ipn.
4. Create an exploded isometric drawing using ANSI (in).idw. Include assembly numbers and a parts list.

See Figure 11-29.

11-10 METRIC KEYS

Figure 11-30 is a partial listing of metric square keys. All dimensions are in millimeters. For other values see reference books such as *Machinery's Handbook* published by Industrial Press, or search the web.

11-11 KEY AND KEYWAY TOLERANCES

The dimensions given in Figures 11-23 and 11-24 are nominal sizes and do not include tolerances. Key to keyway tolerances are generally confined to one of three categories:

Class 1 – Clearance fits
Class 2 – Interference/clearance fits
Class 3 – Interference fits

Figure 11-31 shows how the recommended tolerances would be applied to a nominal shaft and bore of Ø1.000. The square key is .250 × .250.

Shafts of Ø1.000 and less that use square keys that are not tapered but have parallel sides have recommended tolerances shown in Figure 11-32. The notation CL indicates clearance, and INT indicates interference.

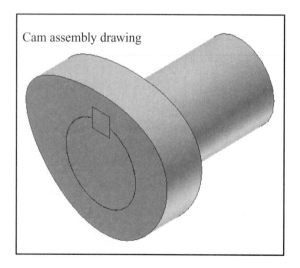

Cam assembly drawing

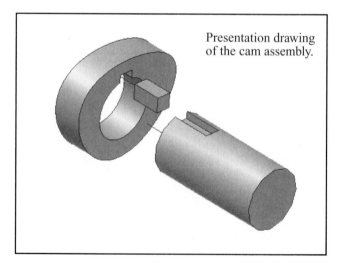

Presentation drawing of the cam assembly.

Figure 11-28

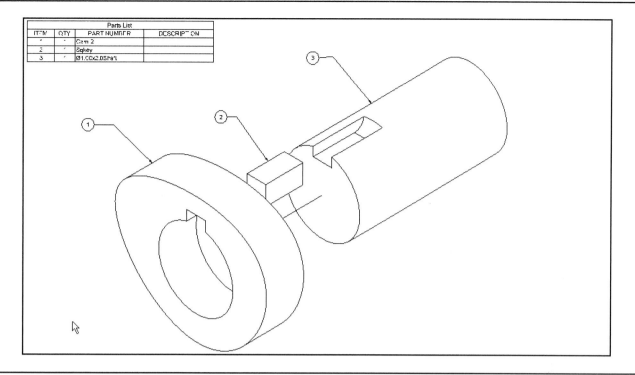

Figure 11-29

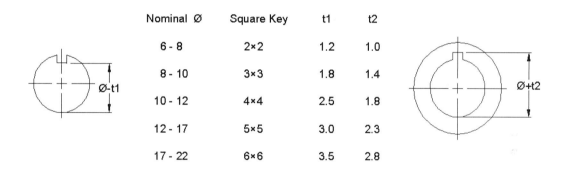

Nominal Ø	Square Key	t1	t2
6 - 8	2×2	1.2	1.0
8 - 10	3×3	1.8	1.4
10 - 12	4×4	2.5	1.8
12 - 17	5×5	3.0	2.3
17 - 22	6×6	3.5	2.8

Key Tolerances: 2 x 2, 3 x 3 Shaft 0.000/+0.025, Hub +0.060/+0.020
Free fit (mm)   4 x 4, 5 x 5, 6 x 6 Shaft 0.000/+0.030, Hub +0.078/+0.030

Figure 11-30

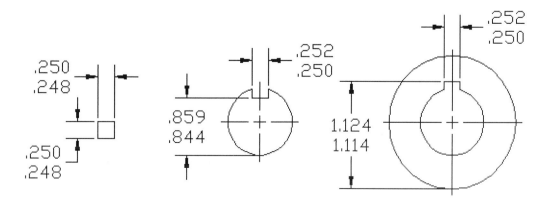

Figure 11-31

CLASS 1

Ø	Width		
	Key	Keyway	Range
.00 - .50	+0.000	+0.002	0.004 CL
	-0.002	+0.000	0.000
.50 - .75	+0.000	+0.003	0.005 CL
	-0.002	+0.000	0.000
.75 - 1.00	+0.000	+0.003	0.006 CL
	-0.003	+0.000	0.000

Depth

	Key	Shaft Keyway	Bore Keyway	Range
.00 - .50	+0.000	+0.000	+0.010	0.032 CL
	-0.002	-0.015	-0.000	0.005
.50 - .75	+0.000	+0.000	+0.010	0.032 CL
	-0.002	-0.015	-0.000	0.005
.75 - 1.00	+0.000	+0.000	+0.010	0.033 CL
	-0.003	-0.015	-0.000	0.005

CLASS 2

Ø	Width		
	Key	Keyway	Range
0.00 - 1.25	+0.001	+0.002	0.002 CL
	-0.000	+0.000	0.001 INT

Depth

	Key	Shaft Keyway	Bore Keyway	Range
0.00 - .500	+0.001	+0.000	+0.010	0.030 CL
	-0.000	-0.015	-0.000	0.004 CL

CLASS 3   There are no standards for Class 3.

Figure 11-32

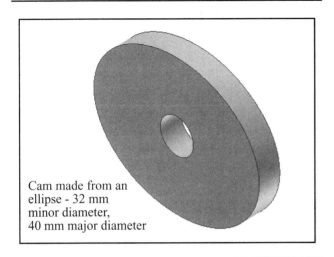

Cam made from an
ellipse - 32 mm
minor diameter,
40 mm major diameter

Figure 11-33

11-12 ANIMATING CAMS AND FOLLOWERS

Cams may be animated, that is, made to rotate on the screen while their followers move linearly. Figure 11-33 shows a cam that was previously created. A new Standard (mm).iam drawing was created, and the Place Component tool was used to add the cam drawing to the assembly drawing. The assembly drawing was saved as Cam Assembly-A.

To create a follower

1. Click the Create Component tool and create a new component named Follower.

   See Figure 11-34. The follower dimensions are .20 × .25 × 2.00.

2. Use the Constraint tool and make the follower tangent to the cam.

To create a follower guide

1. Click the Create Component tool and create a new component named Guide.

   See Figure 11-35.

2. Click the top surface of the follower to create a new sketch plane.
3. Click the Project Geometry tool on the 2D Sketch panel, then click the top surface of the follower. Right-click the mouse and select the Done option.
4. Sketch a rectangle around the projected top surface of the follower, use the Look At tool to create a 2D view of the sketch plane, then use the General Dimension tool to define a 0.05 clearance between the guide and the follower.

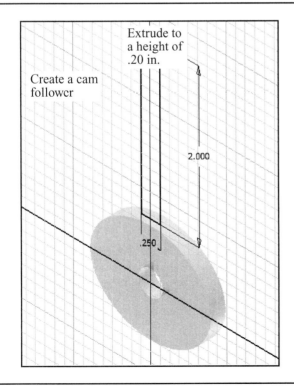

Create a cam
follower

Extrude to
a height of
.20 in.

2.000

.250

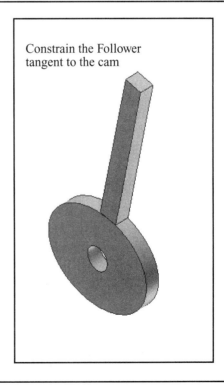

Constrain the Follower
tangent to the cam

Figure 11-34

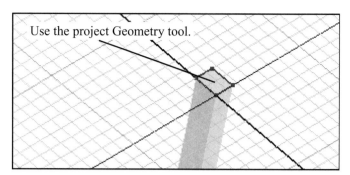

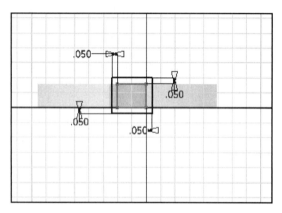

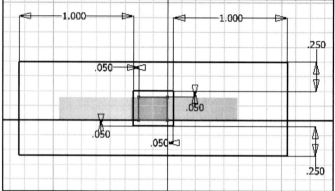

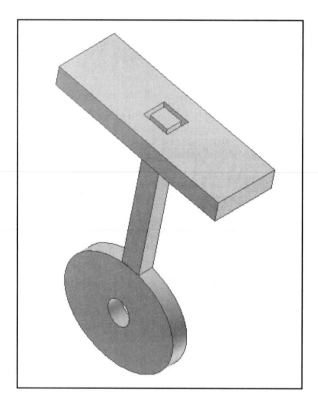

Figure 11-35

5. Sketch a second rectangle around the first rectangle as shown, then right-click the mouse and select the Done option, then the Finish Sketch option. Create an isometric view.
6. Use the Extrude tool to add a 0.25 thickness to the guide.

To animate the cam

1. Go to the browser box and Unground the cam and Ground the guide.

   See Figure 11-36.

2. Add a work axis to the bore of the cam.
3. Activate the cam and add a work plane through the work axis.

   The work plane is part of the cam.

4. Click the Constraint tool, Angle option, and define a 0.00 degree angle between the work plane on the cam and the guide's top surface.
5. Ground the work axis and guide.

   Be sure that nothing else is grounded.

6. Right-click the Angle (0.00 deg) listing in the browser box and select the Drive Constraint option.

   The Drive Constraint dialog box will appear. See Figure 11-37.

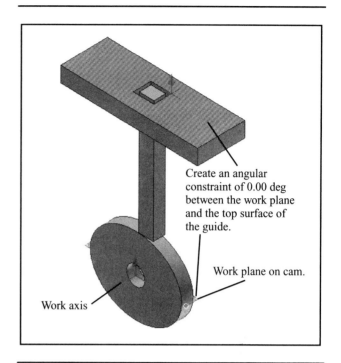

Create an angular constraint of 0.00 deg between the work plane and the top surface of the guide.

Work plane on cam.

Work axis

Figure 11-36

7. Set the Start angle for 0.00 deg and the End angle for 360.00 deg, then click the forward button on the Drive Constraint dialog box.

   The cam should rotate and the follower go up and down within the guide block.

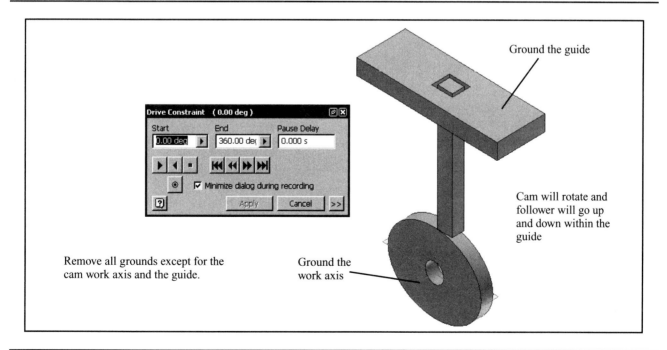

Ground the guide

Cam will rotate and follower will go up and down within the guide

Remove all grounds except for the cam work axis and the guide.

Ground the work axis

Figure 11-37

11-13 EXERCISE PROBLEMS

A. Draw a displacement diagram based on the motion information presented in EX11-1 to EX11-8. Draw a solid model of the cam described.

B. Design a rectangular follower and guide similar to that shown in Section 11-12 and animate the cam and follower.

EX11-1 MILLIMETERS

Motion:
Dwell for 90°
Rise 15 mm over 90° using modified uniform motion
Dwell for 90°
Drop 15 mm over 90° using modified uniform motion

Base circle = Ø120
Bore = Ø20

EX11-2 INCHES

Motion:
Dwell for 90°
Rise .375 in. over 90° using modified uniform motion
Dwell for 90°
Drop .375 in. over 90° using modified uniform motion

Base circle = Ø4.000 in.
Bore = Ø1.00 in. in.

EX11-3 MILLIMETERS

Motion:
Dwell for 30°
Rise 20 mm over 90° using simple harmonic motion
Dwell for 30°
Drop 10 mm over 90° using modified uniform motion
Dwell for 30°
Drop 10 mm over 90° using modified uniform motion

Base circle = Ø100
Bore = Ø20

EX11-4 INCHES

Motion:
Dwell for 30°
Rise .50 in. over 90° using simple harmonic motion
Dwell for 30°
Drop .25 in. over 90° using modified uniform motion
Dwell for 30°
Drop .25 in. over 90° using modified uniform motion

Base circle = Ø3.000
Bore = Ø.75

EX11-5 MILLIMETERS

Motion:
Dwell for 30°
Rise 30 mm over 120° using uniform acceleration and retarded motion
Dwell for 15°
Drop 15 mm over 90° using simple harmonic motion
Dwell for 15°
Drop 15 mm over 90° using modified uniform motion

Base circle = Ø140
Bore = Ø40

EX11-6 INCHES

Motion:
Dwell for 30°
Rise .50 in. over 120° using uniform acceleration and retarded motion
Dwell for 15°
Drop .25 in. over 90° using simple harmonic motion
Dwell for 15°
Drop .2 in. over 90° using modified uniform motion

Base circle = Ø4.50
Bore = Ø.750

EX11-7 MILLIMETERS

Motion:
Dwell for 30°
Rise 40 mm over 120° using cycloidal motion
Dwell for 15°
Drop 20 mm over 90° using simple harmonic motion
Dwell for 15°
Drop 20 mm over 90° using cycloidal motion

Base circle = Ø160
Bore = Ø40

EX11-8 INCHES

Motion:
Dwell for 30°
Rise .75 in. over 120° using cycloidal motion
Dwell for 15°
Drop .375 in. over 90° using simple harmonic motion
Dwell for 15°
Drop .375 in. over 90° using cycloidal motion

Base circle = Ø4.25
Bore = Ø1.00

Design and draw a solid model of a spring that meets the requirements as specified in exercises EX11-9 to EX11-12. Specify the k factor for the spring and the spring wire diameter.

EX11-9 INCHES

Free length = 2.00 in.
Outside diameter = .50 in.
Number of active coils = 12
Material = A228 steel
Diameter of spring wire = .05 in.

k factor = ?
Maximum load?

What is the deflection at maximum load?

EX11-10 MILLIMETERS

Free length = 30 mm
Outside diameter = 20 mm
Number of active coils = 12
Material = A228 steel
Diameter of spring wire = 0.5 mm

k factor = ?
Maximum load = ?

What is the deflection at maximum load?

EX11-11 INCHES

Free length = 1.625 in.
Outside diameter = .25 in.
Number of coils = 18
Material = A228 steel
Diameter of spring wire = .04 in.

k factor = ?
Maximum load = ?

What is the deflection at maximum load?

EX11-12 MILLIMETERS

Free length = 25 mm
Outside diameter = 15 mm
Number of Coils = 20
Material = A228 steel
Diameter of spring wire = 0.8 mm

k factor = ?
Maximum load = ?

What is the deflection at maximum load?

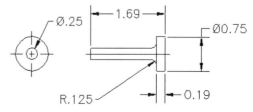

Figure 11-38

EX11-13 INCHES

A. Design a spring that wraps around the follower shown in Figure 11-38. The spring must deflect 0.25 in. and support a load of 5 lb. The spring is made from A228 steel. Determine the springs *k* factor, outside diameter, wire diameter, and number of coils.

B. Draw a solid model assembly of the follower with the spring inserted.

EX11-14

Draw a solid model of the cam defined in EX11-2 and add a shaft and square key according to the following specifications. The cam is .675 in. thick.

A. The shaft has a minimum clearance with the cam bore of .001 in. and a maximum of .005 in. The shaft is 3.00 in. long.

B. Size, dimension, and tolerance the key, keyway in the shaft, and keyway in the cam using the tables given in Figures 11-23, 11-24, and 11-32.

EX11-15

Draw a solid model of the cam defined in EX11-4 and add a shaft and square key according to the following specifications. The cam is 1.000 in. thick.

A. The shaft has a minimum cleararce with the cam bore of .001 in. and a maximum of .005 in. The shaft is 3.25 in. long.

B. Size, dimension, and tolerance the key, keyway in the shaft, and keyway in the cam using the tables given in Figures 11-23, 11-24, and 11-32.

C. Use the follower and spring created for Figure 11-38 and draw a solid model assembly that includes the cam, shaft, and square key. Use the maximum cam height as the maximum spring deformation. Use A228 steel and define the appropriate wire diameter and spring diameter.

D. Draw an exploded isometric drawing of the assembly created in step C.

EX11-16

Draw a solid model of the cam defined in EX11-1 and add a shaft and square key according to the following specifications. The cam is 16 mm thick.

A. The shaft has a minimum clearance with the cam bore of 0.01 mm and a maximum of 0.05 mm. The shaft is 50 mm long.

B. Size, dimension, and tolerance the key, keyway in the shaft, and keyway in the cam using the tables given in Figure 11-30.

C. Convert the follower and spring dimensions created for Figure 11-38 to millimeters and draw a solid model assembly that includes the cam, shaft, and square key. Use the maximum cam height as the maximum spring deformation. Use A228 steel.

D. Draw an exploded isometric drawing of the assembly created in step C.

EX11-17

Section 14-10 presents a cam project that can be manufactured.

Sheet Metal and Weldments

12-1 INTRODUCTION

Inventor 8 can be used to draw both sheet metal parts and weldments. This chapter introduces both types of drawings.

12-2 SHEET METAL DRAWINGS

To create a sheet metal drawing select the New Drawing option, then the Sheet Metal option. The Sheet Metal (mm).ipt option was selected for purposes of demonstration. See Figure 12-1.

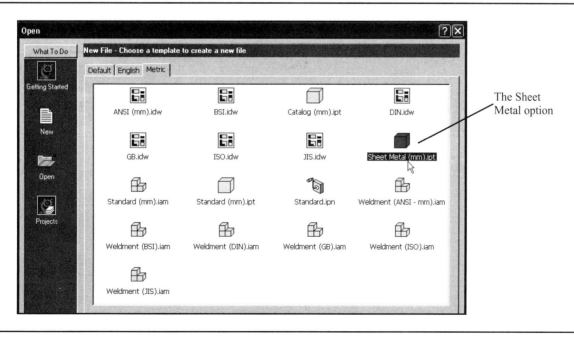

Figure 12-1

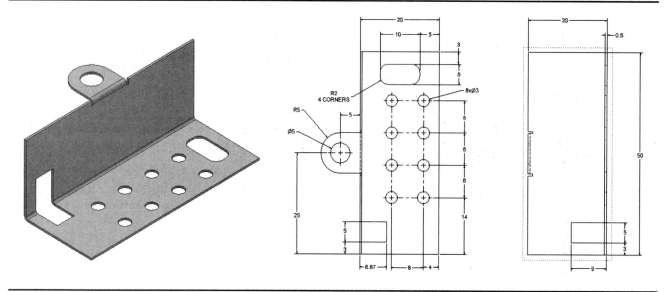

Figure 12-2

When the Sheet Metal option is selected the 2D Sketch screen will appear. Sheet metal drawings are initiated as 2D sketches, then developed using a combination of 2D Sketch and Sheetmetal Features panels. The 2D Sketch Panel was presented in Chapters 1 and 2.

Figure 12-2 shows a sheet metal part. It was created as follows.

The initial sketch

1. Select the Two point rectangle command from the 2D Sketch Panel and sketch a rectangle.

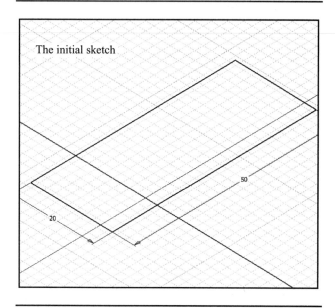

The initial sketch

Figure 12-3

2. Use the General Dimension command to size the rectangle to the given dimensions.
3. Right-click the mouse button and select the Isometric View option.

See Figure 12-3.

To add thickness

1. Right-click the mouse and select the Done option.
2. Either right-click the mouse and select the Finish Sketch option, or click the Return box at the top of the screen.

The Sheetmetal Features panel will appear. See Figure 12-4. Not all commands will be active at this time, but they will become active as the drawing progresses.

3. Select the Styles option from the Sheetmetal Features panel.

The Sheet Metal Styles dialog box will appear. See Figure 12-5. The Sheet Metal Styles box is used to define the thickness and bend characteristics of the part.

Inventor has many default values already in place. Figure 12-5 shows that the default thickness is 0.500 mm. Sheet metal is manufactured in standard thicknesses. Figure 12-6 is a partial listing of available standard sheet metal thicknesses in inches, and Figure 12-7 is a partial listing of sheet metal thicknesses in millimeters.

Figure 12-7 lists 0.500 mm as a standard thickness, so this default value will be used for this example.

4. Accept the 0.500 mm Thickness value and click Done.

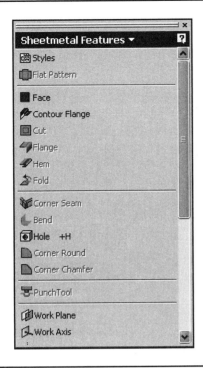

Figure 12-4

Enter thickness value here.

Figure 12-5

Wire and Sheet Metal Gauges			
Gauge	**Thickness (in.)**	**Gauge**	**Thickness (in.)**
000 000	0.5800	18	0.0403
00 000	0.5165	19	0.0359
0 000	0.4600	20	0.0320
000	0.4096	21	0.0285
00	0.3648	22	0.0253
0	0.3249	23	0.0226
1	0.2893	24	0.0201
2	0.2576	25	0.0179
3	0.2294	26	0.0159
4	0.2043	27	0.0142
5	0.1819	28	0.0126
6	0.1620	29	0.0113
7	0.1443	30	0.0100
8	0.1285	31	0.0089
9	0.1144	32	0.0080
10	0.1019	33	0.0071
11	0.0907	34	0.0063
12	0.0808	35	0.0056
13	0.0720	36	0.0050
14	0.0641	37	0.0045
15	0.0571	38	0.0040
16	0.0508	39	0.0035
17	0.0453	40	0.0031

Figure 12-6

Preferred Thickness (mm)		
0.050	0.50	4.0
0.060	0.60	5.0
0.080	0.80	6.0
0.10	1.0	8.0
0.12	1.2	10.0
0.16	1.6	
0.20	2.0	
0.25	2.5	
0.30	3.0	
0.40	3.5	

Figure 12-7

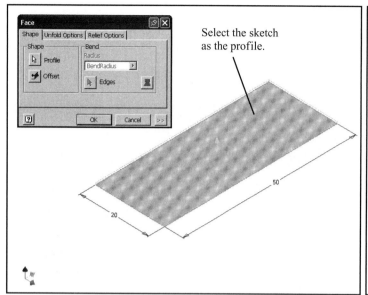

Select the sketch as the profile.

The completed face.

Figure 12-8

5. Select the Face tool from the Sheetmetal Features panel.
6. Select the sketch, then click OK.

   See Figure 12-8.

Bend radius

As sheet metal is bent the inside surface is subject to compression, and the outside surface to tension. These forces cause the material to stretch slightly. Inventor will display bend radii as two concentric arcs.

The default value for the inside bend radius is equal to the sheet metal thickness. Figure 12-9 shows the Sheet Metal Styles dialog box with the Bend tab selected. Note that the Radius box is set to Thickness. The default outside bend radius is equal to twice the sheet metal thickness.

The default values for Radius and Relief Shape will be accepted for this example.

Flange

1. Select the Flange tool from the Sheetmetal Features panel.

   The Flange dialog box will appear. See Figure 12-10.

The inside bend radius is equal to the sheet metal thickness. This is the default value.

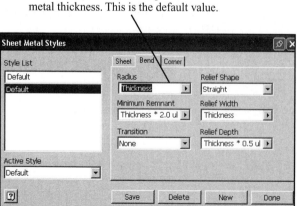

Figure 12-9

Select the lower edge.

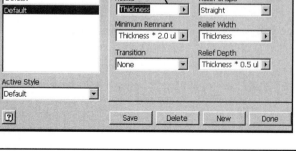

Set the Distance value for 20.

Figure 12-10

If the top edge is selected.

If the bottom edge is selected.

If the bottom edge is selected, then flipped.

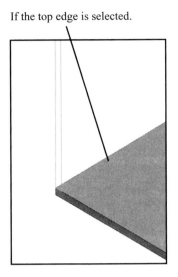 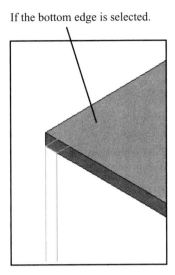 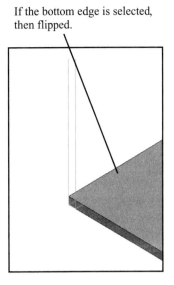

Figure 12-11

2. Set the distance for 20 and select the lower rear edge of the sketch.

The lower edge was chosen because the flange total height is to be 20 mm. If the upper edge were chosen, the total height would be 20.5, the flange height plus the material thickness. Figure 12-11 shows the flange orientation resulting from edge selection.

The bend will be added automatically based on the values defined in the Sheet Metal Styles box.

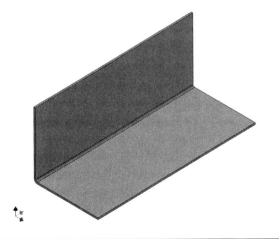

Figure 12-12

3. Use the Flip direction button to change the flange orientation.
4. Click OK.

Figure 12-12 shows the resulting flange.

Tabs

Tabs are similar to flanges, but tabs do not run the entire length of the edge, as flanges do. Tabs are created using a new sketch plane, then the Two point rectangle tool on the 2D Sketch Panel. See Figure 12-13.

1. Select the surface of the vertical flange, right-click the mouse and select the New Sketch option.
2. Use the Two point rectangle tool on the 2D Sketch Panel and draw a rectangle that extends from the edge of the vertical flange as shown.
3. Use the General Dimension tool to size and locate the tab in accordance with the dimensions given in Figure 12-2.
4. Right-click the mouse and select the Done option, then right-click the mouse again and select the Finish Sketch option.
5. Select the Face tool, and define the tab as the Profile.
6. Select OK.

Figure 12-13 shows the resulting tab.

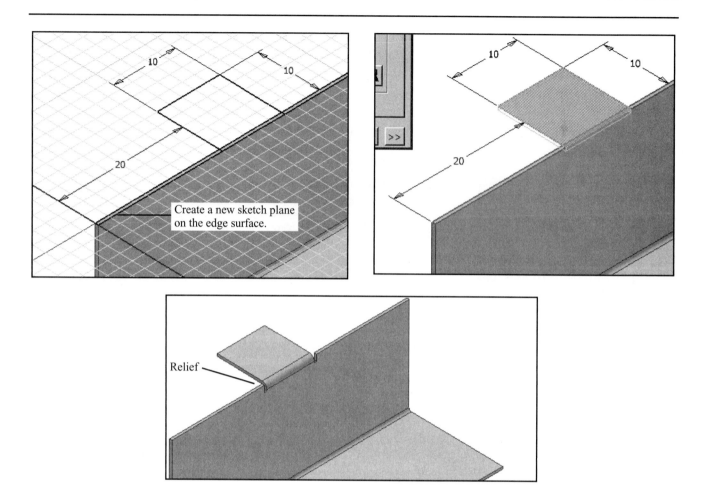

Figure 12-13

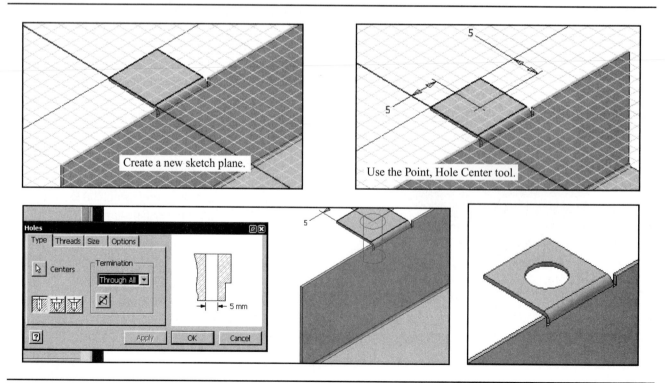

Figure 12-14

Reliefs

Reliefs are cut out in material to allow it to be bent. If the material was not relieved, it would tear uncontrollably as the bend was formed.

Inventor's default relief value is equal to the thickness of the sheet metal material. Figure 12-13 shows the relief that was automatically created as the tab was formed.

Holes

Holes are added to sheet metal parts in the same manner as they are added to solid parts. See Figure 12-14.

1. Create a new sketch plane on the top surface of the tab.
2. Use the Point, Hole Center tool to define the hole's center point.
3. Use the General Dimension tool to dimension the hole's center point location.

4. Right-click the mouse, click the Done option, then click the Finish Sketch option.
5. Use the Hole tool on the Part Features panel to create the hole.

Corners

Both internal and external corners are created using the Corner Round tool found on the Sheetmetal Features panel.

1. Click the Corner Round tool on the Sheetmetal Features panel.

The Corner Round dialog box will appear. See Figure 12-15.

2. Set the Radius value for 5.
3. Select the two outside corners of the tab.
4. Click OK.

Figure 12-15 shows the resulting rounded corners.

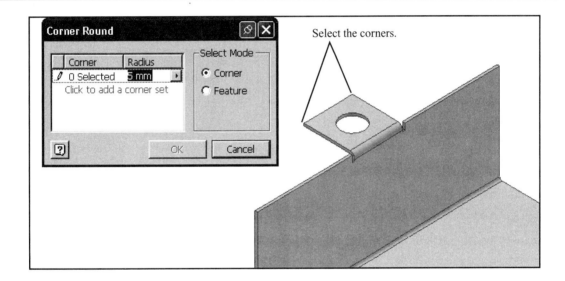

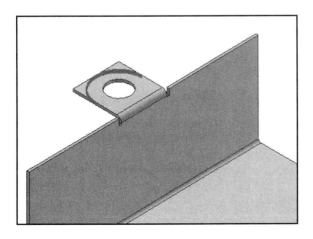

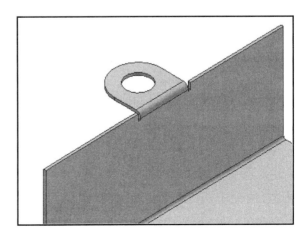

Figure 12-15

Cuts

Cuts may be any shape, other than a hole, that passes through the sheet metal. In this example a rectangular shape is used. See Figure 12-16.

1. Create a new sketch plane and sketch a rectangle as shown. Use the General Dimension tool to size and locate the rectangle.
2. Right-click the mouse and select the Done option, then select the Finish Sketch option.

   The Sheetmetal Features panel will appear.

3. Select the Cut tool.

   The Cut dialog box will appear.

4. Select the rectangle as the Profile.
5. Ensure that the direction of the cut is correct, then click the OK box.

   The rectangular area will be removed.

6. Select the Corner Round tool and set the Radius value for 2 mm.
7. Select the four inside corners of the rectangular cut.
8. Click the OK button on the Corner Round dialog box.

Cuts through normal surfaces

Normal surfaces are surfaces that are perpendicular to each other. Cuts in normal sufaces are made by making intersecting cuts in both surfaces. See Figure 12-17.

1. Create a new sketch plane as shown and sketch a rectangle.

   Ensure that the rectangle extends beyond the rounded edge of the surface.

2. Use the General Dimension tool to locate and size the rectangle.
3. Right-click the mouse and select the Done option, then select the Finish Sketch option.

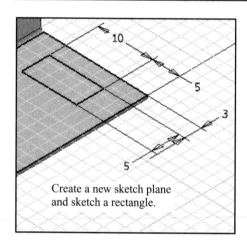

Create a new sketch plane and sketch a rectangle.

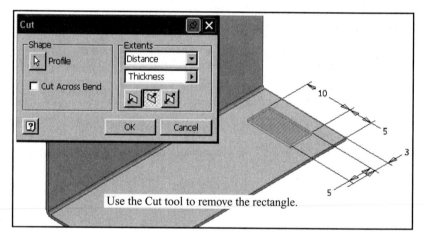

Use the Cut tool to remove the rectangle.

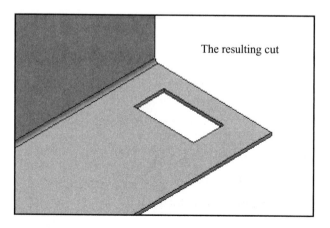

The resulting cut

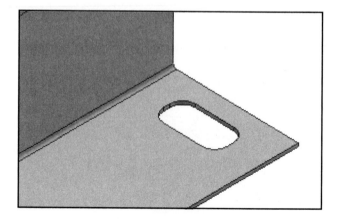

Figure 12-16

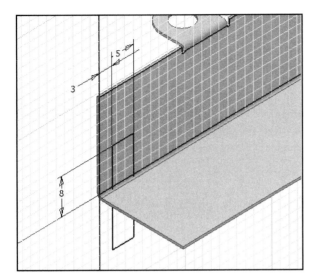
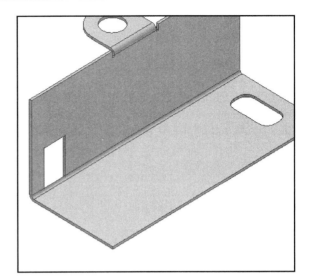
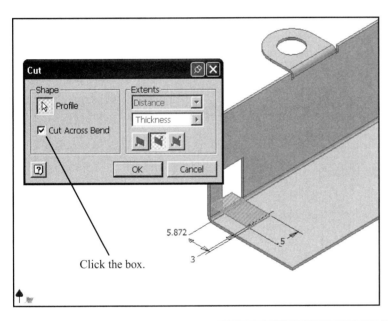

Click the box.

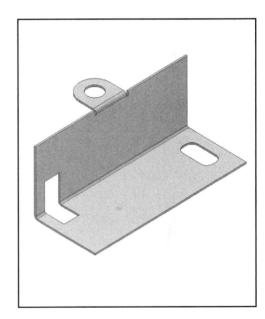

Figure 12-17

4. Use the Cut tool to remove the rectangle.
5. Create another new sketch plane, and use the General Dimension tool to size and locate the rectangle.
6. Right-click the mouse and select the Done option, then the Finish Sketch option.
7. Click the Cut tool.

   The Cut dialog box will appear.

8. Click the Cut Across Bend box, then OK.

Hole patterns

A hole pattern is created from an existing hole. See Figure 12-18.

1. Create a new sketch plane, define a hole center, then add the hole.
2. Click the Rectangular Pattern tool.

   The Rectangular Pattern dialog box will appear.

3. Define the hole as the Feature.

4. Click the arrow under the Direction 1 heading, then the top front edge of the part. Use the Flip button to change directions if necessary.
5. Set the number of holes under Direction 1 for 4 and the spacing for 8 mm.
6. Click the arrow under the Direction 2 heading, then click the left front edge of the part to define the direction.
7. Set the number of holes for 2 and the distance for 8.
8. Click OK.

12-3 WELDMENTS

Weldments are assemblies made from several smaller parts that have been welded together. Weldments are often cheaper to manufacture because they save extensive machining time or replace expensive castings.

Figure 12-19 shows a simple weldment. It was created from two 0.375-in. thick plates. The base plate is 2.00 × 4.00 in. and the vertical plate is 1.25 × 4.00 in. Both parts are made from low-carbon steel.

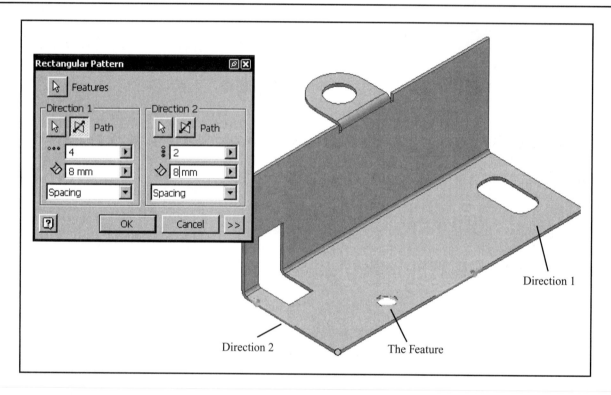

Direction 1

Direction 2   The Feature

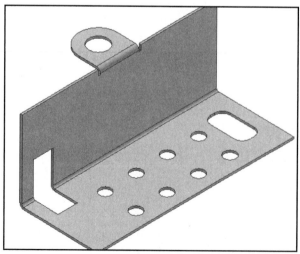

Figure 12-18

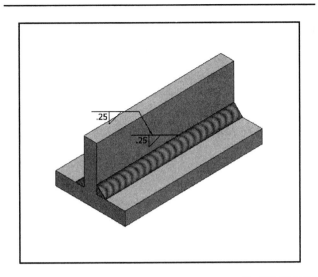

Figure 12-19

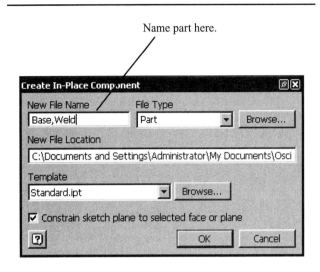

Name part here.

Figure 12-20

To create the components

1. Click the New tool, then the English tab, then Weldment (ANSI).iam.
2. Click the Create Component tool.

The Create In-Place Component dialog box will appear. See Figure 12-20.

3. Define a new component named Base, Weld, then click OK. Switch to an isometric view.

See Figure 12-21.

In the example shown, the sketch plane has been defined as the XZ plane, but any plane may be used. Sketch planes are changed by first selecting the Tools pull-down menu, then Applications Options, then the Parts tab. The sketch plane is selected by choosing one of the radio buttons under the Sketch on New Part Creation heading.

4. Create the Base, Weld part using the commands on the 2D Sketch and the Part Features panels.
5. Create a new sketch plane and create the vertical plate.

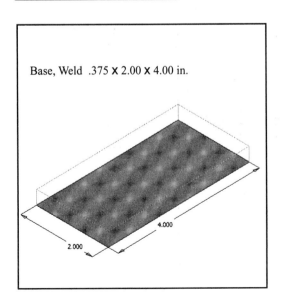

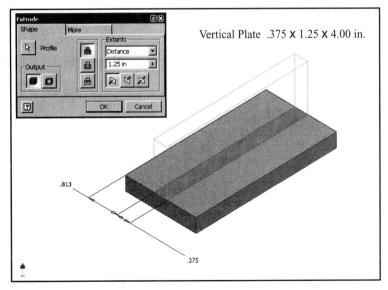

Figure 12-21

To create the welds

1. Access the Weldment Features Panel by double-clicking the Welds tool in the browser.

   See Figure 12-22.

2. Click the Weld tool on the Weldment Features panel.

   The Weld Feature dialog box will appear. See Figure 12-23.

3. Click the Fillet weld option.
4. Define the Arrow Side 1 and Arrow Side 2 surfaces as shown.
5. Click OK.

   Figure 12-24 shows the resulting fillet weld. The .25 value is the default value. In this example all default values were accepted.

6. Use the Rotate tool and rotate the weldment so that the rear surfaces are visible.
7. Add a fillet weld.

   See Figure 12-25.

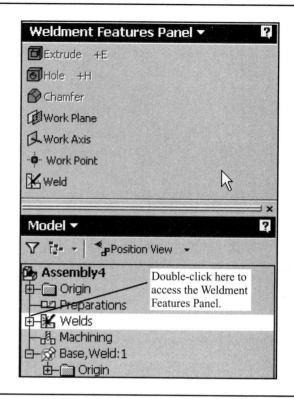

Figure 12-22

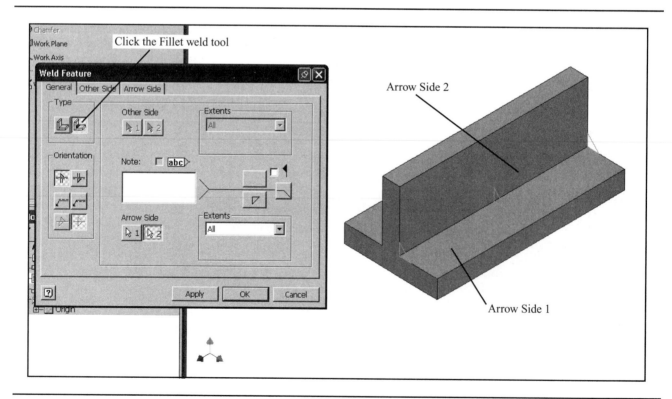

Figure 12-23

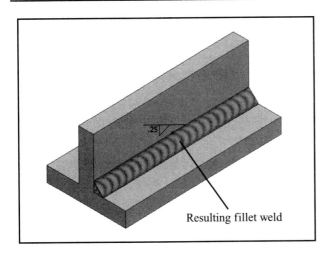

Resulting fillet weld

Figure 12-24

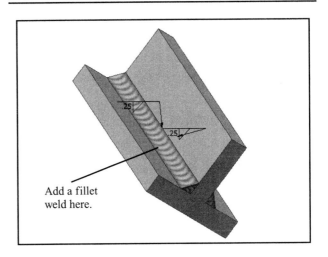

Add a fillet
weld here.

Figure 12-25

Weld symbols

Welds are defined on drawings using symbols. The symbol for a fillet weld is shown in Figure 12-26. Note that the location of the flaglike portion of the symbol defines the location of the weld. It is not always possible to point directly at a weld location, so the Other side symbol is very useful.

The size of the weld is defined as shown. Most fillet welds are created at 45°, although other angles are possible. A fillet weld defined by .25 indicates that the 45° weld is

defined by two sides, both .25 long. Metric values are used to define a weld size in the same manner.

All around

The addition of a circle to the fillet weld symbol indicates that the weld is to placed *all around* the object. Figure 12-27 shows a cylinder welded to a plate. In this example the fillet weld was defined using millimeters. A 5 mm × 5 mm weld is to be created all the way around the cylinder.

Symbols for fillet welds

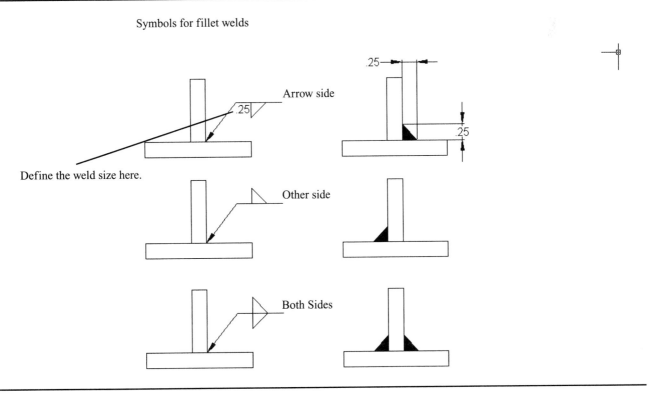

Define the weld size here.

Arrow side

Other side

Both Sides

Figure 12-26

To create an all-around fillet weld

1. Create a new drawing using the Weldment (ANSI-mm).iam format.
2. Create the necessary components.
3. Access the Weldment Features Panel.
4. Click the Weld tool.
5. Select the Arrow side tab on the Weld Feature dialog box.

See Figure 12-27.

6. Set the Small Leg distance for 5 mm, or .19685 in.
7. Select the General tab and click the fillet weld symbol, then define Arrow Side 1 and Arrow Side 2.
8. Click OK.

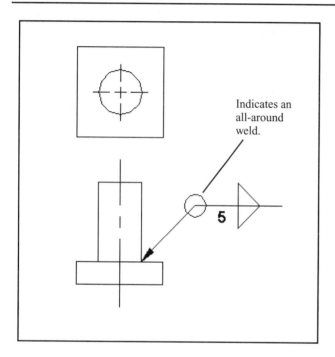

Indicates an all-around weld.

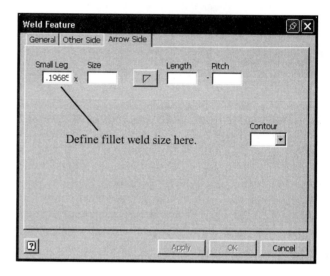

Define fillet weld size here.

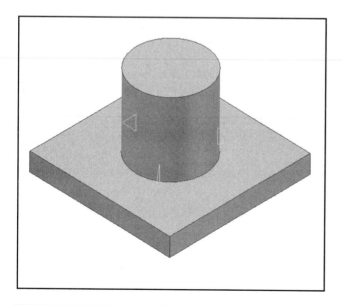

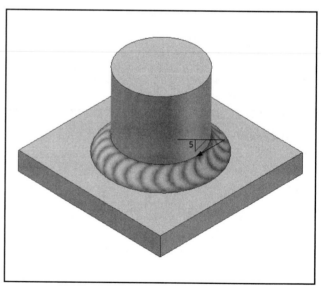

Figure 12-27

To create a V-groove weld

V-groove welds are used when two parts abut. A chamfer is cut in each of the parts, and the weld material is placed into the groove. Figure 12-28 shows an L-bracket manufactured as a weldment. It was created as follows.

1. Draw the horizontal section of the I-bracket .375 × 1.50 × 3.00 in.
2. Create a .19 × .19 chamfer on the upper back edge of the horizontal section.
3. Create a new sketch plane and add the vertical section of the L-bracket as shown.

The 1.125 dimension was derived by subtracting the .375 vertical section thickness from the horizontal section width of 1.500 (1.500 − .375 = .125). The vertical section is .375 × 1.00 × 3.00.

4. Extrude the vertical section to a height of 1.00 in. then use the Rotate tool to turn the L-bracket so that the back surfaces are visible.
5. Add a .19 × .19 chamfer to the lower edge of the vertical section as shown.
6. Access the Weld tool on the Weldment Features panel and add a .25 fillet weld to the groove between the two sections.
7. Return the L-bracket to the isometric orientation and add a second fillet weld as shown.

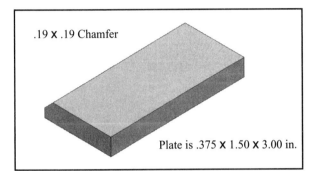

.19 x .19 Chamfer

Plate is .375 x 1.50 x 3.00 in.

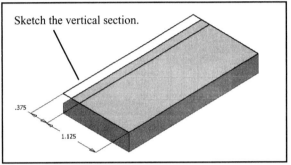

Sketch the vertical section.

.375

1.125

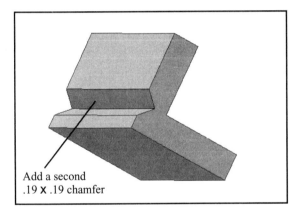

Add a second .19 x .19 chamfer

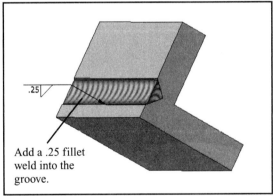

.25

Add a .25 fillet weld into the groove.

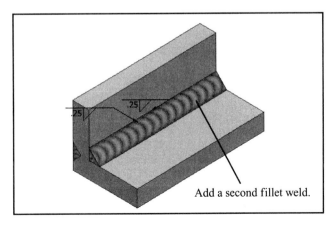

.25

.25

Add a second fillet weld.

Figure 12-28

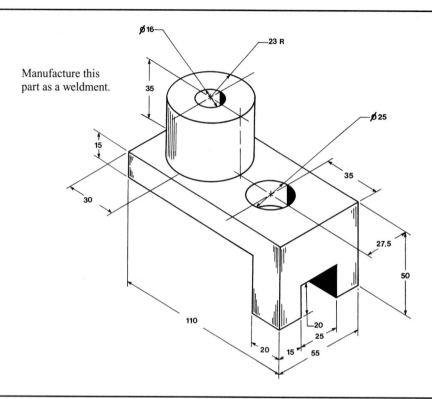

Manufacture this part as a weldment.

Figure 12-29

12-4 SAMPLE PROBLEM SP12-1

Figure 12-29 shows an object that is to be manufactured as a weldment. The object will be created from three parts that are then welded together. The three parts are the barrel, the center plate, and the front plate.

To create the weldment

Use the Weldment (ISO).iam format. The dimensions for each part were derived from those given in Figure 12-29.

1. Use the Create Component tool and draw the center plate. Include a 5 × 5 chamfer as shown.

   See Figure 12-30.

2. Use the Create Component and add the barrel. Add the Ø16 through hole.

3. Use the Rotate tool and position the weldment so that the bottom surface can be seen.

4. Create a new sketch plane and draw the front plate as shown and add the cutout.

   Do not add a chamfer to the front plate. The chamfer on the center plate and the flat edge on the front plate will form a groove to hold the weld.

5. Add a 5-mm fillet weld between the chamfered edge of the center plate and the front plate as shown.

6. Use the Rotate tool and position the weldment so that a 5-mm weld can be added to the inside edge between the center plate and the front plate as shown.

   There is not enough space between the barrel and the edges of the center plate to place 5-mm fillet welds. Two 5-mm fillet welds require 10 mm, and only 9 mm is available.

7. Access the Weld Feature dialog box and change the weld size to 4.00 mm.

8. Create a 4.00-mm weld between the barrel and the center plate.

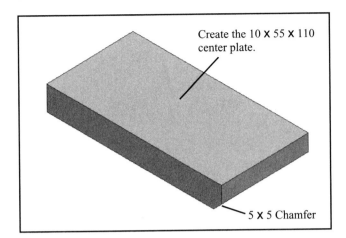

Create the 10 × 55 × 110 center plate.

5 × 5 Chamfer

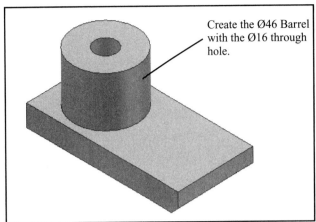

Create the Ø46 Barrel with the Ø16 through hole.

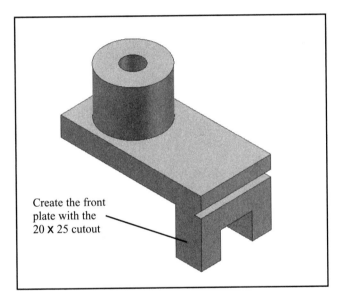

Create the front plate with the 20 × 25 cutout

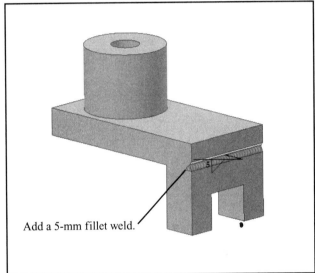

Add a 5-mm fillet weld.

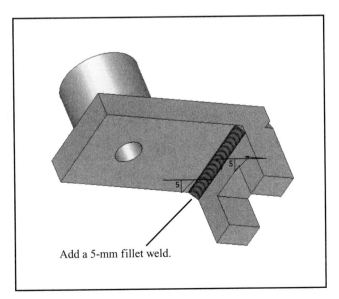

Add a 5-mm fillet weld.

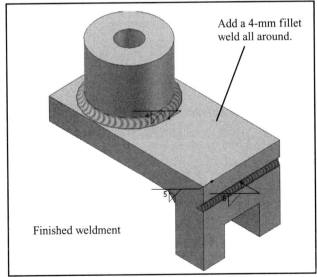

Add a 4-mm fillet weld all around.

Finished weldment

Figure 12-30

12-5 EXERCISE PROBLEMS

Redraw the sheet metal parts in EX12-1 to EX12-6 using the given dimensions. Use the default values for all bend radii and reliefs.

EX12-1 MILLIMETERS

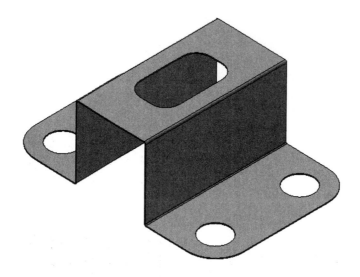

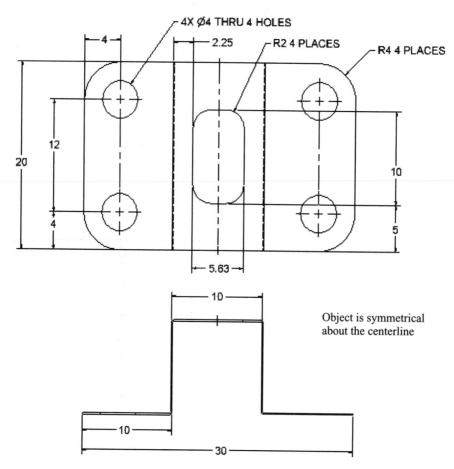

4X Ø4 THRU 4 HOLES

R2 4 PLACES

R4 4 PLACES

Object is symmetrical about the centerline

EX12-2 INCHES

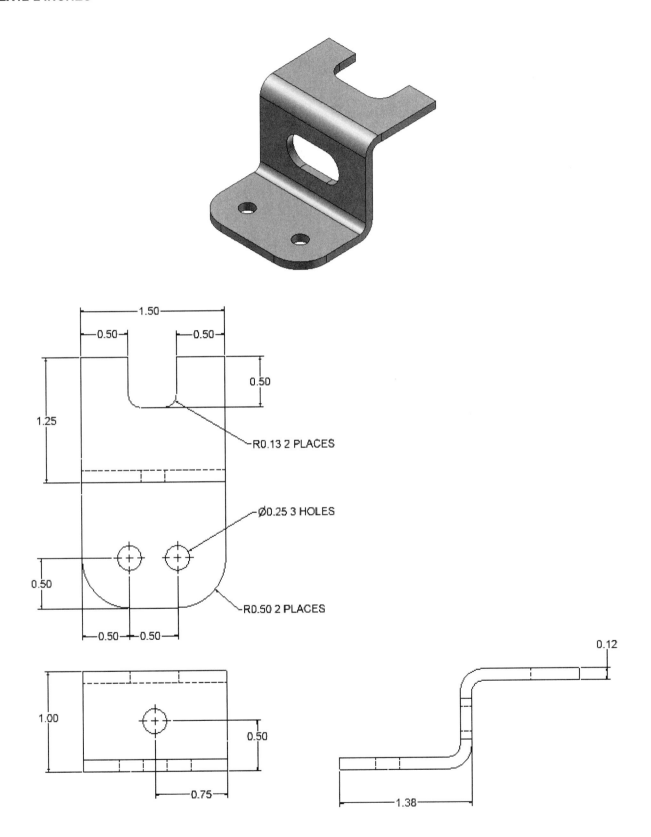

1.50

0.50

0.50

0.50

1.25

R0.13 2 PLACES

Ø0.25 3 HOLES

0.50

R0.50 2 PLACES

0.50 0.50

0.12

1.00

0.50

0.75

1.38

EX12-3 INCHES

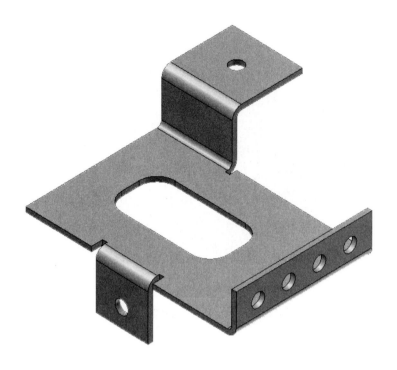

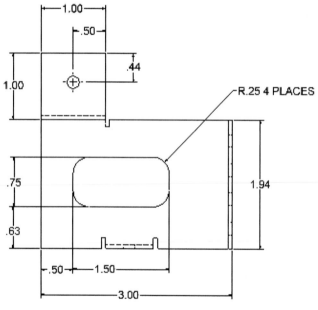

R.25 4 PLACES

1.00

.50

.44

1.00

.75

.63

1.94

.50

1.50

3.00

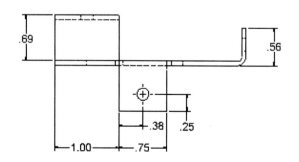

.69

.56

.38

.25

1.00

.75

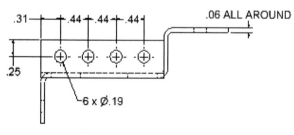

.31

.44

.44

.44

.06 ALL AROUND

.25

6 x Ø.19

EX12-4 MILLIMETERS

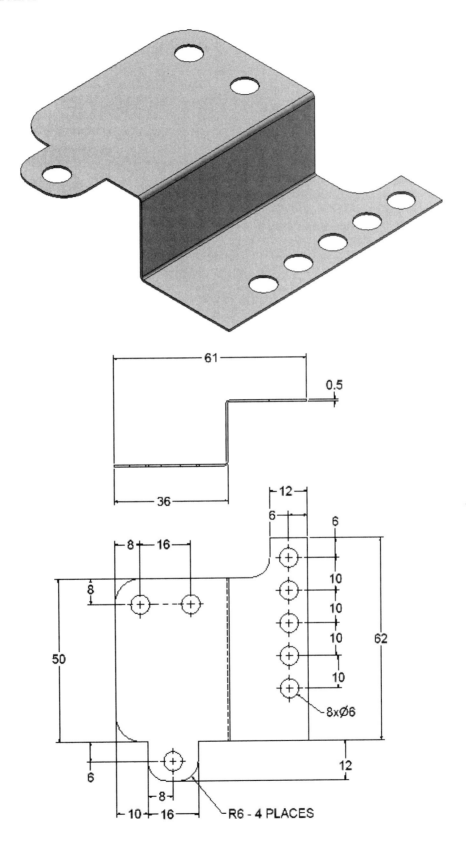

EX12-5 MILLIMETERS

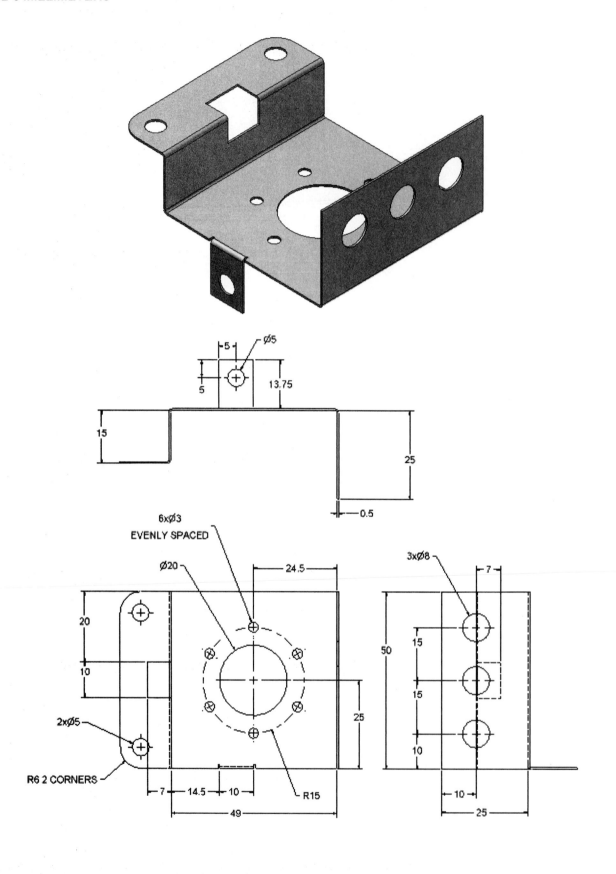

EX12-6 MILLIMETERS

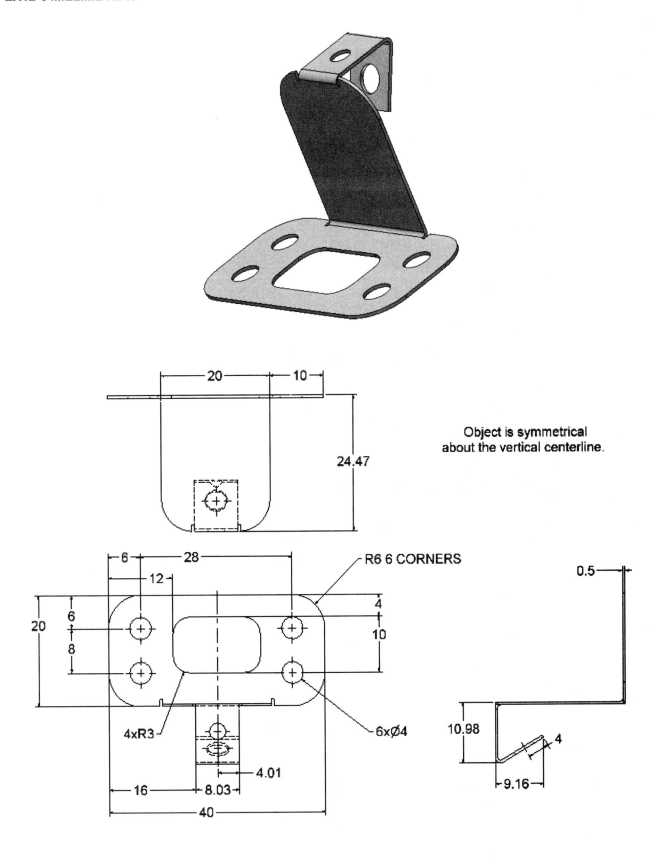

20   10

24.47

Object is symmetrical
about the vertical centerline.

6   28

12

R6 6 CORNERS

20

6

8

4

10

0.5

4xR3

6xØ4

16   8.03

4.01

40

10.98

9.16

4

EX12-7

Design and draw a box similar to that shown that has a capacity of

A. 100 cubic centimeters and is a cube.
B. 4 fluid ounces.
C. 100 cubic centimeters and is rectangular with the length of one side 2 times the length of the other.
D. 125 cubic inches and is a cube.
E. 125 cubic inches and is rectangular with the length of one side 1.5 times the length of the other.
F. 8 fluid ounces.

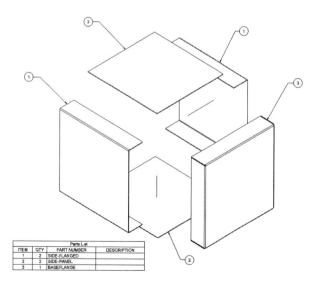

Parts List			
ITEM	QTY	PART NUMBER	DESCRIPTION
1	2	SIDE-FLANGED	
2	2	SIDE-PANEL	
3	1	BASEFLANGE	

EX12-8

Design and draw a sheet metal support assembly as shown. Define all dimensions not given.

> The large center hole is to be Ø1.000 and is to be located 1.50 in. above the base.
> The base is made from Lexan and is .75 thick.
> The two Support Flanges and the Top Plate are made from 0.06-in. 6061 aluminum.
> All screws and nuts are steel.
> Select and add the appropriate nuts.

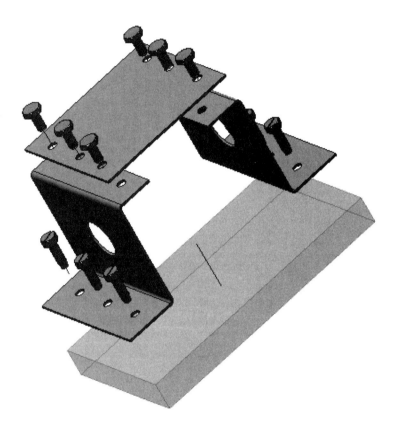

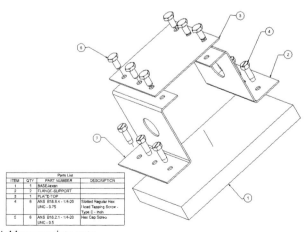

Parts List			
ITEM	QTY	PART NUMBER	DESCRIPTION
1	1	BASE-lexan	
2	2	FLANGE-SUPPORT	
3	1	PLATE-TOP	
4	6	ANS B18.6.4 - 1/4-20 UNC - 0.75	Slotted Regular Hex Head Tapping Screw - Type C - Inch
5	6	ANS B18.2.1 - 1/4-20 UNC - 0.5	Hex Cap Screw

Add appropriate nuts

For exercises EX12-9 to EX12-18, redesign the given parts as weldments. Use either 5-mm or 0.20-in. fillet welds.

EX12-9 MILLIMETERS

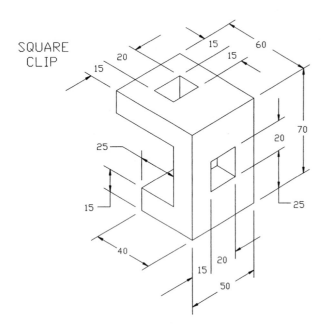

SQUARE CLIP

EX12-11 MILLIMETERS

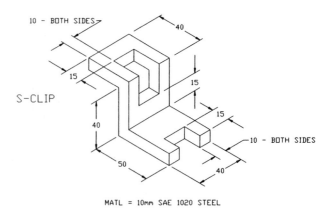

S-CLIP

MATL = 10mm SAE 1020 STEEL

EX12-10 MILLIMETERS

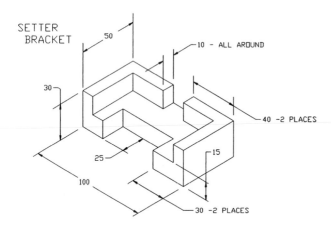

SETTER BRACKET

EX12-12 INCHES

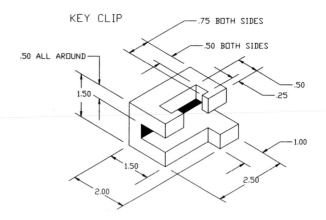

KEY CLIP

EX12-13 MILLIMETERS

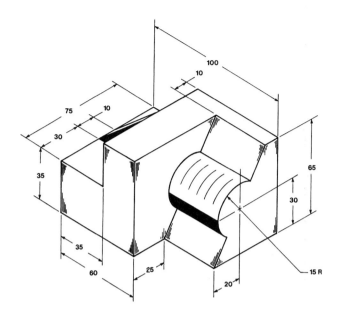

EX12-15 MILLIMETERS

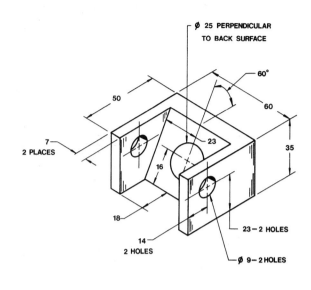

EX12-14 MILLIMETERS

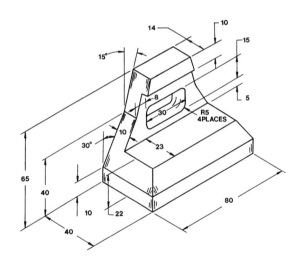

EX12-16 MILLIMETERS

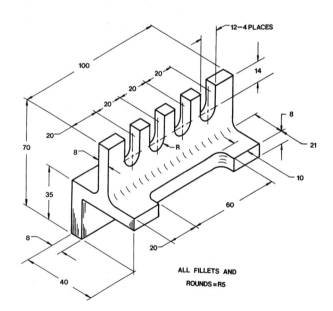

EX12-17 MILLIMETERS

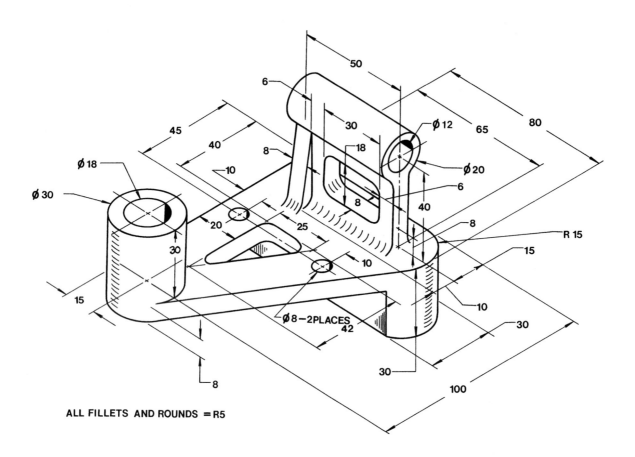

ALL FILLETS AND ROUNDS = R5

EX12-18 MILLIMETERS

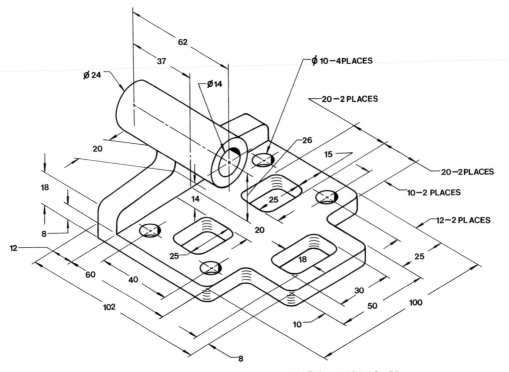

ALL FILLETS AND ROUNDS = R5

C H A P T E R 13

The Design Process

13-1 INTRODUCTION

This chapter explains the design process. Designing is a combination of creative problem solving, applying the principles of math and physics, understanding construction techniques, and much persistent hard work. This chapter demonstrates the design process by outlining the steps used to complete an actual student design project. The chapter ends with a summation and review of the process used to complete the project.

The idea of following a preset sequence of steps to design a new product may seem counterproductive and excessively time consuming. It is tempting to read through the problem, formulate a solution, and then work and rework the solution to fit the problem. This approach often leads to a forced solution that is usually not the optimal solution.

The amount of extra time spent at the beginning of the design process is more than made up for as the design develops. That is not to say that the solution to the problem will be any less difficult. Errors will be made, and ideas will not work out as planned, but overall by *managing* the process, the final result will be achieved faster and in a more satisfactory manner.

13-2 SELECTING A DESIGN GROUP

There has been much study in recent years as to what types of personalities should make up a design group. In a school situation the design groups are usually made up of friends and students who know one another. In industry a design group's personnel may be the random group of people who just happen to be working at a company at a particular time.

The Ned Herrmann Group (thinking@HBDI.com), among others, has devised a test that defines people by the way they think and therefore the way they approach problem solving. The rationale for the test is that if you put together a group of people who all think alike, you will limit the diversity of the solutions the group generates. It is better to have a more diverse group. Each member of the group will look at the problem differently, and the result will be a richer quality of design solutions. An ideal group should have a member who thinks analytically, one with a great imagination, one with good instincts, and one who is a good planner and controller.

One problem with groups composed of people who think differently is communication. As the members think differently, they may also express themselves differently, and

469

this can lead to misunderstanding. It is important that each member of a design group respect the experiences and capabilites of every other member of the group and that differences be thought of as assets and not hindrances.

Pick your design group carefully — look for a variety of experiences and skills.
Respect each member of the group and work toward a common goal.

13-3 THE DESIGN PROJECT

Figure 13-1 presents a design project: the string climber. This chapter will solve the design project using a step-by-step approach that can serve as a model for solving all design project problems.

The string climber project is designed to be a class project in which students compete with one another.

THE STRING CLIMBER

OBJECTIVE:

To build a device that will climb 5 ft up a string, touch the ceiling, and then descend, passing the original starting point. See the following illustration.

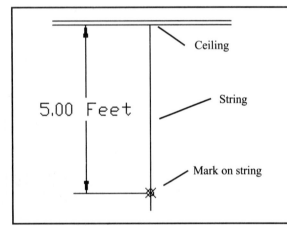

To set up the contest:
Cut a piece of plywood to 2 ft × 2 ft and drill a Ø.50 hole in it.
Remove a 2 ft × 4 ft ceiling tile and replace it with the prepared 2 ft × 2 ft plywood.
Tie the string around something larger than the Ø.50 hole and drop the string through the hole.
Mark the 5-ft mark on the string.
Define the type of string; #21 cabled cotton mason line, industrial quality cabled nylon twine, or the like.

CONSTRAINTS:

(a) 9-V DC battery or maximum voltage (b) 9-V DC, 1200-mA power converter. (Ask the instructor.)
Reversing switch – the change in direction from up to down must be activated by a switch.
You may not control the direction or any other motion of the climber by manipulating the string.
The climber must be activated by an ON/OFF switch.

REQUIREMENTS:

A 3D sketch of your proposed motor setup
A 3D AutoCAD drawing of the motor tester
A working prototype motor tester
A curve of the motor speed (RPM) vs. load (grams)
Show all supporting data and calculations.
3D design sketches for your proposed string climber (three different designs)
A 3D AutoCAD drawing of your string climber
Detail drawings with dimensions of any parts you build or modify
     (continued)

Figure 13-1

A parts list for your string climber including the weight and cost of each part

A center-of-gravity calculation for your design

An electrical schematic for your design

An estimate of the prototype's performance time. Show all supporting data and calculations.

A final report that includes revised drawings and calculations of all the above, the design's performance time, and a conclusion as to why the design was or was not successful.

RULES:

The device must be completely self-contained. No external controls are permitted. There can be no interaction with the device once the device is activated.

The device must start completely below the starting line and climb the string. The device must touch the ceiling. Nothing may be launched into the ceiling and count as touching. A trial will be complete when the device breaks the starting plane. The device may free-fall after touching the top board.

Devices must use a 9-V DC battery or a 1200-mA, 9-V DC power converter (an AC/DC converter is available at Radio Shack) as specified by the instructor. The power source is not considered part of the device. The device must be activated with an ON/OFF switch.

You may attach something to the string to help activate the return up the string. This device may not contain any power but may function only as a locating and rebounding point.

TEAMS:

Students may form teams of up to four members. All members of a team will receive the same grade.

GRADING:

All devices will be timed. The times will be arranged in descending order, starting with the fastest time. Grades will be awarded based on the team's time.

Figure 13-1 (continued)

13-4 UNDERSTANDING THE DESIGN PROBLEM

It is important that all members of the design team have a clear and common understanding of the problem. It is tempting to immediately start to design a solution, but it is better to spend a few minutes reading and discussing the problem.

Have each member of the group read the problem statement out loud.
Discuss the problem.

In this example there is a clear problem statement, but this is not always the case. The problem may be a concept or a goal or a marketing idea. For example, your company's sales and marketing department may feel there is a large market for a "better child's car seat." The design group must first understand what is meant by "better" before any solutions may be formulated.

13-5 CREATING A SOLUTION

Sketch at least three possible solutions to the problem.

Again, the temptation is to create a solution and start creating drawings and a prototype. It is better to create several different solutions and then evaluate each for strengths and weaknesses. One of the solutions may be the final solution, but if something unforeseen occurs, it is good to have other solutions available. It is also possible that the optimal solution is a composite of the various initial solutions.

In this example, three concepts were thought to be possible solutions to the problem. First, the string could be wound around a hub, something like a fishing reel. It was felt that this concept would be very reliable and fairly easy to build. It was also felt that there could be a possible problem with the collection of string. The reel would have to hold at least 5 ft of string. It was also determined that the descent

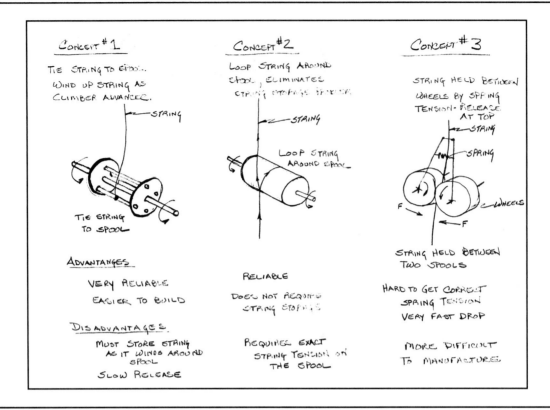

Figure 13-2

speed would depend on how fast the string would unwind from the reel, so the motor would have to reverse to let the string unwind.

The second concept was to loop the string around a spool one time. This would eliminate the problems of collecting and unwinding the string found in the first concept. The disadvantage to the loop concept was that it would require a constant tension on the string. If the string tension was too loose, there would be slippage; if it was too tight, it might cause excess tension on the motor and slow the device down.

The third concept was to hold the string between two wheels. This would completely eliminate any unwinding problems because at the top of the string the wheels would separate and allow the device to drop freely. The disadvantage to the design was the difficulty of manufacture. The tension between the wheels would have to be exact and constant, and the design would require the additional design of a release mechanism.

Figure 13-2 shows sketches of the three possible solutions to the string climber project. It is important to sketch the solutions, as it helps better visualize the prototype.

The next step is to evaluate the differences in the design relative to the objects of the project. This can be done using an *evaluation matrix*. See Figure 13-3.

The purpose of an evaluation matrix is to rate the various design concepts in terms of what is important to the overall design requirements. In this example, the final grade for the project depends on the speed of the device, so speed is very important; however, the device will race more than once, so reliability is also important. Other considerations are ease of manufacture, ease of setup, and stability of the device.

An evaluation matrix is created by assigning a percentage rating to each design criterion in terms of its importance to the success of the design. Next, each of the possible solutions is ranked according to its ability to meet a specific design criterion. The ranks are 1 through 10, with 10 indicating

Evaluation Matrix

Criterion	Weight(%)	Winding		Wrapping		Tension	
		Rating	Weighted	Rating	Weighted	Rating	Weighted
Speed	40	5	2.00	7	2.80	10	4.00
Reliability	30	7	2.10	7	2.10	5	1.50
Ease of manufacture	15	7	1.05	6	0.90	4	0.60
Stability	10	7	0.70	7	0.70	7	0.70
Ease of setup	5	9	0.45	7	0.35	7	0.35
Total score			6.30		6.85		7.15

Figure 13-3

that the solution meets the criterion very well and 0 indicating that the solution will not meet the criterion.

The solution's rating number is then multiplied by the weight percentage to generate a weighted average. The weighted averages are then added. The design with the highest weighted average is considered the best solution for the project.

For this example, the winding concept was selected, although this is not the optimum solution. This selected solution will serve to demonstrate the design process while leaving the optimum solution for the interested student.

13-6 MANAGING THE DESIGN PROCESS

Once a possible solution has been determined, the design and construction of the design must be started. Again, rather than charge ahead, it is better to spend time assigning responsibility for the various aspects of the design and establishing a formal schedule for completion of the various tasks.

Linear responsibility chart

Figure 13-4 shows a *linear responsibility chart*. A linear responsibility chart is used to assign responsibility to individual members of the design team for the various aspects of the design. More than one member may be assigned a task, but only one person should be assigned primary responsibility. Primary responsibility is designated by assigning a 1 to a member. Other types of responsibility may be assigned, such as assistance: 2, must review: 3.

Assigning responsibility helps spread the work evenly among the members of the group and also parcels out accountability among the members.

Team schedule

Once the work is assigned, due dates must be imposed. In this example, the working prototype is due December 6, and the final drawings, December 13. It is assumed that the design is being built during the fall.

It is very difficult to accurately predict how long certain design tasks will take. For this reason, it is recommended that extra time be assigned to each task. If it is thought that the task will take a week, assign at least 10 days.

Linear Responsibility Chart

	Member #1	Member #2	Member #3	Member #4
1.0 Management				
1.1 Overall leadership	1	2	3	3
1.2 Team meeting minutes			1	
1.3 Financial accounts				1
2.0 Procurement				
2.1 Motor	2	1		
2.2 Gears			1	2
2.3 Chassis parts				1
3.0 Drive Train				
3.1 Gear calculations	1	3	2	3
3.2 Chassis supports	3	1	3	2
4.0 Electrical				
4.1 Circuit design		2	1	
4.1.1 Build and test circuit		1	2	
4.2 Add circuit to chassis		1		
5.0 Drawings				
5.1 Concept sketches	1		2	
5.2 Initial 3D drawing	3	2	3	1
5.3 Final drawings				
5.3.1 Individual parts	3	1	3	2
5.3.2 Scenes, 3D isometric	1	2	3	3
5.3.3 Parts list		1	2	
6.0 Construction				
6.1 Build chassis	1	3	3	2
6.2 Assemble drive train	1	2		
6.3 Assemble electronics			1	2
7.0 Testing				
6.1 Test assembled climber	1	2	2	2
6.2 Modify and retest	1	2	2	2
8.0 Final Report	3	3	2	1

Code: 1 = Primary responsibility   2 = Secondary responsibility

3 = Must review

Figure 13-4

September 2005 - Team calendar

Sunday	Monday	Tuesday	Wednesday	Thursday	Friday	Saturday
						1
	Holiday Labor Day 3	Design Class 4	Team Mtg 8:00 Responsibility Concept Skch 5	Design Class 6	7	8
2						
9	10	Design Class 11	Team Mtg 8:00 Formalize Design Sketch 12	Design Class 13	Meet in CAD Lab 2:00 Start drawings 14	Procure Motor Chassis Parts 15
16	17	Design Class 18	Team Mtg 8:00 Design Drive train Electrical 19	Design Class 20	21	Procure Electrical Components 22
23	24	Design Class 25	Team Mtg 8:00 Build & Test ELectric circuit Start chassis 26	Design Class 27	28	Time for more procurement 29
30						

Figure 13-5

Further, it is best to allow for unforeseen events. During a design many things can occur that affect the timing of the design's completion. For example, the team may have planned to use the workshop on a Wednesday, only to find that the shop technician is ill, and the shop will not be open for a week. A part may be delivered later than expected, or a part may break during testing.

Figure 13-5 shows the first part of a team calendar for the month of September. A team calender is created by first listing all known events: class meetings, holidays, due dates for other requirements, and so forth.

Next, a weekly team meeting time is agreed upon. It is critical that a fixed, weekly time be established at the beginning of the project. As other commitments arise, the team members will be able to schedule around the team meeting time and ensure that the design work continues at a constant pace.

List tentative agenda items for each team meeting based on the project's due date requirements. This will serve to give purpose to each meeting and help keep the work on schedule.

Schedule the work realistically

It is recommended that at least 20% be added to the schedule for unforeseen events. In this example the assigned due date is December 6, but the scheduled completion date might be November 15, or just before the traditional Thanksgiving break.

Figure 13-6 shows a Gantt chart. A Gantt chart lists all the project tasks and schedules a date for the work to be done on the task. In addition, certain critical dates are indicated. The Gantt chart gives a visual representation of the project's development and helps organize the work required.

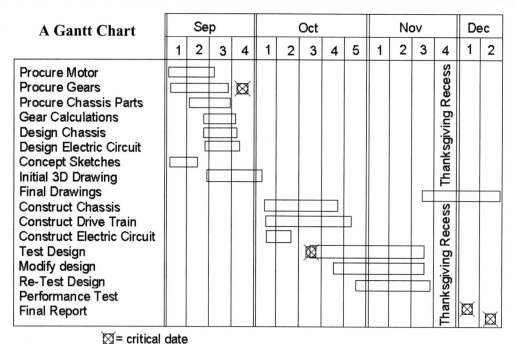

Figure 13-6

13-7 PROCUREMENT

The time required for the gathering of the required parts and tools for the design is often overlooked. A simple trip to a mall to buy a motor or power source can take 2 hours. If parts are ordered off the web, it usually takes several days for the parts to arrive.

Good design teams build procurement times into their schedule. Note that in the team calendar two Saturdays have been set aside for procuring parts. Further procurement time is alloted on the Gantt chart.

13-8 THE MOTOR—THE FIRST UNEXPECTED PROBLEM

In this example, a motor was taken from a Meccano MO set. The motor was designed to operate on power from two AA batteries or 3.0 V DC. It is known that for short periods of time DC motors can be run well above their rated capacity with no damage.

The problem was that nothing was known about the motor's operating characteristics. How fast did it turn, and how much torque did it generate? No data were available.

Literature search (web search)

When confronted with a design problem, it is usually best to take a few minutes and search for any work already done in the particular problem area. Reference books in a library are an excellent source, as is the web.

A web search on the key word Meccano® led to a group called the International Society of Meccanomen, a group of hobbyists who build using Meccano parts. A message was posted on the group's message board, and three responses were received. The responses recommended several different sources. In particular, the article "A Meccano Dynamometer" by Robin Lake, published in the June 1992 issue of *Construction Quarterly,* pages 16 and 17, presented the outline for a small dynamometer. Based on the article a concept sketch was created for a dynamometer that could be used to determine the selected motor characteristics. Figure 13-7 shows the dynamometric concept sketch. Figure 13-8 shows a photograph of the dynamometer built from the concept sketch.

Data are collected from the dynamometer by changing the weight being lifted and recording the change in motor speed. A plot is then created of weight (dependent variable) versus motor speed (dependent variable).

The motor speed was determined by timing the rotation speed of the indicator gear. As the motor lifts the weight

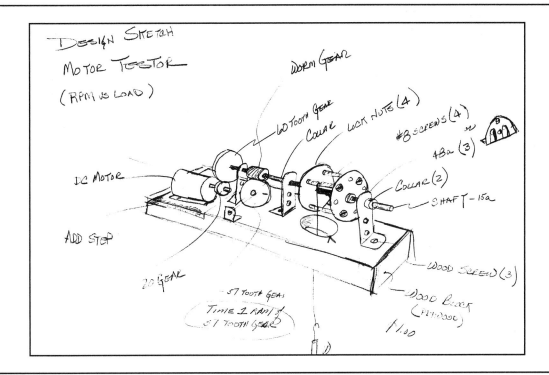

Figure 13-7

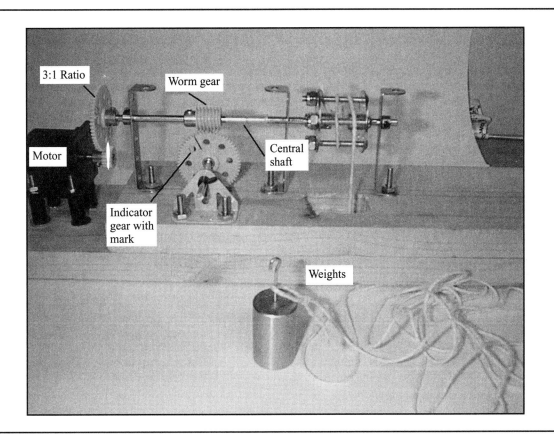

Figure 13-8

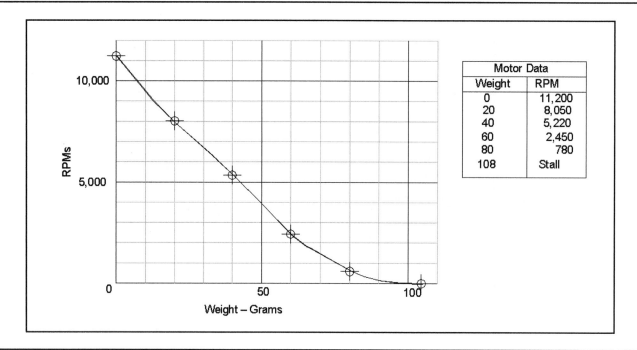

Figure 13-9

the central shaft turns the worm gear, and the worm gear turns the indicator wheel. A mark has to be drawn on the indicator wheel so it can be timed as it rotates. A stopwatch is used to time the indicator wheel's rotation speed.

Say that it takes 1.27 seconds for the indicator wheel to make one revolution for a given weight. This means that the indicator gear is rotating at 1.27 revolutions/second.

The indicator gear has 57 teeth, so the ratio between the indicator gear and the worm gear is 57:1. The worm gear and therefore the central shaft is turning at

$$57(1.27) = 72.39 \text{ revolutions/second}$$

Conversion of this value to RPM yields

$$72.39(60) = 4343 \text{ RPM}$$

The ratio between the 57-tooth gear on the central shaft and the 19-tooth gear attached to the motor is 3:1; therefore, the motor speed is determined to be

$$4343(3) = 13,030 \text{ RPM}$$

Friction

The first value in the plot shown in Figure 13-9 is for no load on the motor. This does not mean that the speed value found is for 0 g load. The motor is driving the dynamometer, which requires some force to turn. There is some friction in the system, and there will also be friction in the climber. It will be assumed that the friction in the climber approximately equals the friction in the dynamometer.

Stall speed

Weights are then added until the motor can no longer lift the load. This is the motor's stall point. In this example the motor stalled at 108 RPM.

13-9 STRING CLIMBER CONCEPT SKETCH

Figure 13-10 shows the first concept sketch of the proposed string climber. It is often difficult to know where to start on a design project. Simply stated, just start sketching, and have a big eraser. It is much easier to change a sketch and resketch than to change formal drawings. An initial concept sketch serves as a starting point, and many potential problems can be addressed at this early stage, making them easier to fix than if they are discovered later.

The word *friction* is included in Figure 13-10. This was felt to be a potential problem, so a question was written directly on the sketch. As a sketch is reviewed any questions and comments should be written directly on the sketch. This will help ensure that all team members are aware of all questions.

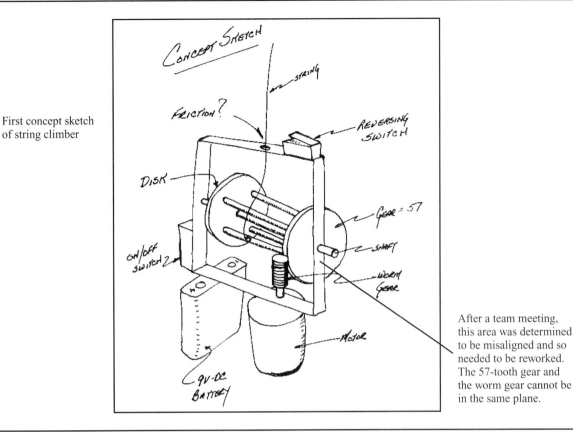

First concept sketch of string climber

After a team meeting, this area was determined to be misaligned and so needed to be reworked. The 57-tooth gear and the worm gear cannot be in the same plane.

Figure 13-10

As the sketch was reviewed during a team meeting a major error was discovered: the worm gear and the gear on the winding mechanism were drawn in the same plane. This is not possible, so a second sketch needed to be done. It is very common for sketches to go through several iterations.

Figure 13-11 shows a second sketch that moved the motor and worm gear into proper alignment with the 57-tooth gear. Changes were also made to the reversing switch, and some of the chassis parts are shown in more detail.

13-10 DRAWING LAYOUT

At this point in the design process sketches become inadequate. The basic concept of the design shape and part location have been established, but more exact information is now needed.

For this example it was determined to use only Meccano erector set parts. This is not the optimum design in terms of speed, as the parts are steel and heavier than, say, plastic or basswood. The solution presented is intended to serve as a guide for a working solution. A better, faster solution can be created.

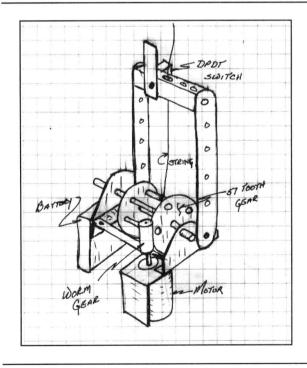

Figure 13-11

String will pass through this hole, so it is assumed to be the climber's centerline.

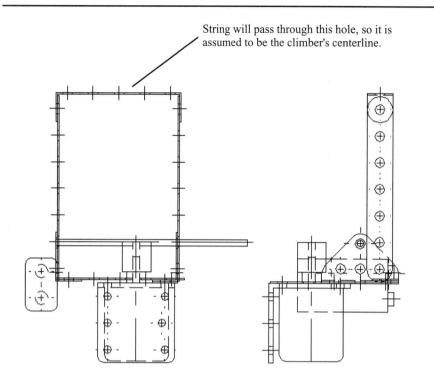

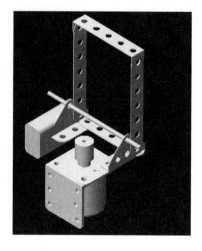

Initial chassis layout

Figure 13-12

Figure 13-12 shows an initial chassis layout. The parts are drawn to scale, although not all parts are included. The fasteners and nuts also are not included. The intent of the layout is to start to establish the actual size of the design and the tentative location of the major parts.

The chassis will be made from Meccano parts. These parts are drawn as individual models, then combined to form an assembly drawing. The sizes of the parts are known, so a chassis shape can be started. The motor and battery sizes are also known, so they too can be modeled.

Weight

The weight of the climber will directly affect its speed. The climber's weight will also determine the gear ratios needed. A running total of the climber's weight should be kept. As parts are added or deleted their weight should be either listed or removed from the list and the total weight revised.

Figure 13-13 is a listing of parts used to build the climber.

Stability

If the climber's weight is not aligned with the string, the climber will flop around during its climb. This will cause the climber to waste power and slow down; however, if the parts are arranged so that they are balanced relative to the string, the climber will move smoothly up and down the string. Balance is achieved by locating the climber's center of gravity in line with the string's axis.

Weights for Climber Parts

Part	Weight (g)	
Motor	54	
Battery	44	
Worm gear	20	
Gear - 57 teeth	16	
Round plate	16	
M16 axle	9	
ON/OFF switch	6	
DPDT switch	6	
M3 [2]	12	
M5 [4]	16	
M48a	6	
M126 [2]	12	
M48—modified	6	
M52 (extension)	2	
M56	8	
Battery clip	2	
Lock nuts [4]	8	
#8x1.50 screws [4]	14	
M111a screw [12]	8	
M37a nut [12]	2	
M36 clip [2]	.5	
Wire (est.)	5	Total = 272.5 g

The prefix M indicates a Meccano part number.

Figure 13-13

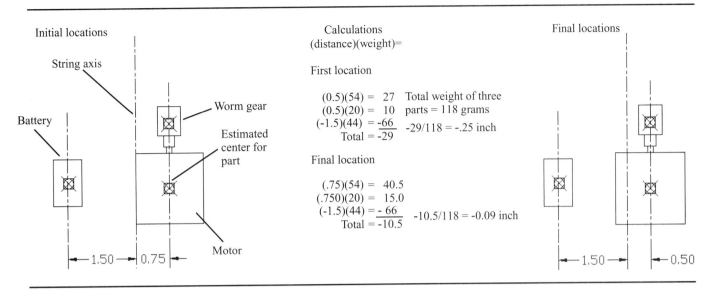

Initial locations

String axis

Battery

Worm gear

Estimated center for part

Motor

|◄— 1.50 —►| 0.75 |◄—

Calculations
(distance)(weight)=

First location

$(0.5)(54) = 27$
$(0.5)(20) = 10$
$(-1.5)(44) = \underline{-66}$
Total = -29

Total weight of three parts = 118 grams

$-29/118 = -.25$ inch

Final location

$(.75)(54) = 40.5$
$(.750)(20) = 15.0$
$(-1.5)(44) = \underline{-66}$
Total = -10.5

$-10.5/118 = -0.09$ inch

Final locations

|◄— 1.50 —►| |◄— 0.50

Figure 13-14

Center of gravity

The center of gravity can be estimated before the drawing is in final form by using the weights and locations of the heaviest parts. From the weights presented in Figure 13-13 the motor, battery and worm gear are the heaviest parts. All the other parts can be arranged symmetrically about the string axis, so they will not be included in this estimate.

The center of gravity is calculated by multiplying the distance from a defined location to the center of the individual part by the weight of the part. The sum of all distance × weight calculations is then divided by the total weight. Distances are calculated using their X, Y, and Z component values.

Figure 13-14 shows some sample calculations for one component of the center of gravity. The string axis is assumed to be 0. Initially the motor and worm gear are located 0.50 in to the right of the string, which is defined as the positive direction. The battery is located 1.50 in. to the left of the string.

The calculation for the initial location indicates that the center of gravity is 0.25 in. to the left of the string. The motor and worm gear are moved .25 in. to the right. This location generates a center-of-gravity value of 0.09 in. This location is deemed acceptable. During final assembly some smaller parts such as the wires, shaft, or 57-tooth gear can be moved slightly as needed to ensure a 0.00 in. value.

Figure 13-15 is a side view of the climber showing the location of the motor and worm gear. The parts were aligned with the string axis by rotating them. The lower portion of the climber is designed so that all the included parts, motor, gears, battery, axle, and support structure can be rotated to establish an exact alignment between the string axis and the climber.

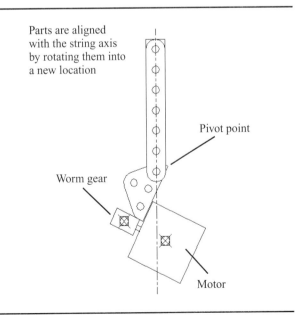

Parts are aligned with the string axis by rotating them into a new location

Worm gear

Pivot point

Motor

Figure 13-15

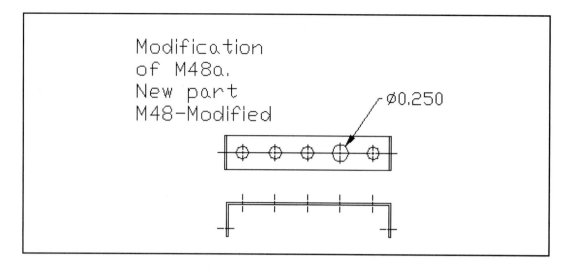

Figure 13-16

13-11 MODIFIED PARTS

Any part that is purchased and used in the *as-purchased* condition need only be listed in the parts list. No drawing is needed. Parts made specifically for the design must be documented by a detail drawing. See Chapter 5. Purchased parts that are modified in any way must also be documented using a detail drawing.

Part M48a was modified to accept the double-pole double-throw (DPDT) switch used to reverse the motor. The existing holes in the part were too small for the switch, so the hole was enlarged from $\varnothing 0.19$ to $\varnothing 0.25$ in. Figure 13-16 shows the drawing.

13-12 ELECTRICAL SCHEMATIC

A DPDT switch was used to reverse the current to the motor and thereby reverse the motor direction. An ON/OFF switch was also included.

Figure 13-17 shows the climber's electrical schematic diagram. The manufacturers' part number for each component should be listed in the climber's parts list.

An extension piece was added to the lever of the switch so that as the climber ascends, the extension will hit the ceiling before any other part of the climber, tripping the DPDT switch and reversing the motor's direction. This will cause the climber to descend the string.

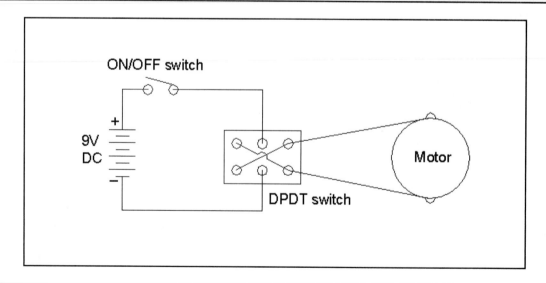

Figure 13-17

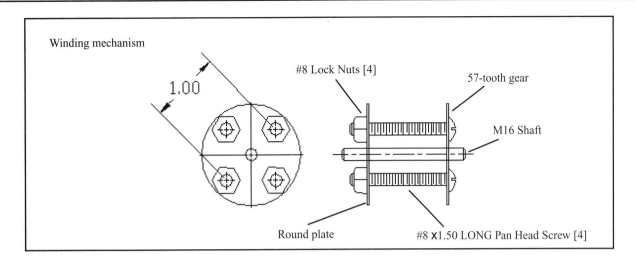

Figure 13-18

13-13 ESTIMATED SPEED

For this example, a worm gear meshed with a 57-tooth spur gear attached to the end of a rotating winding mechanism was used. This is not the optimum solution in terms of speed but it serves to illustrate how to determine a climber's speed based on the motor's power curve and the selected gear ratio.

The string climber weighs 272.5 g. The gear ratio between the worm gear and the 57-tooth gear is 57:1. The weight that the motor must drive is therefore

$$\frac{272.5}{57} = 4.78 \text{ g}$$

At this weight the motor will operate at approximately 9500 RPM, or 158 rev/sec.

$$\left(\frac{9500 \text{ revs}}{\text{min}}\right)\left(\frac{1 \text{ min}}{60 \text{ sec}}\right) = 158.3 \text{ rev/sec}$$

If the motor is turning at 158.3 rev/sec, the winding mechanism driven by the 57-tooth gear is rotating at 2.78 rev/sec:

$$\frac{158.3}{57} = 2.78 \text{ rev/sec}$$

The winding mechanism has a diameter of approximately 1.00 in. See Figure 13-18. The circumference of the winding mechanism is about 3.14 in., so the mechanism winds string at a rate of 8.72 in./sec.

$$(2.78)(3.14) = 8.72 \text{ in./sec}$$

The string is 5 ft, or 60 in., from bottom to top, so the ascent time will be 6.88 seconds:

$$\frac{60}{8.72} = 6.88 \text{ sec}$$

As the climber descends, the motor has little load acting on it, so it rotates faster. It is assumed that this condition is close to the 0 load condition recorded in Figure 13-9. The motor is turning at about 11,100 RPM. Repeating the previous calculations generates a descent time of 5.89 seconds. The total travel time is then $6.88 + 5.89 = 12.77$ seconds.

13-14 FINAL DRAWINGS

Figure 13-19 shows an exploded isometric assembly drawing for the climber. A design should first be built *on paper*; that is, the drawing should be a complete and accurate documentation of the design.

The final drawing is the result of a series of developmental steps. The process starts with a concept sketch. The sketch is then developed as more detail is added. Next, an initial drawing layout is started that starts to size and locate parts. Again, more detail is added.

Calculations are done as needed—for example, to determine the required gear ratio. The number and size of gears required is incorporated into the design by sizing and locating support structures on the chassis. The electrical components are added. The result is a complete drawing that is ready for building.

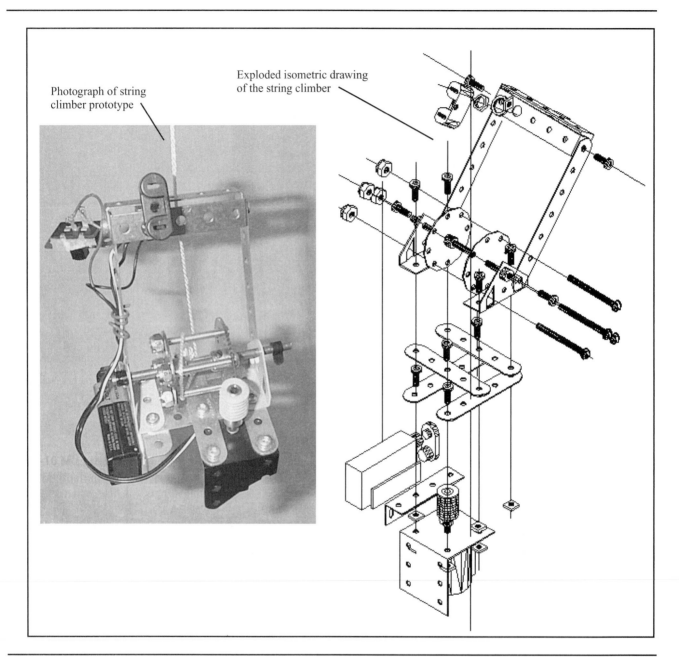

Photograph of string climber prototype

Exploded isometric drawing of the string climber

Figure 13-19

13-15 TEST AND MODIFICATION

The climber prototype should be tested extensively. As problems arise they can be addressed and the prototype modified. The drawing must also be modified and kept up to date.

In this example, it was found during testing that the Meccano nuts located on the winding mechanism constantly came loose. They were exchanged for locknuts that did not come loose.

Adjustments were made on the chassis to ensure that the climber's center of gravity was aligned with the string axis. These final adjustments were added to the drawing, so that the drawing was an exact documentation of the design.

CHAPTER 14

Design Projects

14-1 INTRODUCTION

This chapter presents eight design projects. The purpose of the projects is to provide a hands-on design experience with practical applications for the material covered in this book. The transition from a design solution concept to a working prototype is often more difficult than first imagined. It is analogous to learning to swim. All the concepts may be understood, but until you enter the water you cannot truly appreciate what is required.

The projects are set up to be competitive in nature. Each project presents a set of parameters and challenges designed to achieve a specified goal. Who can go the fastest, who can lift the most weight? The competitiveness serves to make the projects more exciting and more emotional. This helps to create a memorable experience.

Each project is presented as a stand-alone entity. No project is related to any other. All the information needed is presented with the project, and each project has several variations.

Space requirements

Each project can be operated in a classroom, lab, or hallway. No special environment is required. None of the projects requires an outdoor setting.

Tooling

Only hand tools are required. It would be helpful to have a drill press, scroll saw, soldering gun, and vise available.

Parts availability

All projects utilize commonly available parts. The projects have all been built using Meccano® parts, LEGOs, Connects, and other building kits as well as individual parts made from metal or wood.

The use of prefabricated building kits is at the instructor's discretion.

Difficulty rating

Each project includes a difficulty rating. These rating are based on the amount of work required and the difficulty experienced by students actually building the projects.

The ratings run from 1 to 10, with 10 being the hardest and 1 being the easiest.

Grading

Each project has a suggested grading scheme, for performance, and a list of drawing requirements.

14-2 HEAD BANGERS – RATING 2

Object

To design and build a vehicle that will traverse a 6-ft track, touch the wall at the end of the track, then return back down the track, breaking the plane of the original starting line. See Figure 14-1.

The track

The track is made from ¾-in. BD plywood. The side rails are 1 × 4 #2 pine, 6-ft lengths. The wall is made from 2 × 8 #2 pine. All parts are glued and nailed in place.

The start line and the pass line are marked with a permanent marker.

The vehicle

The vehicle must be self-contained, with no external controls. Once activated, the vehicle must advance, turn, and return entirely on its own. If the vehicle is touched during a run, the round will be forfeited.

The vehicle must include wheels and must roll on its wheels while traversing the track. No slingshot-type devices are allowed. A ball is not considered a vehicle. Throwing a ball or other object down the track is not acceptable.

A vehicle must move as a single unit. It is not acceptable to have a stationary object in place that anchors, launches, or in any way controls the vehicle.

All vehicles must be powered by DC electric motors. More than one motor may be used.

All vehicles must derive power from only one standard 9 V DC battery. The term *standard* means the type of battery used to power calculators, transistor radios, and the like, which is available at local convenience stores.

The vehicle must run within the 10-in. walls of the track and cannot touch the top of the walls. The vehicle may not exceed 12 in. in length at any time during the run.

Purchased vehicles may be cannibalized for motors, wheels, gear boxes or other parts but may not be used in the as-purchased condition.

All vehicles must be constructed on their own original chassis.

Teams

Students may work as individuals or in teams of up to three.

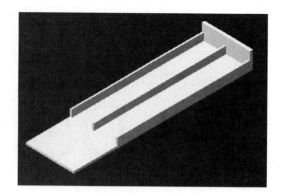

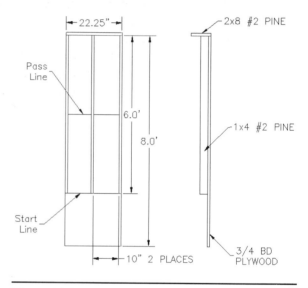

Figure 14-1

Runs

One vehicle will compete against another one. Each vehicle must start from behind the starting line, traverse down the track rolling on its wheels, touch the wall, and return back down the track breaking the plane of the starting line. The first vehicle to break the starting plane will be declared the winner.

A run will end when, in the judgment of the instructor, the vehicles have stopped moving and have ceased to advance down the track.

Rounds

Each vehicle must compete in six runs: three for the semifinals and three for the final. Each win will be worth 1 point. A loss will not earn any points.

If one vehicle returns past the midline and the other vehicle does not, the vehicle that passes the midline will be awarded ½ point.

If neither vehicle passes the midline, both vehicles will be given 0 points.

If the race is a tie, it will be run again. If it is again a tie, both vehicles will receive 1 point.

Grading – performance

Final grades will be awarded at the discretion of the instructor. The following is a possible grading scheme:

Points	Grade
6	A
5	A−
4	B+
3	B
2	B−
1	C+
0 (six runs)	C

Calculations

Time the vehicle and estimate the speed of the motor in RPM.

Drawing requirements

1. An exploded scene of the vehicle
2. An exploded isometric assembly drawing with assembly numbers
3. Dimensioned and toleranced detail drawings of every part not purchased and of any purchased part that has been modified
4. A two-column parts list that gives an estimated price and weight for each part
5. A center-of-gravity calculation. List each part and its X, Y, Z component distances.

Final decisions

The instructor has final decision power on all questions and rulings involving the competition.

14-3 THE STRING CLIMBER – RATING 7

Chapter 13 used the string climber project to explain the design process. The solution presented was not the optimal, that is, not the fastest solution. Use the information presented in Chapter 13 as a guide, and design a faster solution using the same parameters.

If a DC power source is used, the wires may be held away from the climber but may not be used in any way to improve the climber's performance.

See Figure 14-2.

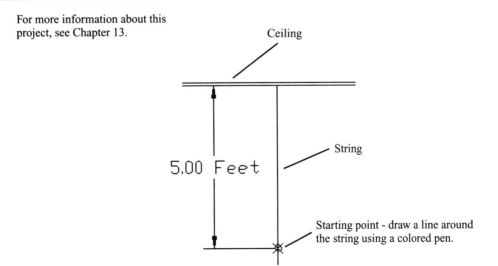

For more information about this project, see Chapter 13.

Ceiling

String

5.00 Feet

Starting point - draw a line around the string using a colored pen.

Figure 14-2

14-4 ROPE BOMBER – RATING 3

Object

To design and build a device (bomber) that will run along a cable, drop an object (bomb) into a specified target area, and then stop within a defined stop zone.

Course

See Figure 14-3. The course will be a plastic-coated Ø0.19 cable (the type of cable used for a dog run and available at pet stores). The cable will be strung across a room as shown. The dimensions given are for a specific room. The dimensions will vary according to the size of the room.

Target area

The target area will be a 3-ft × 3-ft sandbox placed on the floor under the cable. The sand will be approximately 2 in. deep. A 3.5-in. common nail will be located in the center of the sandbox directly under the cable.

Bomb

The bomb must be cylindrically shaped with a diameter between .375 in. and .750 in. and a length between 3.0 in. and 4.0 in. Any material may be used.

Bomber

The bomber must be self-contained. No radio control devices are permitted. No strings or other material may be attached to the string to help control the bomber.

The bomber may be powered only by gravity. No electrical or chemical propellants may be used.

The outside dimensions of the bomber may not exceed 1 ft³ measured as 12 in. × 12 in. × 12 in. At no time during the run may the bomber exceed the volume limits. A string hanging from the bomber is considered a violation of the volume requirement.

The bomber's weight may not exceed 4 lb.

The surfaces of the bomber in contact with the cable may not deteriorate the cable in any way.

Teams

Students may work as individuals or in teams of up to three.

The contest

The bomber must start from behind the start line indicated by a line drawn around the cable. The bomber then must run along the cable, drop its bomb, and continue on, stopping within the stop zone indicated by two lines drawn around the cable.

Each team will be given three runs.

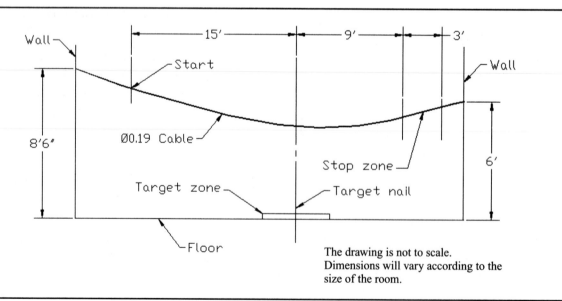

The drawing is not to scale.
Dimensions will vary according to the size of the room.

Figure 14-3

The lanyard

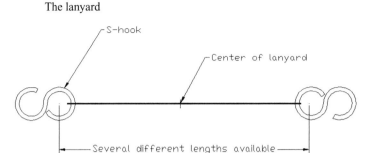

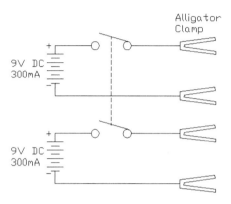

The activating
power source

Wires should be at
least 4' long.

Figure 14-4

Once called, a team will have 3 minutes to set up their bomber on the cable. Failure to be ready within the specified 3 minutes will result in a loss of one run.

Calculations

Measure the slope of the cable after it is in position.

Create a drawing of the cable's slope.

Determine the weight of the bomber and bomb.

Calculate the estimated speed of the bombing run.

Measure the bomber's performance results and compare them with the estimates.

Grades – performance

Each team's bomb drop will be measured from the nail located in the center of the drop zone. The shortest of the three distances will be a team's official distance.

The distances will be arranged in descending order, with the shortest distance first. The instructor will assign grades based on a team's position in the order.

Drawing requirements

1. An exploded scene of the vehicle
2. An exploded isometric assembly drawing with assembly numbers
3. Dimensioned and toleranced detail drawings of every part not purchased and of any purchased part that has been modified
4. A two-column parts list that gives an estimated price and weight for each part
5. A center-of-gravity calculation. List each part and its X, Y, Z component distances.

14-5 TUG OF WAR – RATING 5

Object

To design and build a device capable of pulling a lanyard away from a similar device.

The device

The maximum allowable weight is 3.00 lb.

The maximum volume at the start of the tug is 1 ft³, measured as 12 in. × 12 in. × 12 in. The device may exceed the 1-ft³ requirement during the tug.

The device must be designed to accept electrical input from alligator clamps.

The device must be designed to easily accept the S-hook of the lanyard.

The lanyard connection must be located on the front of the device, that is, the side facing the oponent.

The lanyard

See Figure 14-4. The lanyard may be made from string or wire as determined by the instructor.

An S-hook will be attached to each end of the lanyard. The S-hook will be opened slightly so it can fit over mounting devices on the device.

Several lanyards of different lengths will be available.

The approximate center of the lanyard will be marked in permanent marker.

Pulling power

Power for pulling may come from any one or any combination of the following: DC electric motor, rubber bands, or a mouse trap. The specific limits for each power source are as follows:

Electric motors – Maximum capacity of 6 V DC and a maximum retail price of $5.00.

Rubber bands – Four rubber bands whose relaxed size does not exceed ¼ in. by 4 in. The rubber bands may be doubled over.

Mousetrap – The spring of the mousetrap may not exceed Ø.375 and a length of 1.50 in.

Activation

The device will be activated by an electric current supplied by the instructor. See Figure 14-4.

In this example a 9-V DC 300 mA power source is specified. Other power sources may be used as specified by the instructor.

The end of the activating power source wires will be fitted with alligator clamps. The device must be designed to accept the activating alligator clamps and to use the power to activate the pulling power of the device.

Electrical power will be supplied for 10 seconds.

The activating power may be used to power an electric motor on the device, not just for activating the pulling power of the device but also a pulling electric motor.

The battlefield

The battlefield will be a 4-ft × 4-ft box filled with sand. The sand will be approximately 2 in. deep.

The procedure

Each device will be placed on the sand. The devices may not be pushed down, twisted, or moved so as to gain a positional advantage.

The lanyards will be attached and the devices positioned so that the center mark of the lanyard is off the sand.

The instructor will slide a straightedge (ruler) directly under the center mark of the lanyard.

The alligator clamps will be attached.

Each team will stand back from their device.

Power will be applied for 10 seconds.

At the end of 10 seconds the team with the center mark of the lanyards on their side of the straightedge will be declared the winner.

Teams

Students may work as individuals or in teams of up to three.

Grades – performance

Each team will compete six times. Each win will count as 1 point, each loss 0 points.

If there is a tie, the tug will be done again. If there is a second tie, each team will be awarded 1 point.

It is possible for both teams to receive 0 points if neither device activates properly.

Calculations

Use a pull-scale and measure the pulling power of the mouse trap and the pulling power of one rubber band.

Consider the advantages and disadvantages of each pulling power source.

Calculate any mechanical advantage gained from pulleys, moment arms, or other design features.

Drawing requirements

1. An exploded scene of the vehicle
2. An exploded isometric assembly drawing with assembly numbers
3. Dimensioned and toleranced detail drawings of every part not purchased and of any purchased part that has been modified
4. A two-column parts list that gives an estimated price and weight for each part.

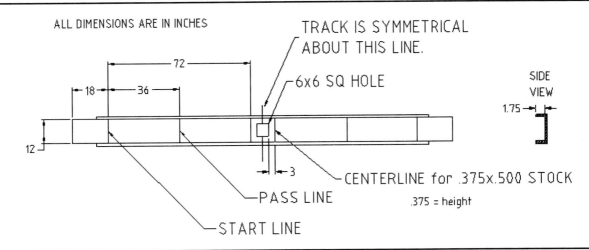

Figure 14-5

14-6 FOOTBAG DROP MADNESS – RATING 3

Objective

To design and build a vehicle that will roll across a track carrying a footbag, drop the footbag into a 6 in. × 6 in. hole and return down the track, passing the starting line in less than 15 seconds.

The vehicle

The vehicle may not exceed 1 ft³ measured as 12 in. × 12 in. × 12 in. The vehicle may change shape during the run, but it may never exceed 1 ft³.

The vehicle may not exceed 4 lb.

The vehicle must have at least three wheels and must roll on its wheels while traversing the track.

The footbag

For purposes of this project, a footbag is a soft object with a knitted outer cover. Its shape is approximately a cylinder 2.50 in. in diameter and 1.50 in. to 1.75 in. high. The instructor has the right to disqualify any footbag that is deemed not to meet the spirit of the project.

Track

Front and side views of the track are shown in Figure 14-5. All dimensions have a tolerance of ± .063 unless otherwise stated. The track is made from ¾-in. BD plywood with the B-side as the track's running surface. The side rails are made from 1 × 3 #2 pine.

Power

Power may be derived only from two "AA" batteries. Additional power may be derived from springs or rubber bands to release the footbag.

The procedure

The vehicle must start from completely behind the start line.

The vehicle must roll down the track staying within the track's walls. The tops of the walls may not be used.

The complete vehicle must pass the "pass line." No part of the vehicle may remain behind the pass line. The vehicle need not travel to the end of the track before dropping its footbag, but the vehicle must be completely beyond the pass line.

The footbag must pass through the hole to the container below. A footbag that hangs on the edge of the hole will not be considered to have dropped into the hole.

The complete vehicle must return down the track breaking the starting line.

The vehicle must complete the course in as close to, but not exceeding, the 15-second time limit.

Points

DROP POINTS: A vehicle that completes a run down the track and successfully drops its footbag into the hole before the opponent does will receive 2 points.

TIME POINTS: An additional point will be awarded to a vehicle that, after successfully dropping its footbag, returns down the track and crosses the starting line closest to, but not exceeding, 15 seconds. For example, if one vehicle crosses the start line at 14.5 seconds, and the second vehicle crosses at 14.7 seconds, the 14.7 time will be awarded 1 point. If a vehicle exceeds the 15-second time limit, no points will be awarded.

Runs

Each team will compete in at least six races. This means that 18 points are available.

Grades – performance

The instructor will arrange each team's total points in descending order with the largest number of points at the top. Grades will be awarded according to relative position on the points list.

Calculations

Determine the vehicle's motor speed, and predict the performance time: the time needed to drop the sack and the return time.

Drawing requirements

1. An exploded scene of the vehicle
2. An exploded isometric assembly drawing with assembly numbers
3. Dimensioned and toleranced detail drawings of every part not purchased and of any purchased part that has been modified

4. A two-column parts list that gives an estimated price and weight for each part
5. A schematic diagram of the vehicle's electrical system

14-7 PEAK PERFORMANCE – RATING 5

Objective

To design and build a vehicle that will climb the "hill" shown in Figure 14-6, stop at the top of the hill, and defend its position against another vehicle.

Hill

See Figure 14-6. The hill is made from ¾-in. plywood covered with carpet. The side rails are made from 1×3 #2 pine. The center tower is made from 2×3 #2 pine topped with a 12 in. $\times$ 12 in. piece of plywood.

Vehicle

The vehicle must be self-contained with no external controls or controlling devices.

The vehicle may be powered mechanically (springs, rubber bands), electrically (limit of 9 V DC as provided by one alkaline 9-V battery), or any combination of the two power sources. No chemicals are permitted (a CO_2 charger

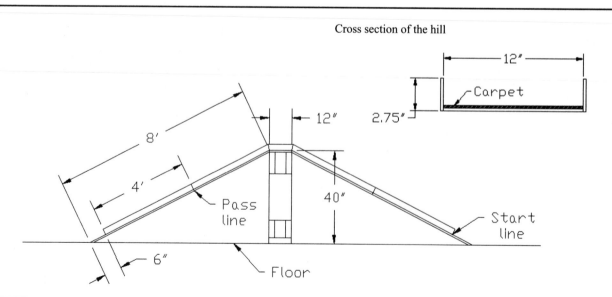

Cross section of the hill

Figure 14-6

is considered a chemical device).

The weight of the vehicle may not exceed 4 lb.

The volume of the vehicle may not exceed 1 ft^3 measured as 12 in. $\times$ 12 in. $\times$ 12 in. at the start of a run. Attacking devices such as a ram may extend beyond the volume limit once the vehicle is activated but cannot be activated before the round begins.

The vehicle must use one activating trigger or switch. The trigger may activate two functions—start the vehicle and launch an attacking device—but only one trigger may be used.

The vehicle must run within the 12-in. track walls and may not ride on top of the side rails.

All vehicles must be constructed on their own original chassis. Purchased vehicles may be cannibalized for motors, wheels, gears but may not be used in their as-purchased condition. An existing chassis that is cut in two and taped back together is not considered an original chassis.

The dot

Each team will locate a clearly visible dot on their vehicle. This dot will be used to determine how close the vehicle is to the hill's centerline. The dot must be located on the main body of the vehicle. It may not be located on any type of extension device.

The procedure

Vehicles must start from completely behind the start line on opposite sides of the hill.

At the instructor's command, the teams will activate their vehicles and stand back. If a vehicle is touched during a run, it will be disqualified from the run.

The complete vehicle must advance beyond the pass line. Nothing can be left behind to serve as an anchor or other type of controlling device.

Vehicles must stay within the designated pass lines. If a vehicle runs up and over the top of the hill and then runs down the other side of the hill past the pass line, it will receive 0 points. It is possible for a vehicle to be awarded a point if it goes beyond the center portion of the hill and stops before crossing the pass line, thereby blocking the other vehicle from getting to the center of the hill.

At the end of 15 seconds, the vehicle whose dot is closest to the center of the hill will be declared the winner and awarded 1 point.

Points

Each team will compete in six runs. The winner of a run will receive 1 point, the loser 0 points.

If there is a tie, the contest will be run again. If it is a tie again, both teams will receive 1 point.

If neither team crosses the pass line, 0 points will be awarded.

If one vehicle crosses the pass line but does not reach the top of the hill, and the other vehicle does not pass the pass line, the vehicle passing the pass line will be awarded 1 point.

Grades – performance

Final grades will be awarded at the discretion of the instructor. The following is a possible grading scheme:

Points	Grade
6	A
5	A−
4	B+
3	B
2	B−
1	C+
0 (six runs)	C

Calculations

Determine the vehicle's motor speed and predict how fast the vehicle will ascend the hill.

Drawing requirements

1. An exploded scene of the vehicle
2. An exploded isometric assembly drawing with assembly numbers
3. Dimensioned and toleranced detail drawings of every part not purchased and of any purchased part that has been modified
4. A two-column parts list that gives an estimated price and weight for each part

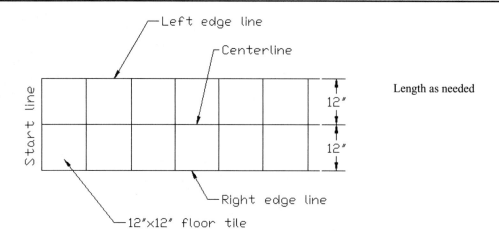

Figure 14-7

14-8 S-CAR – RATING 8

Objective

To design and build a vehicle that will travel in an S-shaped pattern.

Track

The track is defined using standard 12-in. × 12-in. floor tiles. See Figure 14-7. The boundaries of the track may be indicated with masking tape.

The procedure

The vehicle must start from completely behind the start line, approximately centered on the track's centerline.

At the instructor's command, the vehicle will be activated.

The vehicle must first cross the right edge line, run back across the track's centerline, cross the left edge line, then pass back over the centerline.

If two wheels cross an edge line, the vehicle will be considered to have crossed the line.

The vehicle need only break the plane of the centerline to complete a run.

The vehicle

The vehicle may be no larger than 12 in. ×12 in. × 12 in. at any time during a run.

The vehicle must be self-contained with no external controls.

The vehicle must include an ON/OFF switch that is used to activate the vehicle.

The vehicle must have at least three wheels and roll on the wheels throughout a run.

Power

A DC power source may be used. The maximum allowable current is 9 V DC 800 mA. The wires from the power source to the vehicle may be held but may not be used in any way to enhance the vehicle's performance.

Grades – performance

Each vehicle will be timed for three runs. The fastest time will be the team's time. The times will be ranked from fastest to slowest. Grades will be assigned at the discretion of the instructor based on a team's rank in the order.

Calculations

Determine the vehicle's motor speed and predict the time it will take to complete a run.

Drawing requirements

1. An exploded scene of the vehicle
2. An exploded isometric assembly drawing with assembly numbers
3. An isometric assembly drawing of the vehicle's steering mechanism and drive train
4. Dimensioned and toleranced detail drawings of every part not purchased and of any purchased part that has been modified
5. A two-column parts list that gives an estimated price and weight for each part

Side rails may be added at the instructor's discretion.

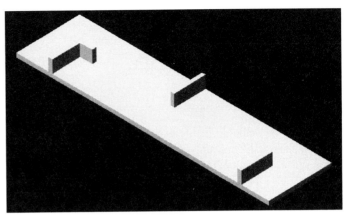

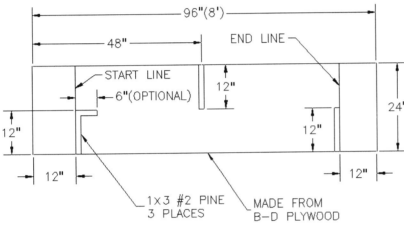

Figure 14-8

14-9 CAM CAR – RATING 9

Objective

To design and build a vehicle that will traverse a defined track using a cam to control the turning.

The track

See Figure 14-8. The track will be constructed on ¾ in. BD plywood with the B-side as the track surface. The rails will be made from 1-in. × 3-in. #2 pine located as shown.

The vehicle

The vehicle may be no larger than 12 in. × 12 in. × 12 in. at any time during a run.

The vehicle must be self-contained with no external controls.

The vehicle must include an ON/OFF switch that is used to activate the vehicle.

The vehicle must have at least three wheels and roll on the wheels throughout a run.

The turning of the vehicle must be controlled exclusively by a cam.

The vehicle may not go over the rails but must travel in a path around the rails.

Power

A DC power source may be used. The maximum allowable current is 9 V DC 800 mA. The wires from the power source to the vehicle may be held but may not be used in any way to enhance the vehicle's performance.

Grades – performance

Each vehicle will be timed for three runs. The fastest time will be the team's time. The times will be ranked from fastest to slowest. Grades will be assigned at the discretion of the instructor based on a team's rank in the order.

Calculations

Determine the vehicle's motor speed and predict the time it will take to complete a run.

Determine the length of the car's potential ideal path.

Drawing requirements

1. An exploded scene of the vehicle
2. An exploded isometric assembly drawing with assembly numbers
3. An isometric assembly drawing of the vehicle's steering mechanism and drive train
4. A displacement diagram for the cam
5. A dimensioned drawing of the cam
6. Dimensional and toleranced detail drawings of every part not purchased and of any purchased part that has been modified
7. A two-column parts list that gives an estimated price and weight for each part

14-10 CAM PROJECT – RATING 7

Objective

Design and build a device with a cam that will lift a dial indicator 10 mm from a defined zero position, hold the indicator at 10 mm for 3 seconds, then return to the zero position. Each lift and return cycle is to be 6 seconds; that is, the cam should turn at 10 RPM.

Design parameters

1. The device must include at least three gears.
2. The device may include a potentiometer, but the potentiometer may be used only to fine-tune the device. It may not be the principal control device.
3. The support structure for the device must be an original design. Kit parts such as LEGO, Itech, and Erector® may be used, but their configuration must be original.

Grades – performance

Grades will be awarded based on the performance ranges shown in Figure 14-9. Each device will start with 100 points.

TEAM MEMBERS: ——————————————   TOTAL POINTS _____

—————————————

—————————————

CAM LIFT – Static test: 10 mm		LIFT TIME: 3 seconds	
9.5–10.5 mm _____	Acceptable range	2.90–3.10 sec Acceptable range	
9–11 mm _____	23	2.75–3.25 sec _____	23
29, + 11 mm _____	28	22.75, + 3.25 sec _____	28
CAM LIFT – Dynamic test: 10 mm		CAM ROTATION: 6 seconds	
9.5–10.5 mm _____	Acceptable range	5.90–6.10 sec Acceptable range	
9–11 mm _____	23	5.75–6.25 sec _____	–3
29, + 11 mm _____	28	25.75, + 6.25 sec _____	–8

Figure 14-9

Power source

Any DC power source may be used. A power source is preferred to batteries because it generates constant power.

Dial indicator

Cam performance will be checked using a dial indicator. See Figure 14-10. A dial indicator requires a force of approximately 0.8 N to lift. Design so that there is a minimum amount of deflection between the cam's edge surface and the dial indicator.

Alternative project

Design and build a device to measure the performance of the cam. Figure 14-11 shows a measuring device designed and built specifically for this project. The measuring device was mounted on a base and post used to support a dial indicator gauge.

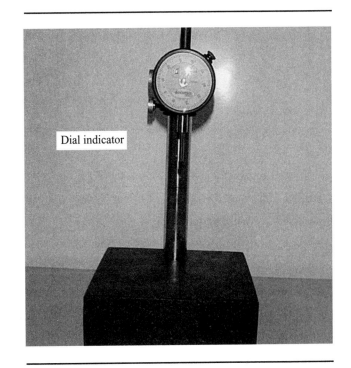

Dial indicator

Figure 14-10

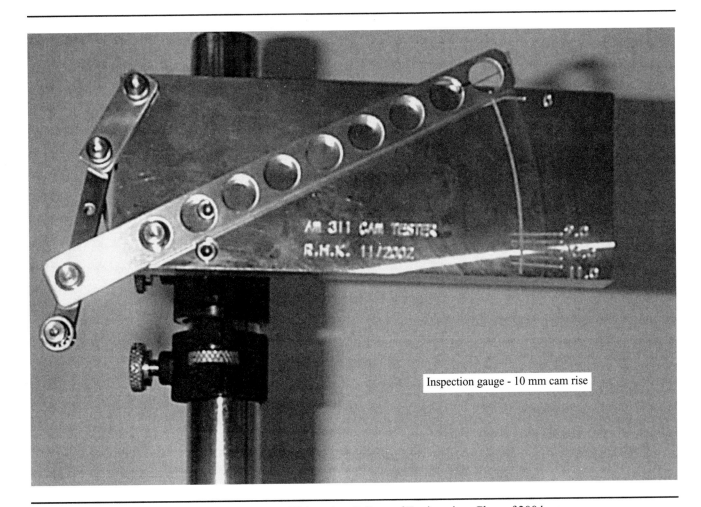

Inspection gauge - 10 mm cam rise

Figure 14-11 Courtesy of Ross Kenyon, Boston University, College of Engineering, Class of 2004.

14-11 GONDOLA PROJECT – RATING 7

Objective

To design and build a gondola that will climb a cable and stop within a defined distance. The gondola must be self-contained with no external controls. An external power adapter (AC to DC) may be used, but the access wires may not be used to control the motion of the gondola in any way.

Cable

The cable is approximately Ø0.19 and is coated with plastic. (The cable used was from a dog run leash.)

The shape of the cable is approximately as shown in Figure 14-12. The cable shape will vary with the amount of tension used when it is mounted.

There are two marks (permanent marker) on the cable, the start mark and the end mark. The two marks are 18 ft apart as measured along the cable.

Performance

The quality of the gondola's performance will be judged by how close it stops to the end mark after starting from behind the start mark. Measurements will be taken from a designated point on the gondola. The measuring mark may be placed anywhere on the gondola so long as it is clearly visible when the gondola stops.

The performance distance will be measured when the gondola has come to a complete stop.

Each team will be allowed three attempts. The closest finish will count toward the team's grade.

Grades

Grades will be earned as follows. All distances are in inches.

±1.00	A
±1.50	A−
±2.00	B+
±2.50	B
±3.00	B−
±3.50	C+
±4.00	C
±4.50	C−
±6.00	D
±12.00	F

Drawing requirements

1. An exploded scene of the vehicle
2. An exploded isometric assembly drawing with assembly numbers
3. Dimensioned and toleranced detail drawings of every part not purchased and of any purchased part that has been modified
4. A two-column parts list that gives an estimated price and weight for each part
5. A center of gravity calculation; list each part and its X, Y, Z component distances. See Figure 13-14.
6. The gear train ratios

Preliminary report

An estimate of the time needed for the gondola to climb the cable must be submitted 2 weeks prior to the performance test. Include all calculations.

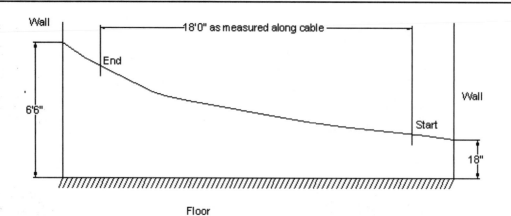

Figure 14-12

Appendix

Wire and Sheet Metal Gauges

Gauge			Thickness	Gauge	Thickness
000	000		0.5800	18	0.0403
00	000		0.5165	19	0.0359
0	000		0.4600	20	0.0320
	000		0.4096	21	0.0285
	00		0.3648	22	0.0253
	0		0.3249	23	0.0226
	1		0.2893	24	0.0201
	2		0.2576	25	0.0179
	3		0.2294	26	0.0159
	4		0.2043	27	0.0142
	5		0.1819	28	0.0126
	6		0.1620	29	0.0113
	7		0.1443	30	0.0100
	8		0.1285	31	0.0089
	9		0.1144	32	0.0080
	10		0.1019	33	0.0071
	11		0.0907	34	0.0063
	12		0.0808	35	0.0056
	13		0.0720	36	0.0050
	14		0.0641	37	0.0045
	15		0.0571	38	0.0040
	16		0.0508	39	0.0035
	17		0.0453	40	0.0031

Figure A-1

American Standard Clearance Locational Fits

Nominal Size Range Inches Over — To	Limits of Clearance	Class LC1 Standard Limits		Limits of Clearance	Class LC2 Standard Limits		Limits of Clearance	Class LC3 Standard Limits		Limits of Clearance	Class LC4 Standard Limits	
		Hole H6	Shaft h5		Hole H7	Shaft h6		Hole H8	Shaft h7		Hole H10	Shaft h9
0 — 0.12	0 / 0.45	+0.25 / 0	0 / -0.2	0 / 0.65	+0.4 / 0	0 / -0.25	0 / 1	+0.6 / 0	0 / -0.4	0 / 2.6	+1.6 / 0	0 / -1.0
0.12 — 0.24	0 / 0.5	+0.3 / 0	0 / -0.2	0 / 0.8	+0.5 / 0	0 / -0.3	0 / 1.2	+0.7 / 0	0 / -0.5	0 / 3.0	+1.8 / 0	0 / -1.2
0.24 — 0.40	0 / 0.65	+0.4 / 0	0 / -0.25	0 / 1.0	+0.6 / 0	0 / -0.4	0 / 1.5	+0.9 / 0	0 / -0.6	0 / 3.6	+2.2 / 0	0 / -1.4
0.40 — 0.71	0 / 0.7	+0.4 / 0	0 / -0.3	0 / 1.1	+0.7 / 0	0 / -0.4	0 / 1.7	+1.0 / 0	0 / -0.7	0 / 4.4	+2.8 / 0	0 / -1.6
0.71 — 1.19	0 / 0.9	+0.5 / 0	0 / -0.4	0 / 1.3	+0.8 / 0	0 / -0.5	0 / 2	+1.2 / 0	0 / -0.8	0 / 5.5	+3.5 / 0	0 / -2.0
1.19 — 1.97	0 / 1.0	+0.6 / 0	0 / -0.4	0 / 1.6	+1.0 / 0	0 / -0.6	0 / 2.6	+1.6 / 0	0 / -1.0	0 / 6.5	+4.0 / 0	0 / -2.5

Figure A-2A

Nominal Size Range Inches Over — To	Limits of Clearance	Class LC5 Standard Limits		Limits of Clearance	Class LC6 Standard Limits		Limits of Clearance	Class LC7 Standard Limits		Limits of Clearance	Class LC8 Standard Limits	
		Hole H7	Shaft g6		Hole H9	Shaft f8		Hole H10	Shaft e9		Hole H10	Shaft d9
0 — 0.12	0.1 / 0.75	+0..4 / 0	-0.1 / -0.35	0.3 / 1.9	+1.0 / 0	-0.3 / -0.9	0.6 / 3.2	+1.6 / 0	-0.6 / -1.6	1.0 / 3.6	+1.6 / 0	-1.0 / -2.0
0.12 — 0.24	0.15 / 0.95	+0.5 / 0	-0.15 / -0.45	0.4 / 2.3	+1.2 / 0	-0.4 / -1.1	0.8 / 3.8	+1.8 / 0	-0.8 / -2.0	1.2 / 4.2	+1.8 / 0	-1.2 / -2.4
0.24 — 0.40	0.2 / 1.2	+0.6 / 0	-0.2 / -0.6	0.5 / 2.8	+1.4 / 0	-0.5 / -1.4	1.0 / 4.6	+2.2 / 0	-1.0 / -2.4	1.6 / 5.2	+2.2 / 0	-1.6 / -3.0
0.40 — 0.71	0.25 / 1.35	+0.7 / 0	-0.25 / -0.65	0.6 / 3.2	+1.6 / 0	-0.6 / -1.6	1.2 / 5.6	+2.8 / 0	-1.2 / -2.8	2.0 / 6.4	+2.8 / 0	-2.0 / -3.6
0.71 — 1.19	0.3 / 1.6	+0.8 / 0	-0.3 / -0.8	0.8 / 4.0	+2.0 / 0	-0.8 / -2.0	1.6 / 7.1	+3.5 / 0	-1.6 / -3.6	2.5 / 8.0	+3.5 / 0	-2.5 / -4.5
1.19 — 1.97	0.4 / 2.0	+1.0 / 0	-0.4 / -1.0	1.0 / 5.1	+2.5 / 0	-1.0 / -2.6	2.0 / 8.5	+4.0 / 0	-2.0 / -4.5	3.0 / 9.5	+4.0 / 0	-3.0 / -5.5

Figure A-2B

American Standard Running and Sliding Fits
(Hole Basis)

Nominal Size Range Inches Over — To	Limits of Clearance	Class RC1 Standard Limits		Limits of Clearance	Class RC2 Standard Limits		Limits of Clearance	Class RC3 Standard Limits		Limits of Clearance	Class RC4 Standard Limits	
		Hole H5	Shaft g4		Hole H6	Shaft g5		Hole H7	Shaft f6		Hole H8	Shaft f7
0 — 0.12	0.1 0.45	+0.2 0	-0.1 -0.25	0.1 0.55	+0.25 0	-0.1 -0.3	0.3 0.95	+0.4 0	-0.3 -0.55	0.3 1.3	+0.6 0	-0.3 -0.7
0.12 — 0.24	0.15 0.5	+0.2 0	-0.15 -0.3	0.15 0.65	+0.3 0	-0.15 -0.35	0.4 1.12	+0.5 0	-0.4 -0.7	0.4 1.5	+0.7 0	-0.4 -0.0
0.24 — 0.40	0.2 0.6	+0.25 0	-0.2 -0.35	0.2 0.85	+0.4 0	-0.2 -0.45	0.5 1.5	+0.6 0	-0.5 -0.9	0.5 2.0	+0.9 0	-0.5 -1.1
0.40 — 0.71	0.25 0.75	+0.3 0	-0.25 -0.45	0.25 0.95	+0.4 0	-0.25 -0.55	0.6 1.7	+0.7 0	-0.6 -1.0	0.6 2.3	+1.0 0	-0.6 -1.3
0.71 — 1.19	0.3 0.95	+0.4 0	-0.3 -0.55	0.3 1.2	+0.5 0	-0.3 -0.7	0.8 2.1	+0.8 0	-0.8 -1.3	0.8 2.8	+1.2 0	-0.8 -1.6
1.19 — 1.97	0.4 1.1	+0.4 0	-0.4 -0.7	0.4 1.4	+0.6 0	-0.4 -0.8	1.0 2.6	+1.0 0	-1.0 -1.6	1.0 3.6	+1.6 0	-1.0 -2.0

Figure A-3A

Nominal Size Range Inches Over — To	Limits of Clearance	Class RC5 Standard Limits		Limits of Clearance	Class RC6 Standard Limits		Limits of Clearance	Class RC7 Standard Limits		Limits of Clearance	Class RC8 Standard Limits	
		Hole H8	Shaft e7		Hole H9	Shaft e8		Hole H9	Shaft d8		Hole H10	Shaft c9
0 — 0.12	0.6 1.6	+0.6 0	-0.6 -1.0	0.6 2.2	+1.0 0	-0.6 -1.2	1.0 2.6	+1.0 0	-1.0 -1.6	2.5 5.1	+1.6 0	-2.5 -3.5
0.12 — 0.24	0.8 2.0	+0.7 0	-0.8 -1.3	0.8 2.7	+1.2 0	-0.8 -1.5	1.2 3.1	+1.2 0	-1.2 -1.9	2.8 5.8	+1.8 0	-2.8 -4.0
0.24 — 0.40	1.0 2.5	+0.9 0	-1.0 -1.6	1.0 3.3	+1.4 0	-1.0 -1.9	1.6 3.9	+1.4 0	-1.6 -2.5	3.0 6.6	+2.2 0	-3.0 -4.4
0.40 — 0.71	1.2 2.9	+1.0 0	-1.2 -1.9	1.2 3.8	+1.6 0	-1.2 -2.2	2.0 4.6	+1.6 0	-2.0 -3.0	3.5 7.9	+2.8 0	-3.5 -5.1
0.71 — 1.19	1.6 3.6	+1.2 0	-1.6 -2.4	1.6 4.8	+2.0 0	-1.6 -2.8	2.5 5.7	+2.0 0	-2.5 -3.7	4.5 10.0	+3.5 0	-4.5 -6.5
1.19 — 1.97	2.0 4.6	+1.6 0	-2.0 -3.0	2.0 6.1	+2.5 0	-2.0 -3.6	3.0 7.1	+2.5 0	-3.0 -4.6	5.0 11.5	+4.0 0	-5.0 -7.5

Figure A-3B

American Standard Transition Locational Fits

Nominal Size Range Inches		Class LT1			Class LT2			Class LT3		
			Standard Limits			Standard Limits			Standard Limits	
		Fit	Hole H7	Shaft js6	Fit	Hole H8	Shaft js7	Fit	Hole H7	Shaft k6
Over	To									
0	0.12	−0.10 +0.50	+0.4 0	+0.10 −0.10	−0.2 +0.8	+0.6 0	+0.2 −0.2			
0.12	0.24	−0.15 −0.65	+0.5 0	+0.15 −0.15	−0.25 +0.95	+0.7 0	+0.25 −0.25			
0.24	0.40	−0.2 +0.5	+0.6 0	+0.2 −0.2	−0.3 +1.2	+0.9 0	+0.3 −0.3	−0.5 +0.5	+0.6 0	+0.5 +0.1
0.40	0.71	−0.2 +0.9	+0.7 0	+0.2 −0.2	−0.35 +1.35	+1.0 0	+0.35 −0.35	−0.5 +0.6	+0.7 0	+0.5 +0.1
0.71	1.19	−0.25 +1.05	+0.8 0	+0.25 −0.25	−0.4 +1.6	+1.2 0	+0.4 −0.4	−0.6 +0.7	+0.8 0	+0.6 +0.1
1.19	1.97	−0.3 +1.3	+1.0 0	+0.3 −0.3	−0.5 +2.1	+1.6 0	+0.5 −0.5	+0.7 +0.1	+1.0 0	+0.7 +0.1

Figure A-4A

Nominal Size Range Inches		Class LT4			Class LT5			Class LT6		
			Standard Limits			Standard Limits			Standard Limits	
		Fit	Hole H8	Shaft k7	Fit	Hole H7	Shaft n6	Fit	Hole H7	Shaft n7
Over	To									
0	0.12				−0.5 +0.15	+0.4 0	+0.5 +0.25	−0.65 +0.15	+0.4 0	+0.65 +0.25
0.12	0.24				−0.6 +0.2	+0.5 0	+0.6 +0.3	−0.8 +0.2	+0.5 0	+0.8 +0.3
0.24	0.40	−0.7 +0.8	+0.9 0	+0.7 +0.1	−0.8 +0.2	+0.6 0	+0.8 +0.4	−1.0 +0.2	+0.6 0	+1.0 +0.4
0.40	0.71	−0.8 +0.9	+1.0 0	+0.8 +0.1	−0.9 +0.2	+0.7 0*	+0.9 +0.5	−1.2 +0.2	+0.7 0	+1.2 +0.5
0.71	1.19	−0.9 +1.1	+1.2 0	+0.9 +0.1	−1.1 +0.2	+0.8 0	+1.1 +0.6	−1.4 +0.2	+0.8 0	+1.4 +0.6
1.19	1.97	−1.1 +1.5	+1.6 0	+1.1 +0.1	−1.3 +0.3	+1.0 0	+1.3 +0.7	−1.7 +0.3	+1.0 0	+1.7 +0.7

Figure A-4B

American Standard Interference Locational Fits

Nominal Size Range Inches Over — To	Limits of Interference	Class LN1 Standard Limits Hole H6	Shaft n5	Limits of Interference	Class LN2 Standard Limits Hole H7	Shaft p6	Limits of Interference	Class LN3 Standard Limits Hole H7	Shaft r6
0 – 0.12	0 / 0.45	+0.25 / 0	+0.45 / +0.25	0 / 0.65	+0.4 / 0	+0.63 / +0.4	0.1 / 0.75	+0.4 / 0	+0.75 / +0.5
0.12 – 0.24	0 / 0.5	+0.3 / 0	+0.5 / +0.3	0 / 0.8	+0.5 / 0	+0.8 / +0.5	0.1 / 0.9	+0.5 / 0	+0.9 / +0.6
0.24 – 0.40	0 / 0.65	+0.4 / 0	+0.65 / +0.4	0 / 1.0	+0.6 / 0	+1.0 / +0.6	0.2 / 1.2	+0.6 / 0	+1.2 / +0.8
0.40 – 0.71	0 / 0.8	+0.4 / 0	+0.8 / +0.4	0 / 1.1	+0.7 / 0	+1.1 / +0.7	0.3 / 1.4	+0.7 / 0	+1.4 / +1.0
0.71 – 1.19	0 / 1.0	+0.5 / 0	+1.0 / +0.5	0 / 1.3	+0.8 / 0	+1.3 / +0.8	0.4 / 1.7	+0.8 / 0	+1.7 / +1.2
1.19 – 1.97	0 / 1.1	+0.6 / 0	+1.1 / +0.6	0 / 1.6	+1.0 / 0	+1.6 / +1.0	0.4 / 2.0	+1.0 / 0	+2.0 / +1.4

Figure A-5

American Standard Force and Shrink Fits

Nominal Size Range Inches Over — To	Limits of Interference	Class FN 1 Standard Limits Hole	Shaft	Limits of Interference	Class FN 2 Standard Limits Hole	Shaft	Limits of Interference	Class FN 3 Standard Limits Hole	Shaft	Limits of Interference	Class FN 4 Standard Limits Hole	Shaft
0 – 0.12	0.05 / 0.5	+0.25 / 0	+0.5 / +0.3	0.2 / 0.85	+0.4 / 0	+0.85 / +0.6				0.3 / 0.95	+0.4 / 0	+0.95 / +0.7
0.12 – 0.24	0.1 / 0.6	+0.3 / 0	+0.6 / +0.4	0.2 / 1.0	+0.5 / 0	+1.0 / +0.7				0.4 / 1.2	+0.5 / 0	+1.2 / +0.9
0.24 – 0.40	0.1 / 0.75	+0.4 / 0	+0.75 / +0.5	0.4 / 1.4	+0.6 / 0	+1.4 / +1.0				0.6 / 1.6	+0.6 / 0	+1.6 / +1.2
0.40 – 0.56	0.1 / 0.8	+0.4 / 0	+0.8 / +0.5	0.5 / 1.6	+0.7 / 0	+1.6 / +1.2				0.7 / 1.8	+0.7 / 0	+1.8 / +1.4
0.56 – 0.71	0.2 / 0.9	+0.4 / 0	+0.9 / +0.6	0.5 / 1.6	+0.7 / 0	+1.6 / +1.2				0.7 / 1.8	+0.7 / 0	+1.8 / +1.4
0.71 – 0.95	0.2 / 1.1	+0.5 / 0	+1.1 / +0.7	0.6 / 1.9	+0.8 / 0	+1.9 / +1.4				0.8 / 2.1	+0.8 / 0	+2.1 / +1.6
0.95 – 1.19	0.3 / 1.2	+0.5 / 0	+1.2 / +0.8	0.6 / 1.9	+0.8 / 0	+1.9 / +1.4	0.8 / 2.1	+0.8 / 0	+2.1 / +1.6	1.0 / 2.3	+0.8 / 0	+2.1 / +1.8
1.19 – 1.58	0.3 / 1.3	+0.6 / 0	+1.3 / +0.9	0.8 / 2.4	+1.0 / 0	+2.4 / +1.8	1.0 / 2.6	+1.0 / 0	+2.6 / +2.0	1.5 / 3.1	+1.0 / 0	+3.1 / +2.5
1.58 – 1.97	0.4 / 1.4	+0.6 / 0	+1.4 / +1.0	0.8 / 2.4	+1.0 / 0	+2.4 / +1.8	1.2 / 2.8	+1.0 / 0	+2.8 / +2.2	1.8 / 3.4	+1.0 / 0	+3.4 / +2.8

Figure A-6

Preferred Clearance Fits — Cylindrical Fits (Hole Basis; ANSI B4.2)

Basic Size		Loose Running			Free Running			Close Running			Sliding			Locational Clear.		
		Hole H11	Shaft c11	Fit	Hole H9	Shaft d9	Fit	Hole H8	Shaft f7	Fit	Hole H7	Shaft g6	Fit	Hole H7	Shaft h6	Fit
4	Max	4.075	3.930	0.220	4.030	3.970	0.090	4.018	3.990	0.040	4.012	3.996	0.024	4.012	4.000	0.020
	Min	4.000	3.855	0.070	4.000	3.940	0.030	4.000	3.978	0.010	4.000	3.988	0.004	4.000	3.992	0.000
5	Max	5.075	4.930	0.220	5.030	4.970	0.090	5.018	4.990	0.040	5.012	4.996	0.024	5.012	5.000	0.020
	Min	5.000	4.855	0.070	5.000	4.940	0.030	5.000	4.978	0.010	5.000	4.988	0.004	5.000	4.992	0.000
6	Max	6.075	5.930	0.220	6.030	5.970	0.090	6.018	5.990	0.040	6.012	5.996	0.024	6.012	6.000	0.020
	Min	6.000	5.885	0.070	6.000	5.940	0.030	6.000	5.978	0.010	6.000	5.988	0.004	6.000	5.992	0.000
8	Max	8.090	7.920	0.260	8.036	7.960	0.112	8.022	7.987	0.050	8.015	7.995	0.029	8.015	8.000	0.024
	Min	8.000	7.830	0.080	8.000	7.924	0.040	8.000	7.972	0.013	8.000	7.986	0.005	8.000	7.991	0.000
10	Max	10.090	9.920	0.260	10.036	9.960	0.112	10.022	9.987	0.050	10.015	9.995	0.029	10.015	10.000	0.024
	Min	10.000	9.830	0.080	10.000	9.924	0.040	10.000	9.972	0.013	10.000	9.986	0.005	10.000	9.991	0.000
12	Max	12.112	11.905	0.315	12.043	11.950	0.136	12.027	11.984	0.061	12.018	11.994	0.035	12.018	12.000	0.029
	Min	12.000	11.795	0.095	12.000	11.907	0.050	12.000	11.966	0.016	12.000	11.983	0.006	12.000	11.989	0.000
16	Max	16.110	15.905	0.315	16.043	15.950	0.136	16.027	15.984	0.061	16.018	15.994	0.035	16.018	16.000	0.029
	Min	16.000	15.795	0.095	16.000	15.907	0.050	16.000	15.966	0.016	16.000	15.983	0.006	16.000	15.989	0.000
20	Max	20.130	19.890	0.370	20.052	19.935	0.169	20.033	19.980	0.074	20.021	19.993	0.041	20.021	20.000	0.034
	Min	20.000	19.760	0.110	20.000	19.883	0.065	20.000	19.959	0.020	20.000	19.980	0.007	20.000	19.987	0.000
25	Max	25.130	24.890	0.370	25.052	24.935	0.169	25.033	24.980	0.074	25.021	24.993	0.041	25.021	25.000	0.034
	Min	25.000	24.760	0.110	25.000	24.883	0.065	25.000	24.959	0.020	25.000	24.980	0.007	25.000	24.987	0.000
30	Max	30.130	29.890	0.370	30.052	29.935	0.169	30.033	29.980	0.074	30.021	29.993	0.041	30.021	30.000	0.034
	Min	30.000	29.760	0.110	30.000	29.883	0.065	30.000	29.959	0.020	30.000	29.980	0.007	30.000	29.987	0.000

Figure A-7

Preferred Transition and Interference Fits — Cylindrical Fits
(Hole Basis, ANSI B4.2)

Basic Size		Locational Trans.			Locational Trans.			Locational Inter.			Medium Drive			Force		
		Hole H7	Shaft k6	Fit	Hole H7	Shaft n6	Fit	Hole H7	Shaft p6	Fit	Hole H7	Shaft s6	Fit	Hole H7	Shaft u6	Fit
4	Max	4.012	4.009	0.011	4.012	4.016	0.004	4.012	4.020	0.000	4.012	4.027	-0.007	4.012	4.031	-0.011
	Min	4.000	4.001	-0.009	4.000	4.008	-0.016	4.000	4.012	-0.020	4.000	4.019	-0.027	4.000	4.023	-0.031
5	Max	5.012	5.009	0.011	5.012	5.016	0.004	5.012	5.020	0.000	5.012	5.027	-0.007	5.012	5.031	-0.011
	Min	5.000	5.001	-0.009	5.000	5.008	-0.016	5.000	5.012	-0.020	5.000	5.019	-0.027	5.000	5.023	-0.031
6	Max	6.012	6.009	0.011	6.012	6.016	0.004	6.012	6.020	0.000	6.012	6.027	-0.007	6.012	6.031	-0.011
	Min	6.000	6.001	-0.009	6.000	6.008	-0.016	6.000	6.012	-0.020	6.000	6.019	-0.027	6.000	6.023	-0.031
8	Max	8.015	8.010	0.014	8.015	8.019	0.005	8.015	8.024	0.000	8.015	8.032	-0.008	8.015	8.037	-0.013
	Min	8.000	8.001	-0.010	8.000	8.010	-0.019	8.000	8.015	-0.024	8.000	8.023	-0.032	8.000	8.028	-0.037
10	Max	10.015	10.010	0.014	10.015	10.019	0.005	10.015	10.024	0.000	10.015	10.032	-0.008	10.015	10.037	-0.013
	Min	10.000	10.001	-0.010	10.000	10.010	-0.019	10.000	10.015	-0.024	10.000	10.023	-0.032	10.000	10.028	-0.037
12	Max	12.018	12.012	0.017	12.018	12.023	0.006	12.018	12.029	0.000	12.018	12.039	-0.010	12.018	12.044	-0.015
	Min	12.000	12.001	-0.012	12.000	12.012	-0.023	12.000	12.018	-0.029	12.000	12.028	-0.039	12.000	12.033	-0.044
16	Max	16.018	16.012	0.017	16.018	16.023	0.006	16.018	16.029	0.000	16.018	16.039	-0.010	16.018	16.044	-0.015
	Min	16.000	16.001	-0.012	16.000	16.012	-0.023	16.000	16.018	-0.029	16.000	16.028	-0.039	16.000	16.033	-0.044
20	Max	20.021	20.015	0.019	20.021	20.028	0.006	20.021	20.035	-0.001	20.021	20.048	-0.014	20.021	20.054	-0.020
	Min	20.000	20.002	-0.015	20.000	20.015	-0.028	20.000	20.022	-0.035	20.000	20.035	-0.048	20.000	20.041	-0.054
25	Max	25.021	25.015	0.019	25.021	25.028	0.006	25.021	25.035	-0.001	25.021	25.048	-0.014	25.021	25.061	-0.027
	Min	25.000	25.002	-0.015	25.000	25.015	-0.028	25.000	25.022	-0.035	25.000	25.035	-0.048	25.000	25.048	-0.061
30	Max	30.021	30.015	0.019	30.021	30.028	0.006	30.021	30.035	-0.001	30.021	30.048	-0.014	30.021	30.061	-0.027
	Min	30.000	30.002	-0.015	30.000	30.015	-0.028	30.000	30.022	-0.035	30.000	30.035	-0.048	30.000	30.048	-0.061

Figure A-8

Preferred Clearance Fits — Cylindrical Fits
(Shaft Basis; ANSI B4.2)

Basic Size		Loose Running			Free Running			Close Running			Sliding			Locational Clear.		
		Hole C11	Shaft h11	Fit	Hole D9	Shaft h9	Fit	Hole F8	Shaft h7	Fit	Hole G7	Shaft h6	Fit	Hole H7	Shaft h6	Fit
4	Max	4.145	4.000	0.220	4.060	4.000	0.090	4.028	4.000	0.040	4.016	4.000	0.024	4.012	4.000	0.020
	Min	4.070	3.925	0.070	4.030	3.970	0.030	4.010	3.988	0.010	4.004	3.992	0.004	4.000	3.992	0.000
5	Max	5.145	5.000	0.220	5.060	5.000	0.090	5.028	5.000	0.040	5.016	5.000	0.024	5.012	5.000	0.020
	Min	5.070	4.925	0.070	5.030	4.970	0.030	5.010	4.988	0.010	5.004	4.992	0.004	5.000	4.992	0.000
6	Max	6.145	6.000	0.220	6.060	6.000	0.090	6.028	6.000	0.040	6.016	6.000	0.024	6.012	6.000	0.020
	Min	6.070	5.925	0.070	6.030	5.970	0.030	6.010	5.988	0.010	6.004	5.992	0.004	6.000	5.992	0.000
8	Max	8.170	8.000	0.260	8.076	8.000	0.112	8.035	8.000	0.050	8.020	8.000	0.029	8.015	8.000	0.024
	Min	8.080	7.910	0.080	8.040	7.964	0.040	8.013	7.985	0.013	8.005	7.991	0.005	8.000	7.991	0.000
10	Max	10.170	10.000	0.260	10.076	10.000	0.112	10.035	10.000	0.050	10.020	10.000	0.029	10.015	10.000	0.024
	Min	10.080	9.910	0.080	10.040	9.964	0.040	10.013	9.985	0.013	10.005	9.991	0.005	10.000	9.991	0.000
12	Max	12.205	12.000	0.315	12.093	12.000	0.136	12.043	12.000	0.061	12.024	12.000	0.035	12.018	12.000	0.029
	Min	12.095	11.890	0.095	12.050	11.957	0.050	12.016	11.982	0.016	12.006	11.989	0.006	12.000	11.989	0.000
16	Max	16.205	16.000	0.315	16.093	16.000	0.136	16.043	16.000	0.061	16.024	16.000	0.035	16.018	16.000	0.029
	Min	16.095	15.890	0.095	16.050	15.957	0.050	16.016	15.982	0.016	06.006	15.989	0.006	16.000	15.989	0.000
20	Max	20.240	20.000	0.370	20.117	20.000	0.169	20.053	20.000	0.074	20.028	20.000	0.041	20.021	20.000	0.034
	Min	20.110	19.870	0.110	20.065	19.948	0.065	20.020	19.979	0.020	20.007	19.987	0.007	20.000	19.987	0.000
25	Max	25.240	25.000	0.370	25.117	25.000	0.169	25.053	25.000	0.074	25.028	25.000	0.041	25.021	25.000	0.034
	Min	25.110	24.870	0.110	25.065	24.948	0.065	25.020	24.979	0.020	25.007	24.987	0.007	25.000	24.987	0.000
30	Max	30.240	30.000	0.370	30.117	30.000	0.169	30.053	30.000	0.074	30.028	30.000	0.041	30.021	30.000	0.034
	Min	30.110	29.870	0.110	30.065	29.948	0.065	30.020	29.979	0.020	30.007	29.987	0.007	30.000	29.987	0.000

Figure A-9

Preferred Transition and Interference Fits — Cylindrical Fits
(Shaft Basis; ANSI B4.2)

Basic Size		Locational Trans.			Locational Trans.			Locational Inter.			Medium Drive			Force		
		Hole K7	Shaft h6	Fit	Hole N7	Shaft h6	Fit	Hole P7	Shaft h6	Fit	Hole S7	Shaft h6	Fit	Hole U7	Shaft h6	Fit
4	Max	4.003	4.000	0.011	3.996	4.000	0.004	3.992	4.000	0.000	3.985	4.000	-0.007	3.981	4.000	-0.011
	Min	3.991	3.992	-0.009	3.984	3.992	-0.016	3.980	3.992	-0.020	3.973	3.992	-0.027	3.969	3.992	-0.031
5	Max	5.003	5.000	0.011	4.996	5.000	0.004	4.992	5.000	0.000	4.985	5.000	-0.007	4.981	5.000	-0.011
	Min	4.991	4.992	-0.009	4.984	4.992	-0.016	4.980	4.992	-0.020	4.973	4.992	-0.027	4.969	4.992	-0.031
6	Max	6.003	6.000	0.011	5.996	6.000	0.004	5.992	6.000	0.000	5.985	6.000	-0.007	5.981	6.000	-0.011
	Min	5.991	5.992	-0.009	5.984	5.992	-0.016	5.980	5.992	-0.020	5.973	5.992	-0.027	5.969	5.992	-0.031
8	Max	8.005	8.000	0.014	7.996	8.000	0.005	7.991	8.000	0.000	7.983	8.000	-0.008	7.978	8.000	-0.013
	Min	7.990	7.991	-0.010	7.981	7.991	-0.019	7.976	7.991	-0.024	7.968	7.991	-0.032	7.963	7.991	-0.037
10	Max	10.005	10.000	0.014	9.996	10.000	0.005	9.991	10.000	0.000	9.983	10.000	-0.008	9.978	10.000	-0.013
	Min	9.990	9.991	-0.010	9.981	9.991	-0.019	9.976	9.991	-0.024	9.968	9.991	-0.032	9.963	9.991	-0.037
12	Max	12.006	12.000	0.017	11.995	12.000	0.006	11.989	12.000	0.000	11.979	12.000	-0.010	11.974	12.000	-0.015
	Min	11.988	11.989	-0.012	11.977	11.989	-0.023	11.971	11.989	-0.029	11.961	11.989	-0.039	11.956	11.989	-0.044
16	Max	16.006	16.000	0.017	15.995	16.000	0.006	15.989	16.000	0.000	15.979	16.000	-0.010	15.974	16.000	-0.015
	Min	15.988	15.989	-0.012	15.977	15.989	-0.023	15.971	15.989	-0.029	15.961	15.989	-0.039	15.956	15.989	-0.044
20	Max	20.006	20.000	0.019	19.993	20.000	0.006	19.986	20.000	-0.001	19.973	20.000	-0.014	19.967	20.000	-0.020
	Min	19.985	19.987	-0.015	19.972	19.987	-0.028	19.965	19.987	-0.035	19.952	19.987	-0.048	19.946	19.987	-0.054
25	Max	25.006	25.000	0.019	24.993	25.000	0.006	24.986	25.000	-0.001	24.973	25.000	-0.014	24.960	25.000	-0.027
	Min	24.985	24.987	-0.015	24.972	24.987	-0.028	24.965	24.987	-0.035	24.952	24.987	-0.048	24.939	24.987	-0.061
30	Max	30.006	30.000	0.019	29.993	30.000	0.006	29.986	30.000	-0.001	29.973	30.000	-0.014	29.960	30.000	-0.027
	Min	29.985	29.987	-0.015	29.972	29.987	-0.028	29.965	29.987	-0.035	29.952	29.987	-0.048	29.939	29.987	-0.061

Figure A-10

Index